T0299082

Random
Processes
First-Passage and Escape

Random
Processes
First-Passage and Escape

Jaume Masoliver
University of Barcelona

 **World Scientific**

JERSEY · LONDON · SINGAPORE · BEIJING · SHANGHAI · HONG KONG · TAIPEI · CHENNAI · TOKYO

Published by

World Scientific Publishing Co. Pte. Ltd.
5 Toh Tuck Link, Singapore 596224
USA office: 27 Warren Street, Suite 401-402, Hackensack, NJ 07601
UK office: 57 Shelton Street, Covent Garden, London WC2H 9HE

British Library Cataloguing-in-Publication Data
A catalogue record for this book is available from the British Library.

Cover image:
Painting: Dylan-1(1995), acrylic on canvass. Artist: Isidre Manils.

RANDOM PROCESSES
First-Passage and Escape

ISBN 978-981-3225-31-2

For any available supplementary material, please visit
https://www.worldscientific.com/worldscibooks/10.1142/10578#t=suppl

Printed in Singapore

To Carme, Miquel and Maria

Preface

The project of this book started in 2014 after receiving an invitation from World Scientific for writing a book on first-passage times and their applications to socio-economic problems. At first I practically declined that kind invitation because I had just finished to write two review articles on the same topic (Masoliver, 2014b; Masoliver and Perelló, 2014b) and thought I had nothing else to say on the subject, at least substantially. Several months passed after I realized that some relevant parts of the first-passage problem, mainly those related to multidimensional problems and fractional processes, had not been included in my previous reports. It occurred to me that, in addition, it would be too a good idea to enlarge the focus of the book and treat not only level crossing problems but also the general theory of random processes to which a great deal of my research and teaching has been devoted during more than three decades of professional activity. The editor gladly agreed with the new project and, after more than three years of discontinuous but intense work, the job is done.

The enlargement of the original project to incorporate a rather complete treatment of random processes, including such topics as stochastic calculus and stochastic differential equations, opened the door to the addition of more applications not only to economics but also to physical systems. In dealing with economic problems the style of the exposition is rigorous, even though being a physicist I have intendedly kept intuition and easiness as much as possible, along the way of what has been lately called *econophysics*.

The book is intended to advanced undergraduate, graduate and PhD students willing to approach, from an intuitive but yet rigorous way, the world of randomness. Selected applications to physical systems and, particularly, to socio-economic systems are also presented. The book can have several levels of reading and more specialized reading is marked in the text.[1]

The first part of this book is devoted to the theory of random processes, an important scientific development of the twentieth century, with significant contributions ranging from pure mathematics (specially in its foundations) to physics, biology, engineering as well as socio-economic sciences. The first chapter consists in a review on probability which is the cornerstone of the whole theory of random processes and the reader familiar with probability may skip this chapter and go directly to subsequent chapters. The second chapter addresses the general properties of random processes and the way of characterizing them. Subsequent chapters describe in detail general classes of random process such as Markov and diffusion processes, including anomalous (or fractional) diffusion processes. The exposition of the subject is rather general and aimed to any sort of application whether to natural sciences, engineering or any other. However, in this part some developments and examples belong to physical sciences. This is, for instance, the case of the exposition in Chapter 2 on the ergodic properties of random processes, which are of prime importance in the foundations of statistical mechanics. Other examples are the description of the Kramers problem, the notion of detailed balance (Chapter 5) or the theory of linear response (Chapter 6) all of them having important applications in statistical physics, chemical physics and thermodynamics.

Part II is dedicated to stochastic calculus which is essentially a mathematical development dealing with the integration of random processes. It constitutes a very technical field which I have tried to explain in the simplest and most intuitive approach that I have been able to, without incurring in too many mathematical inaccuracies. One of the most relevant extensions of stochastic calculus is in stochastic differential equations which, as their name indicates, are differential equations where one or several parameters are random functions of time. A fairly complete exposition of the theme (although by any means not exhaustive) is in Chapter 8. According to the interpretation of stochastic integral we choose, Itô or Stratonovich, we have two different, but otherwise completely equivalent, types of stochastic equations. Itô equations present some advantages such as having simple averages and the fact of dealing with nonanticipating functions, both aspects highly appreciated in mathematical and financial developments. The drawback is a somehow difficult handling of Itô equations because Itô integrals do not follow the usual rules of ordinary calculus. On the other hand,

---

[1] Chapters and Sections marked with an asterisk (*) can be omitted on a first reading.

stochastic equations interpreted in the sense of Stratonovich follow the ordinary rules of calculus which makes their handling easier. Drawbacks are the appearance of clumsier averages and the lack of nonanticipating functions which is conceptually unsatisfactory, specially in finance. Nevertheless, Stratonovich interpretation is the most employed in physical sciences and, besides easiness in handling the calculus, the main reason resides in the fact that, as a consequence of the Wong-Zakai theorem, when modeling real physical systems one usually ends up with Stratonovich equations. The last chapter of this part is dedicated to some financial applications with special emphasis on three important subjects: the process of discounting, the random dynamics of bond prices and a short review on option pricing theory.

In the third and last part of the book I address an important but rather difficult component of randomness: level-crossing (or first-passage) problems. This kind of problems have a long and standing tradition in many parts of natural sciences and engineering but with scarce applications in socio-economic systems. The overall problem consists not only in first-passage and escape problems but also on the theory of the extreme values (maxima and minima) that a random process can attain. The treatment presented here is rather complete, going beyond the two previous review articles mentioned above, specially in the treatment of fractional diffusions, multidimensional systems and inertial processes. The last chapter is devoted to applications to economic systems and, besides the standard treatment of diffusive financial systems (including stochastic volatility models) we also present an account on the first-passage problem in the continuous-time random walk formalism, which is very useful in dealing with tic-by-tic data and not widely known in the literature.

A final word of acknowledgement. This book is based on years of teaching and research which has involved the interaction and enlightening discussions with many colleagues, former students and friends to whom I'm indebted. A necessarily short and incomplete list of them is: Marian Boguná, J. Doyne Farmer, John Geanakoplos, Ryszard Kutner, Katja Lindenberg, Josep Llosa, José Alberto Lobo (†), Miquel Montero, Josep Perelló, Josep M. Pons, Josep M. Porrà, Jaume Puig, Bruce J. West and George H. Weiss (†).

Barcelona, January 2018.

Contents

Stochastic calculus with some applications   169

PART 1
Random processes

Chapter 1

Fundamentals of probability

In this first chapter we will review the basic concepts of probability which constitute the pillar of the theory of random processes and all its subsequent developments and applications.[1] The reader familiar with probability theory may skip this introductory material and go directly to the second chapter.

1.1 Spaces of probability

In many physical phenomena, as well as in biological, technical and socioeconomic matters, just to name a few, the occurrence of a given result cannot be told in advance with certainty. One may give, at most, some numerical estimate, usually a percentage –in fact, a number between 0 and 1– of the likelihood of that result to occur. This numerical estimate is called *probability* and higher probability implies higher likelihood. Different observations of the same phenomenon and under the same circumstances will give different results.

We call *experiment* to any observation that can be repeated under the same conditions an unlimited number of times. The set of all possible outcomes of an experiment is called the *sample space* Ω. It is an abstract set that may or may not contain numerical values. Thus, for instance, the sample space of the experiment consisting in tossing a fair coin consists in two elements: head and tail, i.e., $\Omega = \{H, T\}$, while the sample space of rolling a fair die consists in six (numerical) elements $\Omega = \{1, 2, 3, 4, 5, 6\}$.

---

[1] There are countless excellent books on probability. A very incomplete selection based on my personal taste is: Feller (1991); Gnedenko (1976); Grimmett and Stirzaker (1992); Laha and Rohatgi (1979); Papoulis (1984).

These are examples of finite sample spaces. There are, however, in-finite sample spaces. For example, randomly choosing a positive integer number form the set of all integers or measuring a given voltage in a noisy electric circuit are examples of infinite sample spaces. In the first case the sample space is infinite but countable, the set of all natural numbers $\Omega = \{1, 2, 3, \ldots, n, \ldots\}$ while in the second case, Ω is uncountable.[2]

Any subset $A \subset \Omega$ of the sample space is called *event* and the elements of Ω are *elementary events*. The whole sample space is also an event, often called the *sure event*, and the empty set $\emptyset$ is the *impossible event*. For example, in the experiment of throwing two coins the set of all possible results, the sample space, is $\Omega = \{(H, H); (H, T); (T, H); (T, T)\}$ and the event "the appearance of at least one head" is $A = \{(H, H); (H, T); (T, H)\}$.

Events are subjected to set operations such as unions and intersections. We thus have the union and intersection of two sets A and B defined as

$$A \cup B = \{\omega \in \Omega / \omega \in A \text{ or } \omega \in B\}, \quad A \cap B = \{\omega \in \Omega / \omega \in A \text{ and } \omega \in B\},$$

and the difference of two sets and the complementary of a set, as

$$A - B = \{\omega \in \Omega / \omega \in A \text{ and } \omega \notin B\}, \quad \bar{A} = \{\omega \in \Omega / \omega \notin A\} = \Omega - A.$$

Given a sample space Ω a collection of events, i.e., a collection of subsets of Ω,

$$\mathcal{A} = \{A_1, A_2, \ldots, A_n \ldots\},$$

is called a σ-*algebra* of events if it is closed under unions, intersections and complementations of any of its members. In other words if A_k and A_l belong to $\mathcal{A}$ so do $A_k \cup A_l$, $A_k \cap A_l$ and $\bar{A}_k$ for all k and l.

Axiomatic definition of probability

While almost everybody has an intuitive idea of the concept of probability, a rigorous and precise definition of it, free of paradoxes and misunderstandings, has not been an easy task since the pioneering works of Fermat, Pascal and Laplace in the seventeen and eighteen centuries. It wasn't until rather recently, during the 1930's, when Kolmogorov set the axiomatic approach to probability which is now unanimously accepted.

Suppose we have an experiment with sample space Ω and a σ-algebra of events $\mathcal{A}$. We define the probability of an event $A \in \mathcal{A}$, and denote it by $P(A)$, as any function of A satisfying the so-called *axioms of probability*:

---

[2]If, for instance, the voltage can go from 0 to 10 volts then Ω is the interval of the real line [0, 10] which is uncountable.

(i) $P(A) \geq 0$, for all $A \in \mathcal{A}$.

(ii) $P(\Omega) = 1$.

(iii) If $A_k \in \mathcal{A}$ $(k = 1, 2, 3, \dots)$ is any sequence (finite or infinite) of mutually exclusive events –that is, non-overlapping in the sense that $A_k \cap A_l = \emptyset$ for all $k \neq l$– then

$$P\left(\bigcup_k A_k\right) = \sum_k P(A_k).$$

Note that since A and $\bar{A}$ are mutually exclusive and $A \cup \bar{A} = \Omega$, it follows from (iii) and (ii) that $P(A) + P(\bar{A}) = P(\Omega) = 1$. That is

$$P(\bar{A}) = 1 - P(A),$$

and from (i) and (ii) we conclude that

$$P(A) \leq 1, \quad \text{and} \quad P(\emptyset) = 0.$$

Therefore, *the probability of an event is any real number between 0 and 1.*

Other properties that can be proved from the axioms are:

(a) $P(A) \leq P(B)$ if $A \subset B$.

(b) For an arbitrary sequence of events (not necessarily mutually exclusive) we have *Boole's inequality:*

$$P\left(\bigcup_k A_k\right) \leq \sum_k P(A_k). \tag{1.1}$$

This inequality can be made more explicit. Thus, for instance, for two arbitrary events A and B we have

$$P(A \cup B) = P(A) + P(B) - P(A \cap B), \tag{1.2}$$

which can be extended to any number of arbitrary events although the resulting expressions become intricate very quickly as the number of events involved increase and they are messy and rather impractical.

Let us observe that the axiomatic definition of probability given above is not operative in the sense that it does not provide an algorithm or a procedure to actually calculate the probability of any event. As we will see next, the axiomatic definition provides consistency to any probability measure that we can figure out from empirical observation.

Thus, suppose we have a finite sample space with N elements

$$\Omega = \{\omega_1, \omega_2, \ldots, \omega_N\}.$$

Let $\mathcal{A}$ be the collection of all subsets of Ω ($\mathcal{A}$ is an algebra of events). We define the probability of any event $A \in \mathcal{A}$ as

$$P(A) = \frac{N_A}{N},$$

where N_A is the number of elements (i.e., elementary events) of A and N is the total number of elements of Ω. We left as a simple exercise to the reader to check that $P(A)$ defined in this way satisfies the axioms of probability and hence P is a true measure of probability. This definition is often termed as the *classical definition of probability*.

Another classic example is provided by the *geometric probability*. Suppose we can assign a geometrical representation, a diagram, to any event A and also to the sure event represented by the sample space Ω.[3] Let $S(A)$ and $S(\Omega)$ be the surfaces of the figures corresponding to A and Ω. The probability of A can be represented by

$$P(A) = \frac{S(A)}{S(\Omega)}.$$

This is a probability measure because it satisfies the axioms of probability. Indeed, (i) $P(A) \geq 0$, (ii) $P(\Omega) = 1$ and, since $S(A \cup B) = S(A) + S(B)$ if $A \cap B = \emptyset$, we have (iii) $P(A \cup B) = P(A) + P(B)$.

Let us finally remark that a sample space Ω, resulting from some random experiment or observation, a σ- algebra $\mathcal{A}$ of events defined on Ω, and a probability measure P defined on $\mathcal{A}$ define a *probability space* which is frequently denoted by $(\Omega, \mathcal{A}, P)$.

1.2 Conditional probability and independence

Let A and B be two events belonging to the same probability space and suppose that B has a non-vanishing probability. We define the *conditional probability* of A given B as

$$P(A|B) \equiv \frac{P(A \cap B)}{P(B)}, \tag{1.3}$$

with $P(B) \neq 0$.

This definition is meaningless when $P(B) = 0$ because if the probability that B occurs is zero, then the joint probability, $P(A \cap B)$, that A and B

---

[3] As in Venn diagrams of set theory.

occur simultaneously, is also zero. However, if $P(A \cap B) = 0$ the definition of $P(A|B)$ becomes indeterminate. For finite sample spaces this fact is irrelevant because in these spaces the only event with zero probability is the empty set $\emptyset$. However, this is not the case for infinite sample spaces since in these spaces there are events with zero probability other than the empty set.[4] In such a case the above definition of conditional probability has to be modified giving rise to a rather abstract definition of conditional probability which we will not treat here.[5]

We know that for finite sample spaces, the classical definition of probability of an event is given by the number of elements of such an event divided by the total number of elementary events of the sample space. Therefore,

$$P(A \cap B) = N_{AB}/N, \qquad P(B) = N_B/N,$$

where N_{AB} is the number of common elements of the intersection $A \cap B$, N_B is the number of elements of B and N is the total number of elementary events. Hence

$$P(A|B) = \frac{N_{AB}}{N_B}.$$

The fact that in order to obtain $P(A|B)$ we have to divide the number of elements of the intersection of A and B by the number of elements belonging to B, means that the conditional probability under B is equivalent to choosing B as the new sample space, that is, the new sure event. In other words, $P(A|B)$ *is the probability of A given that B has occurred.*

It is not difficult to show that $P(A|B)$ obeys axioms (i)-(iii) and, hence, the conditional probability is also a true measure of probability.

Independence and the product rule

From an intuitive point of view we can affirm that two experiments are independent if the possible results of one experiment do not affect the results of the other. Thus, if A is any event associated with the first experiment and B is another experiment solely associated with the second experiment, independence means that the occurrence of A does not affect the occurrence of B and vice versa. In this sense we say that A and B are independent and

---

[4]It can be shown, for instance, that the probability of randomly choosing a rational number over the entire real line $\mathbb{R}$ is zero, but the set of rational numbers is obviously not empty.

[5]See, however, the discussion in the next section on conditional probabilities for random variables.

the fact that the knowledge of one of them does not affect the probability of the other leads us to write

$$P(A|B) = P(A), \qquad P(B|A) = P(B).$$

From the definition of conditional probability we can combine both expressions and state that two events A and B are independent if

$$P(A \cap B) = P(A)P(B). \tag{1.4}$$

Let us derive an important result related to the conditional probability. From the definition given in Eq. (1.3) we write

$$P(AB) = P(A|B)P(B),$$

where we have used the shorthand notation

$$AB \equiv A \cap B.$$

We can generalize the above expression to include the product (i.e., the intersection) of three or more events. Indeed, taking $BC \equiv B \cap C$ as conditioning event, we have $P(ABC) = P(A|BC)P(BC)$, but $P(BC) = P(B|C)P(C)$. Hence

$$P(ABC) = P(A|BC)P(B|C)P(C).$$

By means of the induction principle one easily generalizes this expression to include an arbitrary number of events. The final result is called the *product rule*:

$$P(A_1 A_2 \cdots A_n) = P(A_1)P(A_2|A_1) \cdots P(A_n|A_1 \cdots A_{n-1}). \tag{1.5}$$

Let us incidentally note that if $A_1, \cdots, A_n$ are independent events then

$$P(A_n|A_1 \cdots A_{n-1}) = P(A_n)$$

for all $n = 2, 3, \ldots$ and the product rule reduces to

$$P(A_1 \cdots A_n) = P(A_1) \cdots P(A_n), \tag{1.6}$$

which is the generalization of Eq. (1.4) to an arbitrary number of independent events.

Bayes theorem

Let $B_1, B_2, B_3, \ldots$ a collection (finite or infinite) of mutually exclusive events covering the whole sample space. That is

$$\bigcup_k B_k = \Omega, \qquad (B_i \cap B_j = \emptyset, i \neq j).$$

Collections of this kind are called *partitions* of Ω. In such a case the probability of an arbitrary event A can be evaluated by means the *total probability formula*:

$$P(A) = \sum_k P(A|B_k)P(B_k). \qquad (1.7)$$

In effect, from the definition of partition we have

$$A = A \cap \Omega = A \cap \left[\bigcup_k B_k \right] = \bigcup_k [A \cap B_k].$$

However, if the B_k's are mutually exclusive also are the events $A \cap B_k$. Thus from axiom (iii) and the definition of conditional probability we get

$$P(A) = \sum_k P(A \cap B_k) = \sum_k P(A|B_k)P(B_k),$$

which is Eq. (1.7).

On the other hand, from the definition of conditional probability we have

$$P(B_i|A) = \frac{P(B_i \cap A)}{P(A)} = \frac{P(A|B_i)P(B_i)}{P(A)}$$

and substituting for Eq. (1.7) we obtain Bayes theorem

$$P(B_i|A) = \frac{P(A|B_i)P(B_i)}{\sum_k P(A|B_k)P(B_k)}. \qquad (1.8)$$

In many random experiments a given event A only occurs when it holds a series of conditions (or "causes") $B_1, B_2, \ldots$ of which we know their probabilities $P(B_k)$. Suppose that the experiment has taken place and that we know the probabilities, $P(A|B_k)$ of A under conditions B_k ($k = 1, 2, \ldots$). Then Bayes theorem allows us to evaluate $P(B_k|A)$, that is, the probability of the "cause" B_k given the "consequence" A of that cause. This interpretation, formulated by Laplace almost two centuries ago, makes Bayes theorem a very important tool for the calculation of probabilities.

1.3 Random variables. Distribution functions

Let us recall that in a given probability space $(\Omega, \mathcal{A}, P)$ the elements of the sample space Ω are not necessarily numbers. In throwing a coin, the elementary events, heads and tails, are not numerical values even though we can always assign, for instance, heads to 1 and tails to 0. The correspondence between elementary events and numbers leads us to one of the most fundamental concepts in probability theory, that of a random variable.

The fact that to each elementary event $\omega \in \Omega$ it corresponds a numerical value coincides with the mathematical concept of function. Thus, in a first approximation, we can say that a random variable is a function X defined on the sample space Ω such that to each elementary event $\omega \in \Omega$ it corresponds a numerical value $X(\omega)$ which can be a real or complex number or even a vector. In what follows we will generally assume that $X(\omega)$ is a real number.[6]

In order to characterize a random variable we need to know not only the values it can take but, and most important, the probability of taking a particular value. Since from the axiomatic construction of probability only events, that is, the elements of the σ-algebra $\mathcal{A}$, have a defined probability, when we ask for the probability that X takes a given value, we have to relate it to events. In this sense we use the notation $\{X = x\}$, where x is any real number, to denote the subset of elementary events $\omega \in \Omega$ such that $X(\omega)$ takes the value x. In a formal notation we write

$$\{X = x\} = \{\omega \in \Omega | X(\omega) = x\}$$

and, analogously,

$$\{X < x\} = \{\omega \in \Omega | X(\omega) < x\}.$$

In general if M is a subset of the real line, then $\{X \in M\}$ is the subset of the sample space defined by

$$\{X \in M\} = \{\omega \in \Omega | X(\omega) \in M\}.$$

In order to assign probabilities to these sets we have to be sure that all of them are events, that is, elements of the σ-algebra of the probability space $(\Omega, \mathcal{A}, P)$. For discrete sample spaces (finite or infinite) the σ-algebra $\mathcal{A}$ formed by all possible subsets of Ω always contains elements of the form

---

[6]We shall use uppercase letters X, Y, Z, ... for random variables whose actual values are unknown beforehand and subject to the laws of probability. We will generally use lowercase letters x, y, z, ... for the possible values that a random variable can take, that is, for deterministic quantities.

$\{X \in M\}$. However, for continuously infinite sample space one cannot be sure that any subset of the form $\{X \in M\}$ belongs to $\mathcal{A}$. In other words, we cannot always be sure that all the values that X can take have a defined probability. It can be proved, however, that if all subsets of the form $\{\omega \in \Omega | X(\omega) < x\}$ belong to $\mathcal{A}$, then all possible values that the random variable can take have a defined probability. Such a condition necessarily implies some restrictions on the type of random variables, in the sense that not any correspondence between elementary events and numbers cannot automatically be considered as a random variable.[7] Fortunately, these restrictions are unimportant in an overwhelming majority of practical cases. We thus have the following definition:

A (real) random variable is any real function X defined on the sample space Ω such that all subsets of the form $\{\omega \in \Omega | X(\omega) < x\} \in \mathcal{A}$ are events for all $x \in \mathbb{R}$.

Distribution functions

Random variables are characterized by their distribution functions. The *distribution function* of a random variable X is the probability

$$F_X(x) = P\{X < x\}. \qquad (1.9)$$

Along with the distribution function we also have the *probability density function* (PDF) which is the derivative of the distribution function:

$$p_X(x) = \frac{d}{dx} F_X(x). \qquad (1.10)$$

Realizing that

$$dF_X(x) = F_X(x + dx) - F_X(x)$$
$$= P\{X < x + dx\} - P\{X < x\} = P\{x < X < x + dx\}$$

we see that the PDF has the following probabilistic interpretation:

$$p_X(x)dx = P\{x < X < x + dx\}, \qquad (1.11)$$

which is very often taken as the definition of the PDF. From this interpretation we see that $p_X(x)$ can be used to evaluate the probability of general events of the form $\{\omega \in \Omega | X(\omega) \in M\}$, where $M \subset \mathbb{R}$ is any subset of the

---

[7]The reader interested on these rather abstract and formal questions may consult Doob's book (1966).

real line. Indeed, since from the definition of integral one can easily realize
that

$$P\{X \in M\} = \int_M P\{x < X < x + dx\},$$

we have

$$P\{X \in M\} = \int_M p_X(x)dx. \tag{1.12}$$

Let us note that since the distribution function $F_X(x)$ is a probability,
it has the property

$$0 \leq F_X(x) \leq 1 \qquad \forall x \in \mathbb{R}.$$

Moreover, from the definition (1.9) we easily see that

$$\lim_{x \to -\infty} F_X(x) = 0, \qquad \lim_{x \to \infty} F_X(x) = 1.$$

On the other hand, the PDF has the following properties:

   (i) It is non negative

$$p_X(x) \geq 0 \qquad \forall x \in \mathbb{R}.$$

   (ii) It is normalized

$$\int_{-\infty}^{\infty} p_X(x)dx = 1.$$

This latter property coming from the fact that $\{X \in \mathbb{R}\}$ is the sure event
and, hence, $P\{X \in \mathbb{R}\} = 1$.

Suppose that a random variable X can only take finite, or a countably
infinite, different values $x_1, x_2, \ldots, x_k, \ldots$ with probabilities

$$p_k = P\{X = x_k\}$$

$(k = 1, 2, 3, \ldots)$. This is a *discrete random variable* and the PDF is given
by

$$p_X(x) = \sum_k p_k \delta(x - x_k),$$

where $\delta(\cdot)$ is Dirac's delta function.[8] Note that due to the normalization
of $p_X(x)$, the probabilities p_k satisfy

$$\sum_k p_k = 1.$$

On the other hand, random variables that can take any value from a
certain interval of the real line are called *continuous random variables* and
their PDF is a continuous function of x.[9]

---

[8]Let us recall that $\delta(x)$ is a generalized function with defining properties

$$\delta(x) = 0, \quad \forall x \neq 0, \qquad \int_{-\infty}^{\infty} \delta(x)dx = 1, \qquad \text{and} \qquad \int_{-\infty}^{\infty} \delta(x)f(x)dx = f(0),$$

for any sufficiently smooth function $f(x)$ vanishing at infinity (see, for instance
Vladimirov, 1984, for more information).

[9]There can be more general situations in which X can take both continuous and discrete
values. In such a case the PDF is a piecewise continuous function.

Examples

We will next present some examples corresponding to discrete and continuous variables.

(1) *Bernoulli trials.*

These are sequences of independent and identical random experiments with only two possible outcomes, usually termed as "success" (with probability p) and "failure" (with probability $q = 1 - p$). When dealing with $N = 1, 2, 3, \ldots$ consecutive Bernoulli trials, each elementary event can be represented by sequences of 1's (successes) and 0's (failures) randomly distributed:

$$\omega = \overbrace{0100 \cdots 1100}^{N}.$$

Note that the probability of the elementary event ω_k for which there are k successes in N trials is

$$P\{\omega_k\} = p^k (1 - p)^{N-k}.$$

Suppose that X is the random variable giving the number of successes in N trials. $X = 0, 1, 2, \ldots, N$ is discrete and, since the number of k successes in N trials is $\binom{N}{k}$, it obeys the *binomial distribution*:

$$p_k = P\{X = k\} = \binom{N}{k} p^k (1 - p)^{N-k}. \tag{1.13}$$

(2) *Poisson distribution.*

Let X be a discrete random variable taking on any positive integer number, $X = 0, 1, 2, \ldots, n, \ldots$. This variable follows a Poisson distribution if

$$p_n = P\{X = n\} = \frac{\lambda^n}{n!} e^{-\lambda}, \tag{1.14}$$

where $n = 1, 2, 3, \ldots$ and $\lambda > 0$ is a positive parameter characterizing the distribution.

(3) *The Gaussian or normal distribution.*

A continuous random variables taking values on the entire real line is said to be Gaussian if the PDF has the form

$$p_X(x) = \frac{1}{\sigma\sqrt{2\pi}} e^{-(x-m)^2/2\sigma^2}, \tag{1.15}$$

where $-\infty < m < \infty$ and $\sigma > 0$ are parameters of the distribution.

(4) *The uniform distribution over* $[a, b]$.

Suppose now that X is a continuous random variable that can take any real value, $-\infty < X < \infty$. It is said that X follows a uniform distribution over the interval (a, b) if the PDF has the form

$$p_X(x) = \begin{cases} \frac{1}{b-a}, & x \in (a, b), \\ 0, & x \notin (a, b). \end{cases} \tag{1.16}$$

(5) *The Gamma distribution.*

A continuous random variables taking values on the half real line $[0, \infty)$ follows the Gamma distribution if the PDF is given by

$$p_X(x) = \begin{cases} \frac{\beta^{1+\alpha}}{\Gamma(1+\alpha)} x^\alpha e^{-x/\beta}, & x > 0, \\ 0, & x \leq 0, \end{cases} \tag{1.17}$$

where $\alpha > -1$ and $\beta > 0$ are constant parameters.

(6) *The exponential distribution.*

A continuous random variables taking values on $[\beta, \infty)$ follows the exponential distribution if the PDF has the form

$$p_X(x) = \begin{cases} \alpha e^{-\alpha(x-\beta)}, & x > \beta, \\ 0, & x \leq \beta, \end{cases} \tag{1.18}$$

where $\alpha > 0$ and $\beta \in (-\infty, \infty)$ are constants.

(7) *The Laplace distribution.*

A continuous random variables taking values on the entire real line $(-\infty, \infty)$ follows the Laplace distribution if the PDF has the form

$$p_X(x) = \frac{\lambda}{2} e^{-\lambda|x|}, \qquad (\lambda > 0). \tag{1.19}$$

(8) *The Cauchy distribution.*

A continuous random variables taking values on the entire real line $(-\infty, \infty)$ follows the Cauchy distribution if

$$p_X(x) = \frac{1}{\pi(1 + x^2)}. \tag{1.20}$$

All of the above densities are nonnegative and normalized. In fact, any nonnegative function $f(x) \geq 0$ verifying

$$\int_{-\infty}^{\infty} f(x) dx = 1$$

is the density of some probability distribution.

1.4   Multidimensional random variables

We now define the concept of random vector, also called multidimensional random variable. Suppose we have a set of $N = 1, 2, 3, \ldots$ random variables $X_k = X_k(\omega)$, $(k = 1, 2, \ldots, N)$, defined in the same space of probability $(\Omega, \mathcal{A}, P)$, the vector

$$\mathbf{X} = (X_1, \ldots, X_N)$$

is called a N-dimensional random variable.

When all the X_k's are discrete random variables, we deal with a discrete random vector, while if the X_k's are continuous we have a N-dimensional continuous random variable. We can have, however, mixed situations in which some of X_k's are discrete while the rest are continuous although these cases are infrequent.

Let us note that subsets of Ω defined by the intersection of N events of the form $\{\omega \in \Omega | X_k(\omega) < x\}$ $(k = 1, 2, \ldots, N)$, that is,

$$\{X_1 < x_1, \ldots, X_N < x_N\} = \{X_1 < x_1\} \cap \cdots \cap \{X_N < x_N\},$$

belong to the σ-algebra $\mathcal{A}$. They are, therefore, events and have, according to axiomatics, defined probabilities.

Generalizing the one-dimensional case, we define the multivariate distribution function of the random vector $\mathbf{X}$ as:

$$F_N(x_1, \ldots, x_N) = P\{X_1 < x_1, \ldots, X_N < x_N\}. \tag{1.21}$$

The probability density function of $\mathbf{X}$ is then given by

$$p_N(\mathbf{x}) = \frac{\partial^N F_N(\mathbf{x})}{\partial x_1 \cdots \partial x_N}, \tag{1.22}$$

where $\mathbf{x} = (x_1, \ldots, x_N) \in \mathbb{R}^N$.

Properties

Some relevant properties of the multivariate PDF are:

(1)

$$p_N(\mathbf{x}) dx_1 \cdots dx_N = P\{x_1 \leq X_1 < x_1 + dx_1, \ldots, x_N \leq X_N < x_N + dx_N\}.$$

(2)

$$p_N(\mathbf{x}) \geq 0, \qquad \forall \mathbf{x} \in \mathbb{R}^N.$$

(3)
$$F_N(\mathbf{x}) = \int_{-\infty}^{x_1} \cdots \int_{-\infty}^{x_N} p(x_1', \ldots, x_N') dx_1' \cdots dx_N'.$$

(4)
$$\int_{-\infty}^{\infty} \cdots \int_{-\infty}^{\infty} p_N(x_1, \ldots, x_N) dx_1 \cdots dx_N = 1.$$

(5) The probability that $\mathbf{X}(\omega) \in \mathbf{M}$ ($\mathbf{M} \subset \mathbb{R}^N$) is

$$P\{\mathbf{X} \in \mathbf{M}\} = \int \cdots \int_{\mathbf{M}} p(x_1, \ldots, x_N) dx_1 \cdots dx_N,$$

where the N-fold integral extends over the region $\mathbf{M}$ of $\mathbb{R}^N$.

Some examples

(1) Perhaps the most relevant example of a continuous random vector is provided by the *N-dimensional Gaussian distribution*:

$$p_N(\mathbf{x}) = \frac{1}{(2\pi)^{n/2}\sqrt{\det A}} \exp\left[-\frac{1}{2}\sum_{k,l}^{N}(\mathbf{x} - \mathbf{m})_k (\mathbf{A}^{-1})_{kl}(\mathbf{x} - \mathbf{m})_l\right], \quad (1.23)$$

where the N-dimensional vector $\mathbf{m} \in \mathbb{R}^N$ and the $N \times N$ matrix $\mathbf{A} = (A_{kl})$ are parameters of the distribution.

For a two-dimensional random variable $\mathbf{X} = (X, Y)$ we can write this density more explicitly and have the *two-dimensional normal law*:

$$p(x, y) = \frac{1}{2\pi\sigma_1\sigma_2\sqrt{1 - r^2}} \exp\left\{-\frac{1}{2(1 - r^2)}\left[\left(\frac{x - a}{\sigma_1}\right)\right.\right. \quad (1.24)$$
$$\left.\left. -\frac{2r}{\sigma_1\sigma_2}(x - a)(x - b) + \left(\frac{y - b}{\sigma_2}\right)\right]\right\},$$

where $-1 \leq r \leq 1$ is the *correlation coefficient* between X and Y. Note that if $r = 0$ then $p(x, y) = p(x)p(y)$ –being both, $p(x)$ and $p(y)$, Gaussian densities– so that X and Y are independent.

(2) Another example is given by the *N-dimensional uniform distribution*. If V is the N-dimensional volume of the domain $G \subset \mathbb{R}^N$, the uniform density is

$$p_N(\mathbf{x}) = \begin{cases} 1/V & \mathbf{x} \in G, \\ 0, & \mathbf{x} \notin G. \end{cases} \quad (1.25)$$

Marginal distributions

Suppose we have a N-dimensional random variable with distribution function $F_N(x_1, \ldots, x_N)$. Realizing that (i) the event $\{X_k < \infty\} = \Omega$ is the sure event for any $k = 1, \ldots, N$ and (ii) $A \cap \Omega = A$ for any subset $A \subset \Omega$, we easily see from the definition of F_N given in Eq. (1.21) that if, for instance, $x_N \to \infty$ we have

$$\lim_{x_N \to \infty} F_N(x_1, \ldots, x_N) = P\{X_1 < x_1, \ldots, X_{N-1} < x_{N-1}, X_N < \infty\}$$

$$= P\{X_1 < x_1, \ldots, X_{N-1} < x_{N-1}\}.$$

That is,

$$\lim_{x_N \to \infty} F_N(x_1, \ldots, x_N) = F_{N-1}(x_1, \ldots, x_{N-1}),$$

where F_{N-1} is a marginal distribution (of order $N - 1$) of F_N. This procedure can be extended to include several values of the x_k's. Thus if we let a number $m < N$ of the x's go to infinity, we will obtain the marginal distribution function F_{N-m} of the $N - m$ remaining random variables.

As to probability density functions, obtaining marginal densities amounts to integrate with respect some of the x_k's. Indeed from properties (1) and (3) above we see that

$$dx_1 \int_{-\infty}^{\infty} dx_2 \cdots \int_{-\infty}^{\infty} dx_N p_N(x_1, x_2, \ldots, x_N) = P\{x_1 \leq X_1 < x_1 + dx_1\}$$

$$= p_1(x_1) dx_1,$$

so that

$$p_1(x_1) = \int_{-\infty}^{\infty} dx_2 \cdots \int_{-\infty}^{\infty} dx_N p_N(x_1, x_2, \ldots, x_N)$$

is the one-dimensional marginal density of the N-dimensional random variable $(X_1, X_2, \ldots, X_N)$. In an analogous way one can obtain higher-order marginal densities $p_2(x_1, x_2), \ldots, p_{N-1}(x_1, \ldots, x_{N-1})$.

Independence and conditioning

The concept of independence, which we have introduced in the previous section for events, can be extended to random variables. From an intuitive point of view we can say that two random variables X and Y, defined in the same space of probability, are independent when the values taken by X do not affect the values taken by Y and vice versa. Thus, in particular, the event $\{x \leq X < x+dx\}$ must be independent of the event $\{y \leq Y < y+dy\}$, where x and y are arbitrary real numbers and, hence,

$$P\{x \leq X < x + dx, y \leq Y < y + dy\}$$
$$= P\{x \leq X < x + dx\}P\{y \leq Y < y + dy\}.$$

Therefore, the joint density factorizes as

$$p_{XY}(x,y) = p_X(x)p_Y(y).$$

In general, N random variables, $X_1, \ldots, X_N$, defined in the same space of probability are said to be *independent* if their joint probability density function factorizes into a product of individual densities:

$$p_{X_1,\ldots,X_N}(x_1,\ldots,x_N) = p_{X_1}(x_1) \cdots p_{X_N}(x_N).$$

We can also define the *conditional PDF* of a random variable X, knowing that another random variable Y has taken a known value y, as the quotient

$$p(x|y) = \frac{p_{XY}(x,y)}{p_Y(y)}. \tag{1.26}$$

Let us note that if X and Y are independent, then

$$p(x|y) = p_X(x) \quad \text{and} \quad p(y|x) = p_Y(y).$$

It can be easily proved (see Gnedenko, 1976) that the conditional PDF satisfies the *total probability formula*:

$$p_X(x) = \int_{-\infty}^{\infty} p(x|y)p_Y(y)dy$$

and *Bayes' theorem*

$$p(y|x) = \frac{p(x|y)p_Y(y)}{\displaystyle\int_{-\infty}^{\infty} p(x|y)p_Y(y)dy}.$$

Functions of random variables

Consider a random variable X defined in some space of probability and let $g(x)$ be an arbitrary function. The expression $Z = g(X)$ is also a random variable defined in the same space and whose distribution function is given by

$$F_Z(z) \equiv P\{Z < z\} = P\{g(X) < z\}$$
$$= \int_{g(x)<z} p_X(x)dx = \int_{-\infty}^{\infty} \Theta[z - g(x)]p_X(x)dx,$$

where $\Theta(\cdot)$ is the Heaviside step function,

$$\Theta(z) = \begin{cases} 1, & z > 0, \\ 0, & z < 0. \end{cases} \tag{1.27}$$

The probability density function of Z will be given by the derivative of $F_Z(z)$ with respect to z. Taking into account that the Dirac delta function is the derivative of the Heaviside function[10]

$$\delta(z) = \frac{d}{dz}\Theta(z), \tag{1.28}$$

we have

$$p_Z(z) = \int_{-\infty}^{\infty} \delta[z - g(x)] p_X(x) dx. \tag{1.29}$$

This formula can be extended to any number of random variables. Thus if $\mathbf{X} = (X_1, \ldots, X_N)$ is a N-dimensional random variable and $\mathbf{Z} = (Z_1, \ldots, Z_m)$ is an m-dimensional random variable such that

$$Z_1 = g_1(\mathbf{X}), \ldots, Z_m = g_m(\mathbf{X}),$$

then

$$p_{\mathbf{Z}}(z_1, \ldots z_m) \tag{1.30}$$
$$= \int_{-\infty}^{\infty} \cdots \int_{-\infty}^{\infty} \delta[z_1 - g_1(\mathbf{x})] \cdots \delta[z_m - g_m(\mathbf{x})] p_{\mathbf{X}}(x_1, \ldots x_N) dx_1 \cdots dx_N.$$

For example, if $Z = X_1 + X_2$, is the sum of two random variables with joint density $p(x_1, x_2)$, then

$$p_Z(z) = \int_{-\infty}^{\infty} dx_1 \int_{-\infty}^{\infty} \delta(z - x_1 - x_2) p(x_1, x_2) dx_2$$
$$= \int_{-\infty}^{\infty} p(x_1, z - x_1) dx_1.$$

If, in addition, X_1 and X_2 are independent and identically distributed, we have

$$p(x_1, x_2) = p(x_1)p(x_2),$$

and the PDF of the sum is given by the integral:

$$p_Z(z) = \int_{-\infty}^{\infty} p(x_1)p(z - x_1) dx_1, \tag{1.31}$$

which is often called convolution theorem for the sum of two independent random variables.

---

[10]In the sense of generalized functions. (see Gelfand and Shilov, 2016a).

1.5   Averages and moments

Let us start with a discrete random variable taking on values $x_1, x_2, \ldots, x_n, \ldots$ with probabilities $p_k = P\{X = x_k\}$ $(k = 1, 2, \ldots)$. It is clear, from an intuitive point of view, that the *mean* or *average value* (also called *expectation*) of X is

$$\langle X \rangle = \sum_k p_k x_k.$$

Observe that in terms of its probability density function

$$p_X(x) = \sum_k p_k \delta(x - x_k),$$

the expectation can be written as

$$\langle X \rangle = \int_{-\infty}^{\infty} x p_X(x) dx. \qquad (1.32)$$

We will take Eq. (1.32) as the definition of the average of X regardless the discrete or continuous character of the random variable.

Let us now evaluate the mean value of a function $Z = g(X)$ of a random variable. From the definition (1.32) and using (1.29) we write

$$\langle Z \rangle = \int_{-\infty}^{\infty} z p_Z(z) dz = \int_{-\infty}^{\infty} z dz \int_{-\infty}^{\infty} \delta[z - g(x)] p_X(x) dx$$

$$= \int_{-\infty}^{\infty} p_X(x) dx \int_{-\infty}^{\infty} z \delta[z - g(x)] dz = \int_{-\infty}^{\infty} p_X(x) g(x) dx.$$

We have, therefore, proved that the expectation of a function of a random variable is given by

$$\langle g(X) \rangle = \int_{-\infty}^{\infty} g(x) p_X(x) dx. \qquad (1.33)$$

Moments

Particularizing in Eq. (1.33) to the function $g(X) = X^r$, we have the *moment* of order r:

$$\langle X^r \rangle = \int_{-\infty}^{\infty} x^r p_X(x) dx. \qquad (1.34)$$

When $g(X) = (X - \langle X \rangle)^r$, we have the *central moment* of order r:

$$\langle (X - \langle X \rangle)^r \rangle = \int_{-\infty}^{\infty} (x - m)^r p_X(x) dx, \qquad (1.35)$$

where $m = \langle X \rangle$.

One important special case is provided by the *second central moment* or *variance*:

$$\text{Var}\{X\} = \langle [X - \langle X \rangle]^2 \rangle = \langle X^2 \rangle - \langle X \rangle^2. \tag{1.36}$$

The square root of the variance is called *standard deviation*.

Generalizing to random vectors

The above definitions can be easily extended to multidimensional random variables $\mathbf{X} = (X_1, \dots, X_N)$. Thus, for instance, the N-dimensional moments of order r are

$$\langle X_1^{r_1} \cdots X_N^{r_N} \rangle = \int_{-\infty}^{\infty} \cdots \int_{-\infty}^{\infty} x_1^{r_1} \cdots x_N^{r_N} p_{\mathbf{X}}(x_1, \dots, x_N) dx_1 \cdots dx_N, \tag{1.37}$$

where $r_1 + \cdots + r_N = r$. From this definition we easily see that if $X_1, \dots, X_n$ are a set of n independent random variables then

$$\langle X_1^{r_1} \cdots X_n^{r_n} \rangle = \langle X_1^{r_1} \rangle \cdots \langle X_n^{r_n} \rangle.$$

We can also generalize the average formula, Eq. (1.33), to include functions of several random variables. Let $g(\mathbf{X})$ be a function of the N dimensional random vector $\mathbf{X} = (X_1, \dots, X_N)$ having joint PDF $p_{\mathbf{X}}(x_1, \dots, x_N)$. The average of such a function is defined by

$$\langle g(\mathbf{X}) \rangle = \int_{-\infty}^{\infty} \cdots \int_{-\infty}^{\infty} g(x_1, \cdots, x_N) p_{\mathbf{X}}(x_1, \dots, x_N) dx_1 \cdots dx_N. \tag{1.38}$$

It is not difficult to convince ourselves that in the particular case when

$$g(\mathbf{X}) = c_1 X_1 + \cdots + c_2 X_N$$

where $c_1, \dots, c_N$ are arbitrary real (or complex) numbers, Eq. (1.38) yields (recall discussion above on marginal distributions)

$$\langle c_1 X_1 + \cdots c_2 X_N \rangle = c_1 \langle X_1 \rangle + \cdots + c_2 \langle X_N \rangle, \tag{1.39}$$

and averaging is a linear operation.

Examples

(1) Suppose $X = 1, 2, 3, \dots$ follows the *Poisson distribution* [cf. Eq. (1.14)]:

$$p_n = e^{-\lambda} \frac{\lambda^n}{n!}.$$

Then

$$\langle X \rangle = \sum_{n=1}^{\infty} n p_n = e^{-\lambda} \sum_{n=1}^{\infty} n \frac{\lambda^n}{n!} = \lambda e^{-\lambda} \sum_{n=1}^{\infty} \frac{\lambda^{n-1}}{(n-1)!},$$

but making the change $n - 1 \to n$ in the sum we have

$$\sum_{n=1}^{\infty} \frac{\lambda^{n-1}}{(n-1)!} = \sum_{n=0}^{\infty} \frac{\lambda^n}{n!} = e^{\lambda}.$$

Hence

$$\langle X \rangle = \lambda.$$

In an analogous way, the second moment is

$$\langle X^2 \rangle = \sum_{n=1}^{\infty} n^2 p_n = e^{-\lambda} \sum_{n=1}^{\infty} n \frac{\lambda^n}{(n-1)!} = e^{-\lambda} \sum_{n=0}^{\infty} (n+1) \frac{\lambda^{n+1}}{n!}$$

$$= \lambda e^{-\lambda} \left[\sum_{n=0}^{\infty} n \frac{\lambda^n}{n!} + \sum_{n=0}^{\infty} \frac{\lambda^n}{n!} \right] = \lambda e^{-\lambda} \left[\lambda \sum_{n=1}^{\infty} \frac{\lambda^{n-1}}{(n-1)!} + \sum_{n=0}^{\infty} \frac{\lambda^n}{n!} \right]$$

$$= \lambda^2 + \lambda.$$

The variance of the Poisson distribution is thus given by

$$\mathrm{Var}\{X\} = \lambda$$

and equals the mean of the distribution.

(2) Suppose $X \in \mathbb{R}$ has a *normal distribution* of parameters m and σ, that is

$$p_X(x) = \frac{1}{\sqrt{2\pi\sigma^2}} e^{-(x-m)^2/2\sigma^2},$$

then

$$\langle X \rangle = \frac{1}{\sqrt{2\pi\sigma^2}} \int_{-\infty}^{\infty} x e^{-(x-m)^2/2\sigma^2} dx = m$$

and

$$\mathrm{Var}\{X\} = \frac{1}{\sqrt{2\pi\sigma^2}} \int_{-\infty}^{\infty} (x-m)^2 e^{-(x-m)^2/2\sigma^2} dx = \sigma^2.$$

Therefore, m is the mean value and σ is the standard deviation around the mean value.

(3) For the *Cauchy distribution*

$$p_X(x) = \frac{1}{\pi(1+x^2)}, \qquad (-\infty < x < \infty),$$

we have

$$\langle X \rangle = \frac{2}{\pi} \int_0^{\infty} \frac{x}{1+x^2} dx = \frac{1}{\pi} \ln(1+x^2) \Big|_o^{\infty} = \infty$$

and the Cauchy distribution has no expectation. In fact, the Cauchy distribution has no finite moment of any order.

Probability as an expected value

The probability density function of any random variable X can be written as an average over all possible values that X can take. To prove this statement we start from the following identity

$$g(X) = \int_{-\infty}^{\infty} g(x)\delta(X - x)dx,$$

where $\delta(\cdot)$ is the Dirac function.

Suppose now that X is a random variable and that x is any value that X can take. From the above identity we see that the average of $g(X)$ can be written as (recall expectation is a linear operation)

$$\langle g(X) \rangle = \left\langle \int_{-\infty}^{\infty} g(x)\delta(X - x)dx \right\rangle = \int_{-\infty}^{\infty} g(x)\langle \delta(X - x)\rangle dx.$$

However, from the definition of expectation given in (1.33),

$$\langle g(X) \rangle = \int_{-\infty}^{\infty} g(x)p_X(x)dx,$$

we conclude that

$$p_X(x) = \langle \delta(X - x) \rangle, \tag{1.40}$$

and the PDF of X is an average over all possible values X can take.

Let us note that since the PDF is the derivative of the distribution function, $F_X(x)$, and the delta function is the derivative of the Heaviside step function [cf. Eqs. (1.27) and (1.28)] we can write $F_X(x)$ also as an average

$$F_X(x) = \langle \Theta(X - x) \rangle. \tag{1.41}$$

These expressions are readily generalized to include multidimensional variables, $\mathbf{X} = (X_1, \ldots, X_N)$. Thus we have

$$p_{\mathbf{X}}(x_1, \ldots, x_N) = \langle \delta(X_1 - x_1) \cdots \delta(X_N - x_N) \rangle, \tag{1.42}$$

and

$$F_{\mathbf{X}}(x_1, \ldots, x_n) = \langle \Theta(X_1 - x_1) \cdots \Theta(X_N - x_N) \rangle. \tag{1.43}$$

1.6   Characteristic functions

A particular case of expectation, of great importance in the theory of probability and random processes, is the *characteristic function* of a random variable. This is defined as the average value of the complex-valued function

$$g(X) = e^{-i\omega X} = \cos\omega X - i\sin\omega X,$$

where $\omega \in \mathbb{R}$ is an arbitrary real number.

We thus define

$$\varphi_X(\omega) \equiv \langle e^{i\omega X} \rangle = \int_{-\infty}^{\infty} e^{i\omega x} p_X(x)dx. \tag{1.44}$$

Note that the characteristic function of X is the Fourier transform of the probability density function[11]

$$\varphi_X(\omega) = \mathcal{F}\{p_X(x)\}.$$

From the Fourier inversion formula we have

$$p_X(x) = \frac{1}{2\pi}\int_{-\infty}^{\infty} e^{-i\omega x}\varphi_X(\omega)d\omega, \tag{1.45}$$

showing that the characteristic function univocally determines its probability density function and by extension the entire probability distribution of the random variable (Lukacs, 1970).

We next list some relevant properties of the characteristic function:

(1) $\varphi_X(0) = 1$; $|\varphi(\omega)| \le 1 \; \forall \omega \in \mathbb{R}$.
(2) If $\varphi_X^*(\omega)$ denotes the complex conjugate of $\varphi_X(\omega)$, then

$$\varphi_X(-\omega) = \varphi_X^*(\omega).$$

(3) For the linear change $Z = aX + b$ $(a, b \in \mathbb{R})$, we have

$$\varphi_Z(\omega) = e^{i\omega b}\varphi_X(a\omega).$$

(4) The characteristic function of the sum of independent random variables is the product of the individual characteristic functions:

$$\varphi_Z(\omega) = \varphi_{X_1}(\omega)\cdots\varphi_{X_n}(\omega),$$

where $Z = X_1 + \cdots + X_n$.

The proof of these properties is a simple exercise for the reader since they are a direct consequence of the definition of characteristic function (Lukacs, 1970).

---

[11]For this reason it is sometimes used the standard notation for Fourier transforms, $\tilde{p}_X(\omega)$, instead of $\varphi_X(\omega)$.

Examples

(1) Suppose $X = \{x_1, x_2, \ldots, x_n, \ldots\}$ is a discrete random variable with $p_n = \text{Prob}\{X = x_n\}$. We have seen that the probability density function of such a variable can be written as

$$p_X(x) = \sum_n p_n \delta(x - x_n)$$

and we conclude from (1.44) that the characteristic function will be given by the sum

$$\varphi_X(\omega) = \sum_n p_n e^{i\omega x_n}. \tag{1.46}$$

As a particular case, suppose $X = 0, 1, 2, \ldots$ is a Poisson variable for which

$$p_n = e^{-\lambda} \frac{\lambda^n}{n!}.$$

We have

$$\varphi_X(\omega) = e^{-\lambda} \sum_{n=0}^{\infty} e^{i\omega n} \frac{\lambda^n}{n!} = e^{-\lambda} \sum_{n=0}^{\infty} \frac{\left(\lambda e^{i\omega}\right)^n}{n!},$$

that is,

$$\varphi_X(\omega) = \exp\left[\lambda\left(e^{i\omega} - 1\right)\right]. \tag{1.47}$$

(2) It is a simple matter to show that for the Gaussian distribution,

$$p_X(x) = \frac{1}{\sigma\sqrt{2\pi}} e^{-(x-m)^2/2\sigma^2},$$

the characteristic function is also a Gaussian function:

$$\varphi_X(\omega) = \exp\left\{-\frac{1}{2}\sigma^2\omega^2 + im\omega\right\}. \tag{1.48}$$

(3) The characteristic function allows us to easily prove some relevant results concerning averages of exponentials of random variables. Thus, since the characteristic function of X is defined as

$$\varphi_X(\omega) = \left\langle e^{i\omega X} \right\rangle$$

we readily see that

$$\left\langle e^{\pm X} \right\rangle = \varphi_X(\omega = \mp i).$$

Let us apply this useful formula to the case of a *Gaussian variable with zero mean*. From Eq. (1.48) with $m = 0$ we have

$$\left\langle e^{\pm X} \right\rangle = e^{\sigma^2/2}$$

but since $\sigma^2 = \langle X^2 \rangle$ we can write the appealing expression

$$\left\langle e^{\pm X} \right\rangle = e^{\langle X^2 \rangle/2}. \tag{1.49}$$

Moments and cumulants

From a practical point of view the knowledge of the characteristic function is very convenient because it allows for a simple calculation of the integer moments of X by means of the derivatives of $\varphi_X(\omega)$ instead of integrals involving the density $p_X(x)$. Indeed from the definition (1.44), we easily see that

$$\langle X^n \rangle = \frac{1}{i^n} \varphi_X^{(n)}(0), \tag{1.50}$$

($n = 1, 2, 3, \dots$), where $\varphi_X^{(n)}(0)$ is the nth derivative of the characteristic function evaluated at $\omega = 0$.

The expansion of $\varphi_X(\omega)$ around $\omega = 0$ and the use of Eq. (1.50) yield

$$\varphi_X(\omega) = 1 + \sum_{n=1}^{\infty} \frac{i^n \langle X^n \rangle}{n!} \omega^n, \tag{1.51}$$

provided that all integer moments of X exist.

A function closely related to $\varphi_X(\omega)$ is the so-called *cumulant generating function* $\Psi_X(\omega)$. This function is defined as the logarithm of the characteristic function,

$$\Psi_X(\omega) = \ln \varphi_X(\omega). \tag{1.52}$$

The name "cumulant generating function" is due to the fact that the successive derivatives of $\Psi_X(\omega)$ evaluated at $\omega = 0$ are called cumulants (also called semi-invariants) of the random variable. Specifically

$$K_n \equiv \frac{1}{i^n} \Psi_X^{(n)}(0) \tag{1.53}$$

($n = 1, 2, 3, \dots$) is the *cumulant* of order n of the random variable X.

Cumulants and moments are related to each other, although the relationship becomes quite intricate as n grows. However, as we can easily see from Eq. (1.52) for small values of n this relationship is simpler. Thus,

$$K_1 = \langle X \rangle, \qquad K_2 = \langle X^2 \rangle - \langle X \rangle^2.$$

The first cumulant is then the average value while the second cumulant is the variance σ_X^2. For $n = 3$ we have

$$K_3 = \langle X^3 \rangle - 3\langle X^2 \rangle \langle X \rangle + 2\langle X \rangle^2,$$

with increasing complexity as n grows.

Multidimensional variables

Let us briefly comment on the characteristic function of a multidimensional random variable $\mathbf{X} = (X_1, \ldots, X_N)$. As to the one-dimensional case this is a function of N real variables defined by the expectation

$$\varphi_{\mathbf{X}}(\boldsymbol{\omega}) = \left\langle e^{i\boldsymbol{\omega}\cdot\mathbf{X}} \right\rangle,$$

where $\boldsymbol{\omega} = (\omega_1, \ldots, \omega_N) \in \mathbb{R}^N$ and

$$\boldsymbol{\omega} \cdot \mathbf{X} = \omega_1 X_1 + \cdots + \omega_N X_N.$$

Hence, the characteristic function is the N-dimensional Fourier transform of the probability density function

$$\varphi_{\mathbf{X}}(\omega_1, \ldots, \omega_N)$$
$$= \int_{-\infty}^{\infty} \cdots \int_{-\infty}^{\infty} \exp\left(i \sum_{k=1}^{N} \omega_k x_k \right) p_{\mathbf{X}}(x_1, \ldots, x_N) dx_1 \cdots dx_N,$$

and the inverse transform yields the PDF

$$p_{\mathbf{X}}(x_1, \ldots, x_N)$$
$$= \frac{1}{(2\pi)^N} \int_{-\infty}^{\infty} \cdots \int_{-\infty}^{\infty} \exp\left(-i \sum_{k=1}^{N} \omega_k x_k \right) \varphi_{\mathbf{X}}(\omega_1, \ldots, \omega_N) d\omega_1 \cdots d\omega_N.$$

1.7 Limit laws. The central limit theorem

Some of the most important results in the theory of probability refer to the asymptotic behavior of sequences and sums of random variables, that is, to the limit laws of an unlimited number of random variables. In this section we summarize some of these asymptotic results which are known under the collective name of *the law of large numbers*. We will also mention other limit laws known generically as *the central limit theorem*.[12]

Sequences of random variables and their convergence

We define a sequence of random variables as any infinite but countable set, $X_1, X_2, \ldots, X_n, \ldots$, of random variables defined in the same space of probability.

---

[12]We refer the reader to the literature, for detailed proofs (see, for instance Gnedenko, 1976).

Let us recall that a random variable $X = X(\omega)$ is a function of the elementary events. Therefore, for each $\omega \in \Omega$ the sequence of real numbers (we assume real random variables)

$$X_1(\omega), X_2(\omega), \ldots, X_n(\omega), \ldots$$

may converge or diverge depending on the particular elementary event chosen. In other words, the existence of the limit of the sequence will depend of the elementary events. The same sequence of random variables may converge for some ω's but diverge for others. All of this leads to define five types of convergence, which we briefly summarize.

(1) *Sure convergence*

A sequence of random variables, $X_1, X_2, \ldots, X_n, \ldots$, converges *surely* or *everywhere* towards a random variable X, and we denote it by $X_n \overset{\mathrm{s}}{\longrightarrow} X$, if

$$\lim_{n \to \infty} X_n(\omega) = X(\omega), \quad \text{for all } \omega \in \Omega.$$

Note that in this case the subset of Ω for which $X_n \overset{\mathrm{s}}{\longrightarrow} X$ is precisely the whole sample space Ω.

Although this is the strongest kind of convergence for random variables and implies all other types, it is rarely used in practice.

(2) *Almost certain convergence*

The sequence X_n is said to *converge almost certainly*, or *almost everywhere*, or *strongly*, or *with probability* 1 towards X if

$$P\left\{ \lim_{n \to \infty} X_n = X \right\} = 1,$$

that is to say, the subset of Ω for which $X_n \overset{\mathrm{a.s.}}{\longrightarrow} X$ has probability 1.

(3) *Convergence in probability*

We say that a sequence of random variables $X_1, X_2, \ldots, X_n, \ldots$ *converges in probability* towards a random variable X and write

$$X_n \overset{\mathrm{P}}{\longrightarrow} X$$

(or, alternatively, p-lim $X_n = X$) if

$$\lim_{n \to \infty} P\{|X_n - X| > \epsilon\} = 0, \quad \forall \epsilon > 0.$$

In other words, for any fix number $\epsilon > 0$, the probability that X_n and X differ in a quantity greater than ϵ goes to 0 as n increases.

This is a weaker concept of converges than almost sure convergence since a.s. convergence implies convergence in probability but not vice versa (Laha and Rohatgi, 1979).

(4) *Mean square convergence*

The sequence X_n converges in *mean square* to X and write $X_n \xrightarrow{\text{m.s.}} X$ (also, ms-lim $X_n = X$) if $\langle |X_n|^2 \rangle$ and $\langle |X|^2 \rangle$ exist and

$$\lim_{n \to \infty} \langle |X_n - X|^2 \rangle = 0.$$

This definition can be generalized to include the convergence in the *rth mean* $(r \geq 1)$ for which

$$\langle |X_n - X|^r \rangle \longrightarrow 0,$$

as $n \to \infty$.

(5) *Convergence in distribution*

We say that X_n *converge in distribution* to X and write $X_n \xrightarrow{\text{d}} X$ if

$$\lim_{n \to \infty} F_n(x) = F(x),$$

where $F_n(x)$ and $F(x)$ are the distribution functions of X_n and X respectively and the equality holds for any $x \in \mathbb{R}$ where $F(x)$ is continuous. This definition can be extended to include the PDF:

$$\lim_{n \to \infty} p_n(x) = p(x),$$

or the characteristic function:

$$\lim_{n \to \infty} \varphi_n(\omega) = \varphi(\omega).$$

These three cases are subsumed under the name of *convergence in law* and, although they constitute the weakest criteria for convergence, they are most useful because they are instrumental for the central limit theorem.

Relationship between limits

It can be shown the following implications (Laha and Rohatgi, 1979)

Almost certain convergence $\implies$ Convergence in probability
Mean square convergence $\implies$ Convergence in probability
Convergence in probability $\implies$ Convergence in distribution.

The law of large numbers and the central limit theorem

Starting from any sequence $X_1, X_2, \ldots, X_n, \ldots$ of random variables we build the random sequence of partial sums:

$$S_n = \sum_{k=1}^{n} X_k, \qquad (n = 1, 2, 3, \ldots).$$

The generic denomination of "law of large numbers" refers to a collection of theorems dealing with the asymptotic behavior of S_n as $n \to \infty$, that is to say, with infinite sums of random variables. The overall problem consists in obtaining appropriate conditions that establish the existence of limits of the form

$$\lim_{n \to \infty} \left(\frac{S_n}{b_n} - a_n \right),$$

where a_n and b_n are *non-random* sequences of real numbers. We thus see that *the laws of large numbers set the conditions under which sums of random variables converge to non-random limits*. If the existence of such limits is shown to hold in probability we have a "weak law of large numbers", while if S_n/b_n converges almost surely to a_n we have a "strong law of large numbers".

Suppose that $b_n = n$ and since S_n/n is the (random) average of random events represented by $X_1, X_2, \ldots, X_n$, the laws of large numbers ensure stable (i.e., non-random) results for these random averages. We next present two of the most used results (see Gnedenko, 1976, for further details).

Chebyshev's law of large numbers

  If $X_1, X_2, \ldots, X_n, \ldots$ are a sequence of independent random variables having finite variances bounded by one and the same constant $\gamma > 0$,

$$\sigma_n^2 = \left\langle |X_n - \langle X_n \rangle^2 \right\rangle \leq \gamma, \quad \forall n;$$

then, as $n \to \infty$,

$$\frac{S_n - \langle S_n \rangle}{n} \xrightarrow{\text{P}} 0.$$

  A direct consequence of this result is

Bernoulli's classical law of large numbers:

  Let $X_1, X_2, \ldots, X_n, \ldots$ be a sequence of random variables with common mean $\langle X_n \rangle = m$ and finite variances bounded by the same constant $\sigma_n^2 \leq \gamma$ $(n = 1, 2, 3, \ldots)$. Then, as $n \to \infty$

$$\frac{1}{n}(X_1 + X_2 + \cdots + X_n) \xrightarrow{\text{P}} m.$$

This result can also be proved to hold almost everywhere, that is,

$$\frac{1}{n}(X_1 + X_2 + \cdots + X_n) \xrightarrow{\text{a.s.}} m,$$

which constitutes a strong law of large numbers known as *Borel's theorem*.

The laws of large numbers establish the conditions under which sums of random variables converge to some non-random value. They do not, however, specify which are the probabilities of such sums. The so-called "central limit theorem" is the generic name of a collection of results answering this question. We present only one of these results that has a sufficient generality.

The Central limit theorem

If $X_1, X_2, \ldots, X_n, \ldots$ is a sequence of independent random variables with common mean m and finite variance σ^2, then the probability density function $p_n(x)$ of the random variable

$$Z_n = \frac{1}{\sigma\sqrt{n}}(X_1 + \cdots + X_n - nm),$$

converges as $n \to \infty$ to the Gaussian density

$$\lim_{n \to \infty} p_n(x) = \frac{1}{2\sqrt{\pi}}e^{-x^2/2}.$$

This limit theorem can be easily extended to include series of independent (or, even, weakly dependent) random variables each of them with different mean values and variances, m_n and σ_n respectively. The essential condition is the finiteness of these parameters (see Gnedenko, 1976).

The central limit theorem explains the great importance of the Gaussian distribution in Nature. Indeed, many natural phenomena are the result of adding a great number of small and independent random inputs. In these cases, as long as the random inputs have finite variance, the resulting macroscopic phenomena are Gaussian. There are, however, situations in which random inputs do not posses finite variances, in these cases the limiting distribution is not Gaussian. We shall briefly comment on these cases within the theory of stable distributions.

Stable distributions

Let $X_1, \ldots, X_n$ be a collection of independent and identically distributed random variables with common probability density function $p_n(x) = p(x)$ for all $n = 1, 2, \ldots$ and $x \in \mathbb{R}$. We denote by X a random variable whose probability distribution is also $p(x)$. Let us observe that, since the X_n's have a common density given by $p(x)$, they can be seen as independent "identical copies" of the same random variable X.

The probability distribution represented by $p(x)$ is said to be *stable* if there exist constants $c_n > 0$ and $d_n \in \mathbb{R}$ such that the sum $X_1 + \cdots + X_n$ has the same probability distribution than $c_n X + d_n$. This can be symbolically written as

$$X_1 + \cdots + X_n \overset{\mathrm{d}}{=} c_n X + d_n.$$

The distribution is *strictly stable* if $d_n = 0$.

Stable distributions are more easily described by the characteristic function,

$$\varphi(\omega) = \left\langle e^{i\omega X} \right\rangle = \int_{-\infty}^{\infty} e^{i\omega x} p(x) dx.$$

Indeed, recalling that the X_k's are independent, the characteristic function $\varphi_n(\omega)$ of the sum $X_1 + \cdots + X_n$ can be written as

$$\varphi_n(\omega) = \left\langle \exp\left(i\omega \sum_{k=1}^{n} X_k \right) \right\rangle = \left\langle \prod_{k=1}^{n} e^{i\omega X_k} \right\rangle = \prod_{k=1}^{n} \left\langle e^{i\omega X_k} \right\rangle,$$

but the X_k's are identically distributed with common density given by $p(x)$, and

$$\left\langle e^{i\omega X_k} \right\rangle = \left\langle e^{i\omega X} \right\rangle = \varphi(\omega),$$

hence

$$\varphi_n(\omega) = \varphi^n(\omega). \tag{1.54}$$

On the other hand, since $p(x)$ is stable $\varphi_n(\omega)$ is also the characteristic functions of $c_n X + d_n$, so that

$$\varphi_n(\omega) = \left\langle e^{i\omega(c_n X + d_n)} \right\rangle = e^{i\omega d_n} \left\langle e^{i\omega c_n X} \right\rangle = e^{i\omega d_n} \varphi(c_n \omega),$$

Substituting for Eq. (1.54) we get

$$\varphi^n(\omega) = e^{i\omega d_n} \varphi(c_n \omega) \tag{1.55}$$

$(n = 1, 2, 3, \dots)$ which is the condition for a distribution of probability to be stable. Note that for strict stability $d_n = 0$ and (1.56) is simpler and more appealing:

$$\varphi^n(\omega) = \varphi(c_n \omega). \tag{1.56}$$

As an example of stable distribution we consider the *Lévy distribution* which has the following characteristic function

$$\varphi(\omega) = e^{-i\beta\omega^\alpha}, \qquad 1 \le \alpha \le 2. \tag{1.57}$$

In order to prove that the Lévy distribution is (strictly) stable we must show that it satisfies the condition given by Eq. (1.56). In effect, substituting Eq. (1.57) into the logarithmic form of Eq. (1.56),

$$n \ln \varphi(\omega) = \ln \varphi(c_n \omega),$$

we see at once that the Lévy distribution in the form given by Eq. (1.57) is a strictly stable distribution with $c_n = n^{1/\alpha}$ and $d_n = 0$. Note that for $\alpha = 2$ the Lévy distribution reduces to the Gaussian distribution,

$$\varphi(\omega) = e^{-i\beta\omega^2},$$

which shows that the Gaussian distribution is also stable.

We finally observe that the Lévy distribution for $1 \le \alpha < 2$ has no finite moments except the first one. Indeed, we see from Eq. (1.57) that $\varphi'(0) = 0$ and $\varphi^{(n)}(0) = \infty$ for $n = 2, 3, \ldots$ which imply a first moment equal to zero and infinite moments of order 2 and higher.

Chapter 2

Random processes: Definitions and general properties

In the classical theory of probability, whose basics have been outlined in the previous chapter, random variables do not depend of time or other variables. However, in natural sciences and engineering, as well as in socio-economic sciences, one is very often interested on the study of *processes*, that is, phenomena varying with time. The classical theory of probability, mostly developed during the 18th and 19th centuries, was not equipped for dealing with such problems. It was not until well inside the 20th century that the theory of *random* or *stochastic processes* was fully developed. In this chapter we present the most general features of random processes while in subsequent chapters we will develop more specific cases.

2.1 Definitions and examples. The Poisson process

We introduce the concept of random process by means of a simple example borrowed from elementary physics. Suppose a harmonic oscillator such as a simple pendulum, or a spring, whose displacement from its equilibrium position varies with time as

$$x(t) = a\cos(2\pi\nu t + \delta),$$

where a is the amplitude, $\nu > 0$ the frequency and δ the phase. As far as all these parameters are known, this equation allows us to determine the displacement at any time with certainty. If, however, one or several of them are not fully known then $x(t)$ turns out to be uncertain. This is the case, for instance, when the amplitude (or the frequency, or the phase) is a random variable. In such a case $x(t)$ will be also a random variable although *varying with time*. In these situations one says that $x(t)$ is a random process.

We, therefore, define *random processes as random variables varying with time*. Since, as explained in the previous chapter, any random variable is a function defined on some space of probability $(\Omega, \mathcal{A}, P)$, such that to each elementary event $\omega \in \Omega$ of the sample space Ω corresponds a numerical value $X = X(\omega)$, a random process $X(t)$ is defined as a function of two arguments:

$$X(t) = X(\omega, t) \qquad (\omega \in \Omega, t \in \mathbb{R}).$$

The interpretation of a random process $X(t)$ is thus twofold. For a particular value of $t = t_k$, the function $X(\omega, t_k) = X_k(\omega)$ only depends on ω and is a random variable. On the other hand, for every fixed value of the argument ω, that is, for any particular elementary event ω_k, the process $X(\omega_k, t)$ varies only with t and is, therefore, a deterministic (i.e., not random) function of time, $X_k(t)$, which is called *realization, trajectory* or *sample path* of the random process. Let us finally note that $X(\omega_k, t_k)$ is a fixed number.

We can summarize, following Gnedenko (1976), that *a random process may be regarded as either a collection of random variables $X(t)$ that depend on the parameter t, or as a collection of realizations (i.e., deterministic functions of time) $X(t)$ each of them with a given probability.*

The same definition applies to multidimensional random functions taking on values in $\mathbb{R}^N$. In this case we have a vectorial random process $\mathbf{X}(t)$ defined as a collection of N one-dimensional random processes:

$$\mathbf{X}(t) = \big(X_1(t), \ldots, X_N(t)\big).$$

Before proceeding to develop the general theory, we will first present with some detail a particular process of great importance in many theoretical settings and practical applications: the Poisson process.

The Poisson process

The Poisson process is one of the most relevant examples of point processes, that is, random processes consisting of points (usually represented by integer numbers) that occur at random times. Examples are the emission of radioactive particles, the arrival of phone calls, the so-called "shot noise" in electronics circuits, etc.

The Poisson process is a discrete random process taking on integer values and it is habitually represented by $N(t)$ which represents the number of points (i.e., random arrivals) occurring during the time interval $(0, t)$. Let us also denote by $N(t, t + \Delta t)$ the number of points during the time interval $(t, t + \Delta t)$ and note that with this definition $N(t) = N(0, t)$. We say that $N(t)$ is a Poisson process if there exists a positive constant λ, with dimensions $(\text{time})^{-1}$, such that

(i) $p\{N(t, t + \Delta t) = 1\} = \lambda \Delta t + O(\Delta t^2)$

(ii) $p\{N(t, t + \Delta t) = 0\} = 1 - \lambda \Delta t + O(\Delta t^2)$

(iii) $N(t, t + \Delta t)$ is independent of $N(0, t) = N(t)$. That is, the arrival of points between t and $t + \Delta t$ is independent of previous arrivals. As we will see in the next chapter this property means that the Poisson process is Markovian.

We see from (i) and (ii) that

$$p\{N(t, t + \Delta t) > 1\} = O(\Delta t^2). \tag{2.1}$$

In other words, the occurrence of two or more arrivals during a small time interval Δt is highly improbable.

Since the process is discrete, its probability density function is defined as

$$p_n(t) \equiv P\{N(t) = n\}. \tag{2.2}$$

We will now evaluate the value of $p_n(t)$ for $n = 0, 1, 2, \dots$. To this end we first obtain the probability of no arrival during $(0, t)$ which we denote by $p_0(t)$. Notice that if there is no event at time $t + \Delta t$ this necessarily implies that there has been no event before time t as well as between t and $t + \Delta t$. Then, due to assumption (iii), we have

$$\begin{aligned} P\{N(t + \Delta t) = 0\} &= P\{N(0, t) = 0 \text{ and } N(t, t + \Delta t) = 0\} \\ &= P\{N(t) = 0\} P\{N(t, t + \Delta t) = 0\}, \end{aligned}$$

and from Eq. (2.2) and assumption (ii) we have

$$p_0(t + \Delta t) = p_0(t)(1 - \lambda \Delta t) + O(\Delta t^2),$$

that is,

$$\frac{p_0(t + \Delta t) - p_0(t)}{\Delta t} = -\lambda p_0(t) + O(\Delta t),$$

and taking the limit $\Delta t \to 0$ we see that $p_0(t)$ obeys the differential equation

$$\dot{p}_0(t) = -\lambda p_0(t),$$

whose solution reads

$$p_0(t) = p_0(0)e^{-\lambda t}. \tag{2.3}$$

Following an analogous reasoning we see that for $n \geq 1$:

$$
\begin{aligned}
P\{N(t + \Delta t) = n\} &= P\{N(t) = n\}P\{N(t, t + \Delta t) = 0\} \\
&\quad + P\{N(t) = n - 1\}P\{N(t, t + \Delta t) = 1\} \\
&\quad + P\{N(t) < n - 1\}P\{N(t, t + \Delta t) > 1\},
\end{aligned}
$$

and from assumptions (i) and (ii) and Eqs. (2.1) and (2.2) we conclude that

$$
p_n(t + \Delta t) = (1 - \lambda \Delta t)p_n(t) + \lambda \Delta t p_{n-1}(t) + O(\Delta t^2).
$$

That is,

$$
\frac{p_n(t + \Delta t) - p_n(t)}{\Delta t} = -\lambda p_n(t) + \lambda p_{n-1}(t) + O(\Delta t),
$$

taking the limit $\Delta t \to 0$ we see that $p_n(t)$ satisfies the recursive differential equation:

$$
\dot{p}_n(t) = -\lambda p_n(t) + \lambda p_{n-1}(t) \qquad (n = 1, 2, 3 \ldots). \tag{2.4}
$$

As initial condition we assume that at $t = 0$ there is no arrival. That is to say, $p_0(0) = 1$ and $p_n(0) = 0$ $(n = 1, 2, 3 \ldots)$. Hence, form Eq. (2.3) and the integration of Eq. (2.4) we have

$$
p_0(t) = e^{-\lambda t},
$$

$$
p_n(t) = \int_o^{t'} e^{-\lambda(t-t')} p_{n-1}(t'), \qquad (n = 1, 2, 3 \ldots),
$$

which, after iterating, yields the *Poisson distribution*

$$
p_n(t) = \frac{(\lambda t)^n}{n!} e^{-\lambda t}, \qquad (n = 0, 1, 2 \ldots). \tag{2.5}
$$

Let us recall that in the Poisson process the arrival of points takes place at random times. Let us denote by t_n $(n = 1, 2, 3 \ldots)$ the time of the nth arrival. We will show that each t_n is a continuous random variable with the *Erlang density*:

$$
p_{t_n} = \frac{\lambda^n}{(n - 1)!} t^{n-1} e^{-\lambda t}, \tag{2.6}
$$

where $p_{t_n}(t)dt = P\{t < t_n < t + dt\}$. In effect, note that $t_n < t$ implies that there are at least n arrivals during the interval $(0, t)$. In consequence, the distribution function of t_n will be given by

$$
F_{t_n} \equiv P\{t_n < t\} = 1 - \Big[P\{N(t) = 0\} + \cdots + P\{N(t) = n - 1\} \Big],
$$

and from Eq. (2.5) we get

$$F_{t_n}(t) = 1 - e^{-\lambda t} \sum_{k=0}^{n-1} \frac{(\lambda t)^k}{k!}.$$

The probability density function $p_{t_n}(t)$ is the time derivative of $F_{t_n}(t)$ which after some cancellations yields Eq. (2.6), as the reader may easily check.

Another quantity of practical interest is the (random) *time between two consecutive arrivals* defined as the random variable

$$\tau_n = t_n - t_{n-1}$$

(n=1,2,3...). Thus $t_n = t_{n-1} + \tau_n$ is the sum of two independent random variables variables (assumption (iii)) and the convolution theorem for the sum of independent random variables (see Chapter 1, Eq. (1.31)) yields an integral equation for $p_{\tau_n}(t)$:

$$p_{t_n}(t) = \int_0^t p_{t_{n-1}}(t') p_{\tau_n}(t - t') dt'. \qquad (2.7)$$

As is well known (Roberts and Kaufman, 1966), the Laplace transform

$$\hat{p}_{t_n}(s) = \int_0^\infty e^{-st} p_{t_n}(t) dt,$$

transforms Eq. (2.7) into the algebraic equation

$$\hat{p}_{t_n}(s) = \hat{p}_{t_{n-1}}(s) \hat{p}_{\tau_n}(s). \qquad (2.8)$$

Therefore,

$$\hat{p}_{\tau_n}(s) = \frac{\hat{p}_{t_n}(s)}{\hat{p}_{t_{n-1}}(s)}.$$

On the other hand, the Laplace transform of the Erlang density (2.6) is (Roberts and Kaufman, 1966)

$$\hat{p}_{t_n}(s) = \left(\frac{\lambda}{\lambda + s} \right)^n,$$

with a corresponding expression for $\hat{p}_{t_{n-1}}(s)$. Substituting into (Fig. 2.8) yields

$$\hat{p}_{\tau_n}(s) = \frac{\lambda}{\lambda + s},$$

which after Laplace inversion yields

$$p_{\tau_n}(t) = \lambda e^{-\lambda t}. \qquad (2.9)$$

The distribution of the time interval between consecutive arrivals is thus exponential and independent of n, that is to say., the same for all arrivals.

Finally, the mean time between arrivals is

$$\langle \tau_n \rangle = \int_0^\infty t p_{\tau_n}(t) dt = \frac{1}{\lambda}.$$

Therefore, the parameter λ of the Poisson distribution is such that λ^{-1} is the mean time between consecutive arrivals.

2.2   Mathematical functions of random processes

Let us return to the concept of random process introduced above and recall that random processes are collections of random variables $X(t)$ depending on the parameter t. In many situations this parameter can be any positive real number, $0 \leq t < \infty$, as is the case of processes in continuous time. In these cases we have to deal with infinite (and non-countable) collections of random variables. The question is: how can we characterize such collections?

In the classical theory of probability, as explained in Chapter 1, any finite set of random variables $X_1, \ldots, X_n$ is completely characterized by the joint distribution function

$$F_n(x_1, \ldots, x_n) = P\{X_1 < x_1, \ldots, X_n < x_n\}.$$

Thus, when one has infinite sets of random variables, as is the case of random processes, the first step consists in the establishment of the so-called *finite-dimensional distribution functions* of the process defined as

$$F_n(x_1, t_1; \cdots ; x_n, t_n) = P\{X(t_1) < x_1, \ldots, X(t_n) < x_n\}, \qquad (2.10)$$

($n = 1, 2, 3, \cdots$). This infinite collection of positive and bounded functions[1] cannot be completely arbitrary and must satisfy the *compatibility conditions*:

(a)

$$F_{n+r}(x_1, t_1; \ldots ; x_n, t_n; \infty, t_{n+1}, \ldots ; \infty, t_{n+r}) = F_n(x_1, t_1; \ldots ; x_n, t_n).$$

(b)

$$F_n(x_{i_1}, t_{i_1}; \ldots ; x_{i_n}, t_{i_n}) = F_n(x_1, t_1; \ldots ; x_n, t_n),$$

where $i_1, \ldots, i_n$ is any arbitrary permutation of $1, \ldots, n$.

Before proceeding further let us note that these compatibility conditions stem from the fact that the functions F_n are distribution functions for which the question of marginal distributions applies (cf. Sect. 1.4 of Chapter 1) and, hence, the necessity of condition (a). Moreover, condition (b) is the result of the invariance of any probability on the order of its arguments.

We say that a stochastic process is *separable* when its finite-dimensional distribution functions describe almost completely the behavior of the process. In other words, the behavior of the trajectories of $X(t)$ is essentially

---

[1]Indeed, since the F_n's are probabilities they must take values between 0 and 1 and, hence, $0 \leq F_n(x, t) \leq 1$ for all values of x and t.

determined by the F_n's. In what follows we will assume that any random process is separable.[2]

Assuming separability, we can apply to random processes the tools developed in Chapter 1 for the study of random variables. Thus, for processes taking on continuous values, the finite-dimensional probability density functions of a random process are the derivatives of their finite-dimensional distribution functions:

$$p_n(x_1, t_1; \ldots; x_n, t_n) = \frac{\partial^n}{\partial x_1 \cdots \partial x_n} F_n(x_1, t_1, \ldots, x_n, t_n), \qquad (2.11)$$

or, equivalently,

$$F_n(x_1, t_1, \ldots, x_n, t_n) = \int_{-\infty}^{x_1} \cdots \int_{-\infty}^{x_n} p_n(x_1', t_1; \ldots; x_n', t_n) dx_1' \cdots dx_n',$$

from which follows the probabilistic interpretation of the p_n's:

$$p_n(x_1, t_1; \ldots; x_n, t_n) dx_1 \cdots dx_n$$
$$= P\{x_1 \leq X(t_1) < x_1 + dx_1; \ldots; x_n \leq X(t_n) < x_n + dx_n\}.$$

Let us note that for random processes taking on a countable number of discrete values $x_1, \ldots, x_n, \ldots$, the finite-dimensional densities are defined by

$$p_n(x_1, t_1; \ldots; x_n, t_n) = P\{X(t_1) = x_1; \ldots; X(t_n) = x_n\}.$$

From the above definitions we can easily see that the p_n's satisfy the following properties

(1) Positiviness

$$p_n(x_1, t_1; \ldots; x_n, t_n) \geq 0.$$

(2) Normalization

$$\int_{-\infty}^{\infty} \cdots \int_{-\infty}^{\infty} p_n(x_1, t_1; \ldots; x_n, t_n) dx_1 \cdots dx_n = 1.$$

(3) Marginal densities (from compatibility condition (a))

$$p_n(x_1, t_1; \ldots; x_n, t_n)$$
$$= \int_{-\infty}^{\infty} \cdots \int_{-\infty}^{\infty} p_{n+r}(x_1, t_1; \ldots; x_{n+r}, t_{n+r}) dx_{n+1} \cdots dx_{n+r}.$$

(4) Shuffling arguments (from compatibility condition (b))

$$p_n(x_{i_1}, t_{i_1}; \ldots; x_{i_n}, t_{i_n}) = p_n(x_1, t_1; \ldots; x_n, t_n),$$

where $i_1, \ldots, i_n$ is any arbitrary permutation of $1, \ldots, n$.

[2]Separability does not indeed cover all possible mathematical circumstances. However, finite dimensional distributions suffice for an overwhelming majority of practical applications and we refer the interested reader to more specialized and abstract literature on this topic as, for instance, Doob's text (1966).

Conditional densities

For separable processes conditional probability density functions are defined by

$$p_n(x_n, t_n | x_{n-1}, t_{n-1}; \ldots; x_1, t_1) = \frac{p_n(x_n, t_n; \ldots; x_1, t_1)}{p_{n-1}(x_{n-1}, t_{n-1}; \ldots; x_1, t_1)}, \quad (2.12)$$

where $t_n > t_{n-1} > \cdots > t_1$ ($n = 2, 3, \ldots$). They have the probabilistic interpretation

$$p_n(x_n, t_n | x_{n-1}, t_{n-1}; \ldots; x_1, t_1) dx_n$$
$$= P\{x_n \leq X(t_n) < x_n + dx_n | X(t_{n-1}) = x_{n-1}; \ldots; X(t_1) = x_1\}.$$

In other words, the conditional density is the probability density function for the process to take the value x_n at time t_n knowing that at previous instants of time $t_1, \ldots, t_{n-1}$ the process has taken the values $x_1, \ldots, x_{n-1}$.

We easily see from these expressions that the conditional density is non-negative,

$$p_n(x_n, t_n | x_{n-1}, t_{n-1}; \ldots; x_1, t_1) \geq 0,$$

and normalized

$$\int_{-\infty}^{\infty} p_n(x_n, t_n | x_{n-1}, t_{n-1}; \ldots; x_1, t_1) dx_n = 1.$$

2.2.1   *Multidimensional random processes*

Since a stochastic processes is a time-dependent random variable, a vector random process is a N-dimensional random variable evolving with time,

$$\mathbf{X}(t) = (X_1(t), \ldots, X_N(t)) \in \mathbb{R}^N.$$

As in one dimension we suppose that the vector process is separable, so that it is fully described by the finite-dimensional densities $p_n(\mathbf{x}_1, t_1; \ldots; \mathbf{x}_n, t_n)$ ($n = 1, 2, 3, \ldots$), where

$$p_n(\mathbf{x}_1, t_1; \ldots; \mathbf{x}_n, t_n) d\mathbf{x}_1 \cdots d\mathbf{x}_n,$$

is the probability for $\mathbf{X}(t)$ to be found at time t_1 at $\mathbf{x}_1$, at time t_2 at $\mathbf{x}_2$, etc. and $d\mathbf{x}_k$ ($k = 1, 2, \ldots, n$) is the N-dimensional volume element centered at the point $\mathbf{x}_k \in \mathbb{R}^N$.

The finite-dimensional densities of the vector process follow the same properties as those of one-dimensional processes. Thus they are non-negative and normalized

$$\int \cdots \int_{\mathbb{R}^N} p_n(\mathbf{x}_1, t_1; \ldots; \mathbf{x}_n, t_n) d\mathbf{x}_1 \cdots d\mathbf{x}_n = 1.$$

The marginal densities are

$$p_n(\mathbf{x}_1, t_1; \ldots; \mathbf{x}_n, t_n)$$

$$= \int \cdots \int_{\mathbb{R}^N} p_{n+r}(\mathbf{x}_1, t_1; \ldots, \mathbf{x}_{n+r}, t_{n+r}) d\mathbf{x}_{n+1} \cdots d\mathbf{x}_{n+r},$$

and they are invariant under shuffling of their arguments (see above). Finally, conditional densities are defined as in Eq. (2.12):

$$p_n(\mathbf{x}_n, t_n | \mathbf{x}_{n-1}, t_{n-1}; \ldots; \mathbf{x}_1, t_1) = \frac{p_n(\mathbf{x}_n, t_n; \ldots; \mathbf{x}_1, t_1)}{p_{n-1}(\mathbf{x}_{n-1}, t_{n-1}; \ldots; \mathbf{x}_1, t_1)}.$$

2.3 Moments and characteristic functions

In this section we look at other relevant functions of any random process: moments and characteristic function. For the ease of the exposition we will deal with one-dimensional random processes, although, as we have seen, the generalization to include multidimensional random processes is straightforward.

Moments

Suppose we have a random process $X(t)$ with finite-dimensional densities $p_n(x_1, t_1; \ldots; x_n, t_n)$, the average or mean value of a function $f(X(t_1), \ldots, X(t_n))$ is defined as

$$\langle f(X(t_1), \ldots, X(t_n)) \rangle \qquad (2.13)$$

$$\equiv \int_{-\infty}^{\infty} \cdots \int_{-\infty}^{\infty} f(x_1, \ldots, x_n) p_n(x_1, t_1; \ldots; x_n, t_n) dx_1 \cdots dx_n,$$

$(n = 1, 2, 3, \ldots)$.

Special cases of this definition are the *nth-order moments*:

$$m_n(t_1, \ldots, t_n) \equiv \langle X(t_1) \cdots X(t_n) \rangle$$

$$= \int_{-\infty}^{\infty} \cdots \int_{-\infty}^{\infty} x_1 \cdots x_n p_n(x_1, t_1; \ldots; x_n, t_n) dx_1 \cdots dx_n.$$

For $n = 1$ we have the mean value

$$m(t) \equiv m_1(t) = \langle X(t) \rangle = \int_{-\infty}^{\infty} x p_1(x, t) dx,$$

and for $n = 2$ we have the *correlation function*

$$C(t_1, t_2) \equiv m_2(t_1, t_2) = \langle X(t_1) X(t_2) \rangle$$

$$= \int_{-\infty}^{\infty} \int_{-\infty}^{\infty} x_1 x_2 p_2(x_1, t_1; x_2, t_2) dx_1 dx_2.$$

Other special cases are provided by *central moments*:

$$\mu_n(t_1,\ldots,t_n) \equiv \big\langle [X(t_1) - m(t_1)] \cdots [X(t_n) - m(t_n)] \big\rangle.$$

For $n = 2$ we have the *covariance function*

$$K(t_1,t_2) \equiv \mu_2(t_1,t_2)$$
$$= \int_{-\infty}^{\infty} \int_{-\infty}^{\infty} [x_1 - m(t_1)][x_2 - m(t_2)] p_2(x_1,t_1;x_2,t_2) dx_1 dx_2.$$

2.3.1 *Characteristic functions and cumulants*

Given a separable random process $X(t)$ their *finite-dimensional characteristic functions* are (Lukacs, 1970)

$$\varphi_n(\omega_1,t_1;\ldots,\omega_n,t_n) = \big\langle \exp[i\omega_1 X(t_1) + \cdots + i\omega_n X(t_n)] \big\rangle. \qquad (2.14)$$

That is [cf Eq. (2.13)],

$$\varphi_n(\omega_1,t_1;\ldots,\omega_n,t_n) \qquad\qquad (2.15)$$
$$= \int_{-\infty}^{\infty} \cdots \int_{-\infty}^{\infty} e^{i(\omega_1 x_1 + \cdots + \omega_n x_n)} p_n(x_1,t_1;\ldots;x_n,t_n) dx_1 \ldots dx_n.$$

In other words, the finite-dimensional characteristic functions are the multidimensional Fourier transforms of the finite-dimensional densities. From this equation we can see that moments are derivatives of the characteristic functions:

$$m_n(t_1,\ldots,t_n) = \frac{1}{i^n} \left. \frac{\partial^n \varphi_n(\omega_1,t_1;\ldots;\omega_n,t_n)}{\partial\omega_1 \cdots \partial\omega_n} \right|_{\omega_k=0}. \qquad (2.16)$$

Closely related to moments are the *cumulants* defined as

$$K_n(t_1,\ldots,t_n) \equiv \frac{1}{i^n} \left. \frac{\partial^n \Psi_n(\omega_1,t_1;\ldots;\omega_n,t_n)}{\partial\omega_1 \cdots \partial\omega_n} \right|_{\omega_k=0}, \qquad (2.17)$$

where Ψ_n is the logarithm of the characteristic function,

$$\Psi_n(\omega_1,t_1;\ldots;\omega_n,t_n) \equiv \ln \varphi_n(\omega_1,t_1;\ldots;\omega_n,t_n), \qquad (2.18)$$

and it is called the *cumulant generating function*.

It can be shown after some simple algebra that the relation between the first three moments and cumulants is

$$K_1(t_1) = m(t_1)$$

$$K_2(t_1,t_2) = m_2(t_1,t_2) - m(t_1)m(t_2)$$

$$K_3(t_1, t_2, t_3) = m_3(t_1, t_2, t_3) - m_2(t_1, t_2)m(t_3) - m_2(t_1, t_3)m(t_2)$$
$$-m_2(t_2, t_3)m(t_1) + 2m(t_1)m(t_2)m(t_3).$$

Unfortunately this relation becomes rather intricate and less practical as n increases.

One of the chief advantages of cumulants is that they provide a clear indication of the correlation between the values taken by the process at different times, this is clearly seen by the following example on independent processes.

Independent processes

A random process $X(t)$ is said to be independent when $X(t_i)$ is independent of $X(t_j)$ for any $t_i \neq t_j$. In such a case all finite-dimensional densities factorize

$$p_n(x_1, t_1; \ldots; x_n, t_n) = p_1(x_1, t_1) \cdots p_n(x_n, t_n)$$

$(n = 1, 2, 3, \ldots)$, which is equivalent to say that conditional densities (2.12) do not depend on conditions:

$$p_n(x_n, t_n | x_{n-1}, t_{n-1}; \ldots; x_1, t_1) = p_1(x_n, t_n).$$

Characteristic functions also factorize:

$$\varphi_n(\omega_1, t_1; \ldots; \omega_n, t_n) = \varphi_1(\omega_1, t_1) \cdots \varphi_1(\omega_n, t_n)$$

and, hence

$$\ln \varphi_n(\omega_1, t_1; \ldots; \omega_n, t_n) = \ln \varphi_1(\omega_1, t_1) + \cdots + \ln \varphi_1(\omega_n, t_n).$$

The first cumulant is

$$K_1(t_1) = \frac{1}{i} \left. \frac{\partial \ln \varphi_n}{\partial \omega_1} \right|_{\omega_1 = 0} = \frac{1}{i} \left. \frac{\partial \ln \varphi_1}{\partial \omega_1} \right|_{\omega_1 = 0} = \langle X(t_1) \rangle,$$

while higher cumulants vanish (cf. Eq. (2.17))

$$K_n(t_1, \ldots, t_n) = 0, \qquad (n = 2, 3, \ldots).$$

Therefore, the independent processes have all cumulants equal to zero except the first one. This clearly shows that cumulants indicate the degree of correlation among the different values taken by a random process.

2.4   Gaussian processes

We next present the main properties of one of the most important random processes in nature, the Gaussian process. A quite convenient way of introducing the process is through the expansion of the cumulant generating function $\Psi_n(\omega_1, t_1; \ldots; \omega_n, t_n)$ (cf. Eq. (2.18)) for small values of ω_k.

Expanding Ψ_n and using the definition of cumulants given in Eq. (2.17) we have

$$\Psi_n(\omega_1, t_1; \ldots; \omega_n, t_n) \qquad\qquad (2.19)$$
$$= \sum_{m=1}^{\infty} \left\{ \frac{i^m}{m!} \sum_{\alpha_1=1}^{n} \cdots \sum_{\alpha_n=1}^{n} K_m(t_{\alpha_1}, \ldots, t_{\alpha_m}) \omega_{\alpha_1} \cdots \omega_{\alpha_m} \right\}.$$

Recall that the cumulant generating function is the logarithm of the characteristic function. In other words, the characteristic function is the exponential of the cumulant generating function,

$$\varphi_n(\omega_1, t_1; \ldots; \omega_n, t_n) = \exp\{\Psi_n(\omega_1, t_1; \ldots; \omega_n, t_n)\},$$

which after substituting for Eq. (2.19) yields the following expansion for the characteristic function

$$\varphi_n(\omega_1, t_1; \ldots; \omega_n, t_n) \qquad\qquad (2.20)$$
$$= \exp\left\{ \sum_{m=1}^{\infty} \frac{i^m}{m!} \sum_{\alpha_1=1}^{n} \cdots \sum_{\alpha_n=1}^{n} K_m(t_{\alpha_1}, \ldots, t_{\alpha_m}) \omega_{\alpha_1} \cdots \omega_{\alpha_m} \right\}.$$

Particularizing in the first terms of this expansion we write

$$\varphi_n(\omega_1, t_1; \ldots; \omega_n, t_n) = \exp\left\{ \sum_{j=1}^{n} K_1(t_j)\omega_j - \frac{1}{2} \sum_{j,k=1}^{n} K_2(t_j, t_k)\omega_j\omega_k \right.$$
$$\left. - \frac{i}{3!} \sum_{j,k,l=1}^{n} K_3(t_j, t_k, t_l)\omega_j\omega_k\omega_l + \ldots \right\}.$$

As we have seen we can define an independent process as any process having all cumulants equal to 0 except the first one. For independent processes we then have

$$\varphi_n(\omega_1, t_1; \ldots; \omega_n, t_n) = \exp\left\{ \sum_{j=1}^{n} K_1(t_j)\omega_j \right\}.$$

Following this way the next process we could define would be that of having different from zero the first and second cumulants. This is precisely the case of Gaussian processes. Thus, and by definition, $X(t)$ is a *Gaussian* or *normal process* if their finite-dimensional characteristic functions have the form

$$\varphi_n(\omega_1, t_1; \ldots; \omega_n, t_n) = \exp\left\{\sum_{j=1}^{n} K_1(t_j)\omega_j - \frac{1}{2}\sum_{j,k=1}^{n} K_2(t_j, t_k)\omega_j\omega_k\right\},$$

(2.21)

$(n = 1, 2, 3, \ldots)$. Since the characteristic function is the Fourier transform of the probability density function, the Fourier inversion of Eq. (2.21) will give the finite-dimensional densities of the Gaussian process as (see, for instance, Stratonovich, 1967, for more details)

$$p_n(x_1, t_1; \ldots; x_n, t_n)$$

(2.22)

$$= \frac{1}{(2\pi)^{n/2}\Delta_n^{1/2}}\exp\left\{-\frac{1}{2}\sum_{j,k=1}^{n} a_{jk}^{(n)}\left[x_j - K_1(t_j)\right]\left[x_k - K_1(t_k)\right]\right\},$$

where $K_1(t) = \langle X(t)\rangle$ is the mean value of the process,

$$\Delta_n = \det\big(K_2(t_j, t_k)\big)_n$$

is the determinant of the *correlation matrix*

$$\big(K_2(t_j, t_k)\big)_n = \begin{pmatrix} K_2(t_1, t_1) & K_2(t_1, t_2) & \ldots & K_2(t_1, t_n) \\ K_2(t_2, t_1) & K_2(t_2, t_2) & \ldots & K_2(t_2, t_n) \\ \vdots & \vdots & & \vdots \\ K_2(t_n, t_1) & K_2(t_n, t_2) & \ldots & K_2(t_n, t_n) \end{pmatrix}$$

(2.23)

and

$$a_{jk}^{(n)} = \big(K_2(t_j, t_k)\big)_n^{-1}$$

is the inverse of the correlation matrix. Let us observe that *any Gaussian process is completely determined by first and second moments.*

Gaussian processes play a very important role in natural and social sciences. This is essentially due to the *central limit theorem* which, as we have seen in Chapter 1, states that under rather general conditions, the sum of a great number of random variables with finite variances (but otherwise arbitrary probability distributions) is approximately a Gaussian random function. Since many natural phenomena are the effect of the addition of a great number of small random causes, hence the significance of Gaussian processes.

2.4.1   The characteristic functional *

Perhaps one of the most general and complete descriptions of a random process $X(t)$ is through the *characteristic functional*. This is a functional defined as the average [3]

$$\varphi[\omega] \equiv \left\langle \exp\left[i \int_{-\infty}^{\infty} \omega(t) X(t) dt \right] \right\rangle, \tag{2.24}$$

where $\omega(t)$ is an arbitrary function conveniently behaved at infinity so that the integral in Eq. (2.24) converges. Note that in the special case when

$$\omega(t) = \sum_{k=1}^{n} \omega_k \delta(t - t_k),$$

where $\delta(\cdot)$ is the Dirac function and ω_k are real numbers, the characteristic functional reduces to the sequence of characteristic functions

$$\varphi_n(\omega_1, t_1; \ldots; \omega_n, t_n) = \exp\langle \exp\{ i\omega_1 X(t_1) + \cdots + \omega_n X(t_n) \} \rangle.$$

On the other hand, by expanding the exponential in (2.24),

$$\exp\left[i \int_{-\infty}^{\infty} \omega(t) X(t) dt \right] = \sum_{n=0}^{\infty} \frac{i^n}{n!} \left[\int_{-\infty}^{\infty} \omega(t) X(t) dt \right]^n$$

$$= 1 + \sum_{n=1}^{\infty} \frac{i^n}{n!} \int_{-\infty}^{\infty} \cdots \int_{-\infty}^{\infty} \omega(t_1) \cdots \omega(t_n) X(t_1) \cdots X(t_n) dt_1 \cdots dt_n,$$

we see that

$$\varphi[\omega] = 1 + \sum_{n=1}^{\infty} \frac{i^n}{n!} \int_{-\infty}^{\infty} \cdots \int_{-\infty}^{\infty} \omega(t_1) \cdots \omega(t_n) m_n(t_1, \ldots, t_n) dt_1 \cdots dt_n, \tag{2.25}$$

where $m_n(t_1, \ldots, t_n)$ are the nth-order moments. Comparison of Eqs. (2.20) and (2.25) suggests that cumulants $K_n(t_1, \ldots, t_n)$ can be defined by

$$\ln \varphi[\omega] = \sum_{n=1}^{\infty} \frac{i^n}{n!} \int_{-\infty}^{\infty} \cdots \int_{-\infty}^{\infty} \omega(t_1) \cdots \omega(t_n) K_n(t_1, \ldots, t_n) dt_1 \cdots dt_n.$$

That is,

$$\varphi[\omega] = \exp\left\{ \sum_{n=1}^{\infty} \frac{i^n}{n!} \int_{-\infty}^{\infty} \cdots \int_{-\infty}^{\infty} \omega(t_1) \cdots \omega(t_n) K_n(t_1, \ldots, t_n) dt_1 \cdots dt_n \right\}. \tag{2.26}$$

---

[3] Equation (2.24) defines a functional, not an ordinary function, since it does not depend on any independent variable, as ordinary functions, because time is integrated. It does, however, depend on kind of function $\omega(t)$ chosen.

We have seen that for an *independent process*, $K_1(t) = \langle X(t) \rangle$ and $K_n(t_1, \ldots, t_n) = 0$ for $n = 2, 3, \ldots$. In this case the characteristic functional reads

$$\varphi[\omega] = \exp\left\{ i \int_{-\infty}^{\infty} \omega(t) \langle X(t) \rangle dt \right\}. \tag{2.27}$$

For a *Gaussian process* $K_3 = K_4 = \cdots = 0$ and the characteristic functional is

$$\varphi[\omega] = \exp\left\{ i \int_{-\infty}^{\infty} \omega(t_1) \langle X(t_1) \rangle dt_1 \right.$$
$$\left. -\frac{1}{2} \int_{-\infty}^{\infty} \int_{-\infty}^{\infty} \omega(t_1) \omega(t_2) K_2(t_1, t_2) dt_1 dt_2 \right\}. \tag{2.28}$$

2.5 Stationary processes

The concept of stationary random process is of great importance in the theory of stochastic systems. This is essentially due to the fact that many phenomena are stationary, at least in an approximate way, which means that the properties of the process are invariant under time translations. In particular this implies that the time evolution of the process is independent of the initial time at which we start observing it. Roughly speaking, the time evolution of the system does not depend on whether we begin to measure the process yesterday, today or tomorrow.

We will first define the concept and then apply it to the spectral representation which constitutes a very useful technique for practical applications such as the theory of signal analysis and the handling of random data.

2.5.1 *Definitions*

A stochastic process $X(t)$ is *strictly stationary* –also termed *strongly stationary*– if all finite-dimensional probability distributions are invariant under time translations; that is, the probability distributions of the random variables $X(t_1 + h), X(t_2 + h), \ldots, X(t_n + h)$ are independent of $h \in \mathbb{R}$ for all n:

$$p_n(x_1, t_1; x_2, t_2; \ldots; x_n, t_n) \tag{2.29}$$
$$= p_n(x_1, t_1 + h; x_2, t_2 + h; \ldots; x_n, t_n + h).$$

Since h is an arbitrary real number we can choose $h = -t_1$ and condition (2.29) can be written as

$$p_n(x_1, t_1; x_2, t_2; \ldots; x_n, t_n) \tag{2.30}$$
$$= p_n(x_1, 0; x_2, t_2 - t_1; \ldots; x_n, t_n - t_1).$$

We thus see that under strict stationarity, the finite-dimensional distributions depend on time differences $t_n - t_1$ $(n = 1, 2, \ldots)$. This means, for instance, that the one time PDF, $p_1(x, t_1) = p(x)$, is independent of time, while the two-time PDF, $p_2(x_1, t_1; x_2, t_2) = p(x_1, x_2; t_2 - t_1)$, only depends on the time difference.[4]

There is a milder concept than that of strict stationarity that is very frequently used in data analysis. A random process $X(t)$ is *wide-sense stationary*, or *weakly stationary*, if their first and second moments are invariants under time translations:

$$\langle X(t) \rangle = \langle X(t+h) \rangle, \qquad \langle X(t_1)X(t_2) \rangle = \langle X(t_1 + h)X(t_2 + h) \rangle, \quad (2.31)$$

for all $h \in \mathbb{R}$. For wide-sense stationarity, and contrary to strict stationarity, it is not necessary invariance under time translations of the complete sequence of probability densities but only that of the first two moments. Hence, if a process is strongly stationary, it will also be weakly stationary. However, the reverse statement is not generally true except for Gaussian processes. Indeed, as we have seen above, Gaussian processes are completely determined by their first and second moments [see Eqs. (2.21) and (2.22)], therefore, the time invariance of the later will imply the time invariance of the entire probability distribution. In other words, if a Gaussian process is stationary in wide sense so is in strict sense and we do not have to distinguish between both kinds of stationarity.

Note that definition (2.31) is valid for any possible value of $h \in \mathbb{R}$. Thus, choosing $h = -t$ for the first moment and $h = -t_1$ for the second, we have

$$\langle X(t) \rangle = \langle X(0) \rangle \equiv m, \quad (2.32)$$

and

$$\langle X(t_1)X(t_2) \rangle = \langle X(0)X(t_2 - t_1) \rangle \equiv C(t_2 - t_1). \quad (2.33)$$

For stationary processes (both weak and strong) their average value is constant and the correlation function only depends on the time difference.

Up to now we have considered real random process. However, In some applications, as for instance in signal theory, the random process $X(t)$ can take complex values. For such cases the correlation function is defined as

$$C(t_1, t_2) \equiv \langle X(t_1) \rangle X^*(t_2) \rangle, \quad (2.34)$$

where $X^*(t)$ is the complex conjugate of $X(t)$. In this case the conditions for wide-sense stationarity are given by Eq. (2.32) but with Eq. (2.33) replaced by

$$C(t_2 - t_1) = \langle X(t_1)X^*(t_2) \rangle. \quad (2.35)$$

---

[4]The same applies to sequences of distribution functions F_n and characteristic functions φ_n.

2.5.2 *Spectral representation* *

The notion of spectral representation of a wide-sense stationary process is twofold. For one hand, we have the *spectral density* $f(\lambda)$ which is the Fourier transform of the correlation function $C(\tau)$ (note that both functions $C(\tau)$ and $f(\lambda)$ are nonrandom). On the other hand, it is called *spectrum* to the Fourier transform of the process itself.[5] In this case the spectrum is random. We will here present both representations and their connection. We begin with an example.

Example

Suppose we have a (finite) collection $X_1, \ldots, X_n$ of orthogonal random variables with zero mean, $\langle X_k \rangle = 0$. The orthogonality condition implies that

$$\langle X_k X_l^* \rangle = b_k \delta_{kl}, \tag{2.36}$$

where $b_k = \langle |X_k|^2 \rangle > 0$ and δ_{kl} is the Kronecker delta $(k, l = 1, \ldots, n)$. Note that condition (2.36) indicates that this collection is a set of uncorrelated random variables. With such random variables we construct the random process:

$$X(t) = \sum_{k=1}^{n} X_k e^{i\lambda_k t}, \qquad (\lambda_k \in \mathbb{R}), \tag{2.37}$$

which is the superposition of n harmonic oscillations with frequencies λ_k and random amplitudes X_k.

It is a simple matter to show that $X(t)$ is a wide-sense stationary process. Indeed

$$\langle X(t) \rangle = \sum_{k=1}^{n} \langle X_k \rangle e^{i\lambda_k t} = 0.$$

Moreover, since the complex conjugate of $X(t)$ is

$$X^*(t) = \sum_{l=1}^{n} X_l^* e^{-i\lambda_l t},$$

by virtue of the orthogonality condition (2.36) we have

$$C(t, t+\tau) \equiv \langle X(t+\tau) X^*(t) \rangle = \sum_{k,l=1}^{n} \langle X_k X_l^* \rangle e^{i\lambda_k (t+\tau)} e^{-\lambda_l t}$$

$$= \sum_{k=1}^{\infty} b_k e^{i\lambda_k \tau} \equiv C(\tau).$$

---

[5]Fourier series in case of discrete processes.

We thus see that $X(t)$ has zero (i.e., constant) mean and correlation function depending on time differences.

These relations are valid also when there are infinitely many X_k's and λ_k's provided that

$$\sum_{k=1}^{\infty} \langle |X_k|^2 \rangle = \sum_{k=1}^{\infty} b_k < \infty.$$

In this case we can write

$$X(t) = \sum_{k=1}^{\infty} X_k e^{i\lambda_k t} \tag{2.38}$$

and

$$C(\tau) = \sum_{k=1}^{\infty} b_k e^{i\lambda_k \tau}. \tag{2.39}$$

A stationary process of the form (2.37) or (2.38) is called *process with a discrete spectrum*, and the set of numbers $\lambda_1, \ldots, \lambda_n$ is called the *spectrum of frequencies* of the process. The following theorem tells us that in a certain sense any (wide-sense) stationary process can be approximated by processes of this simple type.

Theorem

Let $X(t)$ be a wide-sense stationary process, then $X(t)$ has the spectral representation

$$X(t) = \int_{-\infty}^{\infty} e^{i\lambda t} dZ(\lambda), \tag{2.40}$$

where $Z(\lambda)$ is a random process with zero mean

$$\langle Z(\lambda) \rangle = 0, \tag{2.41}$$

and orthogonal increments. That is,

$$\left\langle [Z(\lambda_2) - Z(\lambda_1)][Z(\mu_2) - Z(\mu_1)]^* \right\rangle = 0, \tag{2.42}$$

provided that the intervals (λ_1, λ_2) and (μ_1, μ_2) do not overlap.

The integral in Eq. (2.40) is a particular case of *stochastic integral* which we will study in Chapter 7. It is defined in the sense of Stieljes as

$$\int_{-\infty}^{\infty} e^{i\lambda t} dZ(\lambda) = \lim_{\Lambda \to \infty} \left\{ \lim_{\delta_k \to 0} \sum_{k=1}^{\infty} e^{i\lambda_k' t} [Z(\lambda_k) - Z(\lambda_{k-1})] \right\}, \tag{2.43}$$

where $-\Lambda = \lambda_0 < \lambda_1 < \cdots < \lambda_n = \Lambda$ is a partition of the real interval $[-\Lambda, \Lambda] \in \mathbb{R}$, $\delta_k = \max |\lambda_k - \lambda_{k-1}|$ and λ'_k is an arbitrary point of the interval $[\lambda_{k-1}, \lambda_k]$.

The proof of this result is based on the fact that the series (2.38) is a discretization of the continuous representation (2.40), which means that every stationary process can be approximated arbitrarily closely by linear combinations of harmonic oscillations of the form $X_k e^{i\lambda_k t}$ (see Yaglom, 1973, for a detailed discussion).

As we will see below, the random process $Z(\lambda)$ is not differentiable in the sense of ordinary functions. It can be shown, however, that it is differentiable in the sense of generalized functions (Gelfand and Shilov, 2016b). We thus call *spectrum* $\tilde{X}(\lambda)$ of stationary process $X(t)$ to the derivative of the random process $Z(\lambda)$

$$\tilde{X}(\lambda) = \frac{dZ(\lambda)}{d\lambda}. \tag{2.44}$$

From (2.40) we have

$$X(t) = \int_{-\infty}^{\infty} e^{i\lambda t} \tilde{X}(\lambda) d\lambda, \tag{2.45}$$

which means *the spectrum $\tilde{X}(\lambda)$ is the Fourier transform of $X(t)$.* That is,

$$\tilde{X}(\lambda) = \frac{1}{2\pi} \int_{-\infty}^{\infty} e^{-i\lambda t} X(t) dt. \tag{2.46}$$

In the special case when $Z(\lambda)$ is a random "jump function", then

$$\tilde{X}(\lambda) = \sum_k A_k \delta(\lambda - \lambda_k),$$

where A_k is a (finite or infinite) sequence of orthogonal random variables with zero mean. In this case the spectral representation of $X(t)$ reduces to [cf. (2.45)]

$$X(t) = \sum_k A_k e^{i\lambda_k t}$$

and we recover (2.37) or (2.38).

Up to this point we have presented the interpretation of the spectral representation involving the process itself and resulting in another random process, the spectrum $\tilde{X}(\lambda)$. We now introduce a second interpretation of the spectral representation which is nonrandom and consists in the continuous analog of Eq. (2.39) for the correlation function. This is given by the following theorem.

Theorem (Wiener-Khinchin)

For any wide-sense stationary process with zero mean and finite variance, the correlation function can be represented in the form of an integral

$$C(\tau) = \int_{-\infty}^{\infty} e^{i\lambda t} dF(\lambda), \tag{2.47}$$

where $F(\lambda)$ is a real nondecreasing function of bounded variation called spectral distribution (or measure) of the process.

Proof. Using the spectral representation (2.40) and the definition of the Stieljes integral (2.43) we write

$$C(\tau) = \left\langle X(t+\tau)X^*(t) \right\rangle = \left\langle \left[\int_{-\infty}^{\infty} e^{i\lambda(t+\tau)} dZ(\lambda) \right] \left[\int_{-\infty}^{\infty} e^{-i\mu t} dZ^*(\mu) \right] \right\rangle$$

$$= \lim_{\Lambda, M \to \infty} \left\{ \lim_{\delta_k, \nu_l \to 0} \sum_{k,l=1}^{n} e^{i\lambda'_k(t+\tau)-i\mu'_k t} \left\langle \Delta Z(\lambda_k)\Delta Z^*(\mu_l) \right\rangle \right\}, \tag{2.48}$$

where $-\Lambda = \lambda_0 < \lambda_1 < \cdots < \lambda_n = \Lambda$ and $-M = \mu_0 < \mu_1 < \cdots < \mu_n = M$ are two different partitions of $[-\Lambda, \Lambda] \in \mathbb{R}$ and $[-M, M] \in \mathbb{R}$ respectively; $\lambda'_k \in [\lambda_{k-1}, \lambda_k]$ and $\mu'_k \in [\mu_{k-1}, \mu_k]$ are intermediate points of these partitions;

$$\delta_k = \max |\lambda_k - \lambda_{k-1}|, \qquad \nu_l = \max |\mu_l - \mu_{l-1}|$$

and

$$\Delta Z(\lambda_k) \equiv Z(\lambda_k) - Z(\lambda_{k-1}), \qquad \Delta Z^*(\mu_l) \equiv Z^*(\mu_l) - Z^*(\mu_{l-1}).$$

But, from the orthogonality condition on the increments of $Z(\lambda)$ given in (2.42), we have

$$\left\langle \Delta Z(\lambda_k)\Delta Z^*(\mu_l) \right\rangle = 0, \qquad \forall \lambda_k \neq \mu_l.$$

Hence,

$$C(\tau) = \lim_{\Lambda \to \infty} \left\{ \lim_{\delta_k \to 0} \sum_{k=1}^{n} e^{i\lambda'_k \tau} \left\langle |\Delta Z(\lambda_k)|^2 \right\rangle \right\}$$

$$\equiv \int_{-\infty}^{\infty} e^{i\lambda \tau} dF(\lambda), \tag{2.49}$$

where $F(\lambda)$ is such that

$$F(\lambda + \Delta\lambda) - F(\lambda) = \left\langle |Z(\lambda + \Delta\lambda) - Z(\lambda)|^2 \right\rangle \geq 0, \tag{2.50}$$

showing that $F(\lambda)$ is real and nondecreasing. This equation determines $F(\lambda)$ except for an additive constant, which we choose such that

$$F(-\infty) = 0.$$

On the other hand, from (2.49) we write

$$C(0) = \langle |X(t)|^2 \rangle = \int_{-\infty}^{\infty} dF(\lambda),$$

and since, by assumption $X(t)$ has zero mean and finite variance, we have that $C(0)$ is finite, hence

$$\int_{-\infty}^{\infty} dF(\lambda) < \infty,$$

which proves that $F(\lambda)$ has bounded variation.

□

We write Eq. (2.47) in the form

$$C(\tau) = \int_{-\infty}^{\infty} e^{i\lambda\tau} f(\lambda) d\lambda, \qquad (2.51)$$

where $f(\lambda)$ is the derivative of the spectral distribution

$$f(\lambda) = \frac{dF(\lambda)}{d\lambda},$$

which always exists in the sense of generalized functions and it is called *spectral density*[6]. Note that since $F(\lambda)$ is nondecreasing then the spectral density is nonnegative:,

$$f(\lambda) \geq 0 \qquad \forall \lambda \in \mathbb{R}.$$

and since $F(-\infty) = 0$, we have

$$F(\lambda) = \int_{-\infty}^{\lambda} f(\lambda') d\lambda'. \qquad (2.52)$$

---

[6]In fact, $f(\lambda)$ exits as an ordinary function as long as $C(\tau)$ is absolutely integrable (Yaglom, 1973), that is

$$\int_{-\infty}^{\infty} |C(\tau)| d\tau < \infty.$$

Differentiability of $Z(\lambda)$

From Eq. (2.52) we write

$$F(\lambda + \Delta\lambda) - F(\lambda) = \int_{\lambda}^{\lambda+\Delta\lambda} f(\lambda)d\lambda,$$

from which we see that for a continuous and non vanishing $f(\lambda)$

$$F(\lambda + \Delta\lambda) - F(\lambda) = O(\Delta\lambda).$$

Combining this with (2.50) we have

$$\langle|\Delta Z(\lambda)|^2\rangle = O(\Delta\lambda),$$

that is, the quadratic mean square increment of process $Z(\lambda)$ is of order $\Delta\lambda$. This implies that the increment of Z is of order $\sqrt{\Delta\lambda}$

$$\Delta Z(\lambda) = O(\sqrt{\Delta\lambda}).$$

Hence

$$\frac{\Delta Z(\lambda)}{\Delta\lambda} = O\left(\frac{1}{\Delta\lambda}\right),$$

and $|\Delta Z/\Delta\lambda| \to \infty$ as $\Delta\lambda \to 0$ and the process $Z(\lambda)$ is not differentiable (in the sense of an ordinary function). As we have remarked above $Z(\lambda)$ is differentiable in the sense of generalized function and its derivative, itself a generalized function, is the spectrum $\tilde{X}(\lambda)$ of process $X(t)$ (Gelfand and Shilov, 2016b).

Example (white noise)

Suppose that a stationary process $X(t)$ has constant spectral density independent of the frequency λ:

$$f(\lambda) = \gamma = \text{constant}.$$

Substituting into (2.51) we have

$$C(\tau) = \gamma \int_{-\infty}^{\infty} e^{i\lambda\tau}d\lambda,$$

but, from the representation of Dirac delta function (Vladimirov, 1984)

$$\delta(\tau) = \frac{1}{2\pi} \int_{-\infty}^{\infty} e^{i\lambda\tau}d\lambda, \qquad (2.53)$$

we obtain

$$C(\tau) = 2\pi\gamma\delta(\tau)$$

and $X(t)$ is delta-correlated

$$\langle X(t + \tau)X(t) \rangle = k\delta(\tau),$$

where $k = 2\pi\gamma$. Note that in particular

$$C(0) = \langle |X(0)|^2 \rangle = \infty,$$

On the other hand, since $\delta(\tau) = 0$ for $\tau \neq 0$ we see that $C(\tau) = 0$ $(\tau \neq 0)$ and the values of $X(t)$ and $X(t + \tau)$ are uncorrelated even for arbitrarily small values of the time interval τ.

This process is called *white noise* (when $f(\lambda)$ is not constant $X(t)$ is called "colored noise").[7] From the above discussion we see that white noise is a "purely random process" which is a convenient mathematical idealization. We shall discuss this important point at several places of this book.

One aspect we want to comment on the spectral representation of a stationary process is the relationship, mentioned at the beginning of this section, between the two interpretations of spectral representation, one of them based on the spectrum $\tilde{X}(\lambda)$ and the other on the spectral density $f(\lambda)$. The link is given by the following result.

Theorem

The spectrum $\tilde{X}(\lambda)$ of any wide-sense stationary process is white noise, i.e., a delta-correlated random process:

$$\langle \tilde{X}(\lambda)\tilde{X}^*(\lambda') \rangle = f(\lambda)\delta(\lambda - \lambda'), \tag{2.54}$$

where $f(\lambda)$ is the spectral density.

Proof. Using Eq. (2.46) we write

$$\langle \tilde{X}(\lambda)\tilde{X}^*(\lambda') \rangle = \frac{1}{4\pi^2} \int_{-\infty}^{\infty} dt \int_{-\infty}^{\infty} e^{-i\lambda t + i\lambda' t'} C(t - t')dt'.$$

The change of variables

$$\tau = t - t', \qquad s = \frac{1}{2}(t + t'),$$

yields

$$\langle \tilde{X}(\lambda)\tilde{X}^*(\lambda') \rangle = \frac{1}{4\pi^2} \int_{-\infty}^{\infty} e^{-i(\lambda+\lambda')\tau/2}C(\tau)d\tau \int_{-\infty}^{\infty} e^{-i(\lambda-i\lambda')s}ds.$$

---

[7]This comes from the analogy of white light for which the spectral density density does not depend of any frequency, by contrast to colored light where the spectral density depends on a frequency (that of the particular color).

Recalling that [cf. (2.53)]

$$\int_{-\infty}^{\infty} e^{-i(\lambda - i\lambda')s} ds = 2\pi\delta(\lambda - \lambda'),$$

we write

$$\langle \tilde{X}(\lambda)\tilde{X}^*(\lambda')\rangle = \frac{1}{2\pi} \int_{-\infty}^{\infty} d\tau e^{-i\lambda\tau} C(\tau)\delta(\lambda - \lambda').$$

From (2.51) we see that the correlation function $C(\tau)$ is the Fourier transform of the spectral density $f(\lambda)$ and by inverting we have

$$f(\lambda) = \frac{1}{2\pi} \int_{-\infty}^{\infty} e^{-i\lambda\tau} C(\tau) d\tau.$$

Therefore

$$\langle \tilde{X}(\lambda)\tilde{X}^*(\lambda')\rangle = f(\lambda)\delta(\lambda - \lambda'),$$

which is Eq. (2.54).

□

2.5.3   *Filters* *

The real physical meaning of the spectral representation is explained by the possibility of separating spectral components corresponding to different parts of the spectrum by using suitable designed *filters* which are devices that pass periodic oscillations in a certain frequency range (the *pass band*) while suppressing oscillations with different frequencies. We will here present a short summary on this topic of which there is an immense literature specially in communication theory and radio engineering (Middleton, 1960; Papoulis, 1977; Hamilton, 1994).

We shall here restrict ourselves to *linear filters*. From an analytical point of view, a linear filter is characterized by a *transfer function* $h(s)$ which relates the output of the filter $X_2(t)$ in terms of the input signal $X_1(t)$ as

$$X_2(t) = \int_{-\infty}^{\infty} X_1(t-s)h(s)ds. \tag{2.55}$$

Let us recall that either input and output are random processes (usually complex valued). The transfer function may be also a complex function although deterministic.

Suppose next that the input signal is a wide-sense stationary process with zero mean and spectral density $f_1(\lambda)$,

$$C_1(\tau) = \langle X_1(t+\tau)X_1(t)\rangle = \int_{-\infty}^{\infty} e^{i\lambda\tau} f_1(\lambda)d\lambda. \tag{2.56}$$

Let us now show that *the output signal is also a wide-sense stationary process*. In effect, from (2.55) we see that $\langle X_2(t)\rangle = 0$ if $\langle X_1(t)\rangle = 0$. On the other hand, using (2.55) we write

$$\langle X_2(t+\tau)X_2^*(t)\rangle = \int_{-\infty}^{\infty} h(s)ds \int_{-\infty}^{\infty} h^*(r)\langle X_1(t+\tau-s)X_1^*(t-r)\rangle dr,$$

and since $X_1(t)$ is stationary, its correlation function is only a function of time differences, that is

$$\langle X_1(t+\tau-s)X_1^*(t-r)\rangle \equiv C_1(\tau-s+r),$$

does not depend on t. Hence, the output correlation,

$$\langle X_2(t+\tau)X_2^*(t)\rangle = \int_{-\infty}^{\infty} h(s)ds \int_{-\infty}^{\infty} h^*(r)C_1(\tau-s+r)dr \equiv C_2(\tau), \quad (2.57)$$

only depends on τ proving the stationarity of the output.

Let us now obtain the spectral density of the output, $f_2(\lambda)$, in terms of that of the input, $f_1(\lambda)$. Plugging (2.56) into (2.57) we have

$$C_2(\tau) = \int_{-\infty}^{\infty} e^{i\lambda\tau}f_1(\lambda)d\lambda \int_{-\infty}^{\infty} e^{-i\lambda s}h(s)ds \int_{-\infty}^{\infty} e^{i\lambda r}h^*(r)dr. \quad (2.58)$$

The Fourier transform of the transfer function is

$$\tilde{h}(\lambda) = \frac{1}{2\pi}\int_{-\infty}^{\infty} e^{-i\lambda s}h(s)ds \quad \text{and} \quad \tilde{h}^*(\lambda) = \frac{1}{2\pi}\int_{-\infty}^{\infty} e^{i\lambda r}h^*(r)dr. \quad (2.59)$$

Substituting (2.59) into (2.58) yields

$$C_2(\tau) = 4\pi^2 \int_{-\infty}^{\infty} e^{i\lambda\tau}f_1(\lambda)\big|\tilde{h}(\lambda)\big|^2 d\lambda. \quad (2.60)$$

Comparing (2.60) with (2.51) we conclude that the output spectral density, $f_2(\lambda)$, in terms of the input density, $f_1(\lambda)$, is

$$f_2(\lambda) = 4\pi^2\big|\tilde{h}(\lambda)\big|^2 f_1(\lambda). \quad (2.61)$$

Likewise we can obtain the link between spectra of input, $\tilde{X}_1(\lambda)$, and output, $\tilde{X}_2(\lambda)$. Thus, if $X_1(t)$ is stationary with spectrum $\tilde{X}_1(\lambda)$ we have [cf. Eq. (2.45)]

$$X_1(\tau) = \int_{-\infty}^{\infty} e^{i\lambda\tau}\tilde{X}_1(\lambda)d\lambda.$$

Substituting for Eq. (2.55) and taking into account (2.59), we get

$$X_2(t) = \int_{-\infty}^{\infty} d\lambda e^{i\lambda t}\tilde{X}_1(\lambda)\int_{-\infty}^{\infty} ds e^{-i\lambda s}h(s) = 2\pi\int_{-\infty}^{\infty} e^{i\lambda t}\tilde{h}(\lambda)\tilde{X}_1(\lambda)d\lambda.$$

That is,

$$\tilde{X}_2(\lambda) = 2\pi\tilde{h}(\lambda)\tilde{X}_1(\lambda), \quad (2.62)$$

which is the connection sought [compare with Eq. (2.61)].

Example: Moving averages

A classical example of filter, intensively used in time series analysis, is the moving average, $X_2(t)$, of a stationary process, $X_1(t)$, inside the interval $(t - T, t + T)$ $(T > 0)$. For continuous processes we define this average by the integral [8]

$$X_2(t) = \frac{1}{2T} \int_{t-T}^{t+T} X_1(s)ds. \qquad (2.63)$$

We thus see that the moving average is a linear filter with transfer function

$$h(s) = \begin{cases} 1/2T & \text{if } -T \leq s \leq T, \\ 0 & \text{otherwise.} \end{cases} \qquad (2.64)$$

The Fourier transform of the transfer function is

$$\tilde{h}(\lambda) = \frac{1}{2\pi} \int_{-\infty}^{\infty} e^{-i\lambda s} h(s)ds = \frac{1}{4\pi T} \int_{-T}^{T} e^{-i\lambda s} ds = \frac{1}{2\pi} \frac{\sin \lambda T}{\lambda T},$$

and, hence, the correlation function of the moving average is [see (2.60)]

$$C_2(\tau) = \int_{-\infty}^{\infty} e^{i\lambda \tau} \frac{\sin^2 \lambda T}{\lambda^2 T^2} f_1(\lambda)d\lambda, \qquad (2.65)$$

where $f_1(\lambda)$ is the spectral density of the input signal.

Suppose we are now filtering white noise, that is to say

$$C_1(\tau) = \delta(\tau) \qquad \text{and} \qquad f_1(\lambda) = \frac{1}{2\pi}.$$

From (2.65) we have

$$\begin{aligned}
C_2(\tau) &= \frac{1}{2\pi T^2} \int_{-\infty}^{\infty} e^{i\lambda \tau} \frac{\sin^2 \lambda T}{\lambda^2} d\lambda \\
&= \frac{1}{2\pi T^2} \left[\int_{-\infty}^{\infty} \cos \lambda \tau \frac{\sin^2 \lambda T}{\lambda^2} d\lambda + i \int_{-\infty}^{\infty} \sin \lambda \tau \frac{\sin^2 \lambda T}{\lambda^2} d\lambda \right] \\
&= \frac{1}{\pi T^2} \int_{0}^{\infty} \cos \lambda \tau \frac{\sin^2 \lambda T}{\lambda^2} d\lambda.
\end{aligned}$$

---

[8] In the discrete version (mostly used in practical applications) the integral is replaced by the sum:

$$X_2(t) = \sum_{k=-n}^{n} a_k X_1(t-k) \qquad (t = \ldots, -2, -1, 0, 1, 2, \ldots),$$

where a_k are given complex numbers.

Using the Fourier inversion formula (Erdelyi, 1954)

$$\int_0^\infty \cos(xy)\frac{\sin^2 ax}{x^2}dx = \begin{cases} (\pi/2)(a - y/2) & y < 2a, \\ 0 & y > 2a, \end{cases}$$

$(y > 0)$ and recalling that $C_2(-\tau) = C_2(\tau)$, we finally get the following correlation function for the moving average of white noise

$$C_2(\tau) = \begin{cases} (1/4T^2)(2T - \tau) & -2T \leq \tau \leq 2T, \\ 0 & \text{elsewhere.} \end{cases} \tag{2.66}$$

2.6 Ergodicity of random processes *

Let $X(t)$ be a stochastic processes representing some physical observable as, for instance, the position of a Brownian particle or the intensity of a noisy electronic circuit, etc. In order to evaluate the mean value

$$m(t) = \langle X(t) \rangle \equiv \int_{-\infty}^\infty x p_1(x, t) dx,$$

or the correlation function

$$C(t_1, t_2) = \langle X(t_1)X(t_2) \rangle \equiv \int_{-\infty}^\infty \int_{-\infty}^\infty x_1 x_2 p_2(x_1, t_1; x_2, t_2),$$

we should know the densities $p_1(x, t)$ and $p_2(x_1, t_1; x_2, t_2)$ but unfortunately this is beyond reach in most real situations. An alternative way of evaluating averages consists in having a great number of realizations of the process:

$$X^{(1)}(t), X^{(2)}(t), \ldots, X^{(n)}(t) \qquad (n \gg 1),$$

where $X^{(k)}(t) \equiv X(\omega_k, t)$, and evaluate the averages, which we can identify with the mean value and correlation, as:

$$\langle X(t) \rangle = \lim_{n \to \infty} \left[\frac{1}{n} \sum_{k=1}^n X^{(k)}(t) \right],$$

$$\langle X(t_1)X(t_2) \rangle = \lim_{n \to \infty} \left[\frac{1}{n} \sum_{k=1}^n X^{(k)}(t_1)X^{(k)}(t_2) \right].$$

Averages of this kind are called *ensemble averages*.

However, this second procedure is also unrealistic because the observation of a great number of $X^{(k)}(t)$'s is just about impossible[9]. During

---

[9] One usually has a single experimental setting which results in the observation of only one realization of $X(t)$, but not 10^6 or 10^9 identical settings which provide many different realizations.

a certain period of time $(0, T)$ an observer can only obtain a single realization which allows to get *time averages* of the process using this unique realization. We thus define

$$\overline{X(t)} \equiv \lim_{T \to \infty} \frac{1}{T} \int_0^T X(t)dt,$$

as the time average of $X(t)$ and

$$\overline{X(t+\tau)X(t)} \equiv \lim_{T \to \infty} \frac{1}{T} \int_0^T X(t+\tau)X(t)dt.$$

as the time correlation function. This kind of averages are called *time averages*.

We say that a random process is *ergodic* if ensemble averages equal time averages. That is to say, when the statistical properties of the process can be obtained from a single trajectory (i.e., realization) of the process provided that we follow it for a sufficiently long time.

In stationary processes the conditions for being ergodic are rather general as it is shown by the following theorem:

Theorem.

Let $X(t)$ a weakly stationary random process, then the time average converges to the ensemble average (in mean square sense), that is

$$\frac{1}{T} \int_0^T X(t)dt \xrightarrow[T \to \infty]{} \langle X(t) \rangle = m, \qquad (2.67)$$

if and only if the so-called Slutsky condition holds

$$\lim_{T \to \infty} \frac{1}{T} \int_0^T k(\tau)d\tau = 0, \qquad (2.68)$$

where $k(\tau) = \langle X(t+\tau)X(t) \rangle - m^2$ is the correlation function of $X(t)$.

Proof. Define the random variable

$$X_T = \frac{1}{T} \int_0^T X(t)dt$$

$(T > 0)$, so that the time average of $X(t)$ is the limit

$$\overline{X(t)} = \lim_{T \to \infty} X_T$$

and the mean-square convergence of Eq. (2.67) is equivalent to

$$\lim_{T \to \infty} \left\langle [X_T - m]^2 \right\rangle = 0. \qquad (2.69)$$

Let us first see that due to the assumed weakly stationarity $\langle X(t) \rangle = m$ is time independent and

$$\langle X_T \rangle = \frac{1}{T} \int_0^T \langle X(t) \rangle dt = \frac{1}{T} m \int_0^T dt = m.$$

Hence, the average of X_T equals that of $X(t)$:

$$\langle X_T \rangle = \langle X(t) \rangle = m.$$

From the definition of X_T and the wide-sense stationary character of $X(t)$ we see that the variance of X_T is

$$\left\langle [X_T - \langle X_T \rangle]^2 \right\rangle = \langle X_T^2 \rangle - \langle X_T \rangle^2 = \langle X_T^2 \rangle - \langle X(t) \rangle^2$$

$$= \frac{1}{T^2} \int_0^T dt \int_0^T \left[\langle X(t)X(t') \rangle - \langle X(t) \rangle \langle X(t') \rangle \right] dt'$$

$$= \frac{1}{T^2} \int_0^T dt \int_0^T k(t - t') dt'.$$

A further manipulation involving the even character of the correlation function [i.e., $k(\tau) = k(-\tau)$] yields

$$\left\langle [X_T - \langle X_T \rangle]^2 \right\rangle = \frac{2}{T^2} \int_0^T dt \int_0^t k(\tau) d\tau.$$

It can be shown that (Rytov *et al.*, 1987)

$$\lim_{T \to \infty} \frac{2}{T^2} \int_0^T dt \int_0^t k(\tau) d\tau = 0 \quad \text{iff} \quad \lim_{T \to \infty} \frac{1}{T} \int_0^T k(\tau) d\tau = 0.$$

Therefore, the variance of X_T goes to 0 as $T \to \infty$ if and only if the Slutsky condition (2.68) holds, which proves the theorem.

$\square$

Remark 1

The ergodic theorem can be generalized to include functions $f\big(X(t)\big)$ of the stationary process $X(t)$. Thus, if we define the time average

$$\overline{f(X(t))} = \lim_{T \to \infty} \frac{1}{T} \int_0^T f\big(X(t)\big) dt,$$

the ensemble average,

$$\langle f\big(X(t)\big) \rangle = \int_{-\infty}^{\infty} f(x)p(x) dx,$$

equals the time average in mean square sense. We thus have the following generalized ergodic theorem (see Rytov *et al.*, 1987, for more details).

Theorem.

The time average of any function of a strictly stationary random process converges in mean square sense to the ensemble average,

$$\frac{1}{T}\int_0^T f\big(X(t)\big)dt \;\xrightarrow[T\to\infty]{}\; \big\langle f\big(X(t)\big)\big\rangle, \tag{2.70}$$

if and only if

$$\lim_{T\to\infty}\frac{1}{T}\int_0^T k_f(\tau)d\tau = 0, \tag{2.71}$$

where

$$k_f(\tau) \equiv \big\langle f(X(t+\tau))f\big(X(t)\big)\big\rangle - \big\langle f(X(t+\tau))\big\rangle\big\langle f\big(X(t)\big)\big\rangle. \tag{2.72}$$

Remark 2

Let $X(t)$ be a strict sense stationary random process which is also ergodic. Since the process is strongly stationary the univariate probability density function,

$$p_1(x,t) \equiv p(x),$$

is time independent and the bivariate density,

$$p_2(x,t+\tau;x',t) \equiv p(x,\tau;x'),$$

only depends on the time difference.

We define the *relative time for $X(t)$ to fall into the infinitesimal interval* $(x, x+dx)$ *during* $(0,T)$ as the ratio

$$\frac{T(x,x+dx)}{T},$$

where

$$T(x,x+dx) = \Big\{t \in (0,T)|x < X(t) < x+dx\Big\}$$

is the time spent by the process in the infinitesimal interval. Clearly $T(x, x+dx)$ is a random variable because it depends on the particular trajectory taken by $X(t)$. It also depends on the parameter T which measures the observation time on the random phenomenon described by $X(t)$.

For strictly stationary processes this relative time is related to the univariate probability density by

$$\frac{T(x,x+dx)}{T} \;\xrightarrow[T\to\infty]{}\; p(x)dx \tag{2.73}$$

in mean square sense.

In the language of physics, Eq. (2.73) affirms that *the stationary probability of a given state equals the relative time spent by the system in such a state provided that the system has been under observation during a sufficiently long time (i.e., $T \to \infty$).* In other words, statistical averages equal time averages taken over sufficiently long times which is, precisely, the definition of ergodicity.

To prove the above statement we define the following function

$$f(X(t), y) = \begin{cases} 1 & \text{if } y \leq X(t) < y + dy \\ 0 & \text{otherwise.} \end{cases} \tag{2.74}$$

Note that $T(y, y + dy)$ is given by the time integral of f and

$$\frac{T(x, x + dx)}{T} = \frac{1}{T} \int_0^T f(X(t), y) dt.$$

On the other hand, the ensemble average of f is

$$\langle f(X(t), y) \rangle = \int_{-\infty}^{\infty} f(X(t), y) P\{x \leq X(t) < x + dx\}$$

$$= P\{y \leq X(t) < y + dy\} = p_1(y, t) dy,$$

and due to the strict stationarity of the process $p_1(y, t) = p(y)$ and we have

$$\langle f(X(t), y) \rangle = p(y) dy.$$

The ergodic theorem (2.70)

$$\frac{1}{T} \int_0^T f(X(t), y) dt \xrightarrow[T \to \infty]{} \langle f(X(t) y) \rangle,$$

implies Eq. (2.73). That is

$$\lim_{T \to \infty} \left\langle \left[\frac{T(y, y + dy)}{T} - p(y) dy \right]^2 \right\rangle = 0.$$

This results holds as long as the function $f(X(t), y)$ meets the generalized Slutsky condition Eq. (2.71). The necessary and sufficient condition (2.71) can be replaced by more strict requirements which are sufficient for Eq. (2.73) to hold and that are met in many practical applications. Thus writing $k_f(\tau)$ defined in Eq. (2.72) as

$$k_f(\tau) = \int_{-\infty}^{\infty} \int_{-\infty}^{\infty} f(x, y) f(x', y') [p(x, \tau; x') - p(x) p(x')] dx dx',$$

and, since $f(x, y) = 1$ if $y < x < y + dy$ and $f(x, y) = 0$ otherwise, we get

$$k_f(\tau) = p(y, \tau; y') - p(y)p(y').$$

It is, therefore, sufficient for the Slutsky condition (2.71),

$$\lim_{T \to \infty} \frac{1}{T} \int_0^T k_f(\tau) d\tau = 0,$$

to hold that the covariance $k_f(\tau)$ is bounded everywhere and decreases as $|\tau|^{-\alpha}$ where $\alpha > 0$ when $|\tau| \to \infty$. Another sufficient condition for Eq. (2.73) to hold, and that is most important from a physical point of view, is provided by the assumption that as τ increases the process $X(t + \tau)$ is independent of $X(t)$ in such a case the bivariate density factorizes

$$p_2(x, \tau; x') \underset{\tau \to \infty}{\longrightarrow} p(x)p(x'),$$

which ensures the validity of the ergodic theorem.

We finally address the interested reader to Rytov et al (1987) for more information and some examples.

Chapter 3

Markov processes

We now from the previous chapter that any separable random process $X(t)$ is fully characterized by their finite-dimensional densities $p_n(x_1, t_1; \ldots; x_n, t_n)$ $(n = 1, 2, 3, \ldots)$. This characterization is so general that we can barely go ahead with any further analysis, specially for practical purposes. We have also seen in Chapter 2 alternative descriptions of $X(t)$, based on characteristic functions and cumulants, which allow us to define simple and operative forms for independent and Gaussian processes. In this chapter we return to the general description of stochastic processes given by finite-dimensional densities and obtain one of the most important idealizations, the Markov process.

3.1 Definitions and general properties

In order to introduce Markov processes let us briefly go back to independent processes. Recall that $X(t)$ is an independent process if the value of the process at time t does not depend of any previous value. This means that $X(t_1), X(t_2), \ldots, X(t_n)$ are independent random variables for all $n = 1, 2, 3, \ldots$ and arbitrary $t_1 < t_2 \cdots < t_n$.. In other words, all finite-dimensional densities factorize:

$$p_n(x_1, t_1; x_2, t_2; \cdots; x_n, t_n) = p_1(x_1, t_1) p_2(x_2, t_2) \ldots p_n(x_n, t_n),$$

which, in terms of the conditional density

$$p_n(x_n, t_n | x_{n-1}, t_{n-1}; \ldots; x_1, t_1) = \frac{p_n(x_n, t_n; \ldots; x_1, t_1)}{p_{n-1}(x_{n-1} t_{n-1}; \ldots; x_1, t_1)}, \quad (3.1)$$

implies

$$p(x_n, t_n | x_{n-1}, t_{n-1}; \ldots; x_1, t_1) = p_1(x_n, t_n),$$

showing that the values taken by independent processes do not depend on the previous history.

For more than three centuries the theory of probability essentially consisted in the study of independent processes occurring at discrete times such as, for instance, coin tossing and dice throwing. However, many natural phenomena occur on continuous time scales and with them it is practically impossible to have truly independent processes in which $X(t + \tau)$ does not depend on $X(t)$ no matter how small $\tau > 0$ is.

It is, nonetheless, possible to consider that physical phenomena may be described by independent processes on discrete scales of time. This is rather accurate if time intervals between consecutive observations are much larger than any individual time scale governing elementary events. Think, for instance, about the motion of a Brownian particle through a fluid in equilibrium. The position of the particle may be considered an independent random process over time scales such that the time intervals between observations are large compared to the times between consecutive collisions of the particle with the molecules of the surrounding fluid.

It is, however, clear that independent processes are not sufficient to describe many aspects of physical reality. We should consider some sort of "temporal correlation" or "memory", specially in continuous time scales. A decisive step along this direction was taken by Andrei Markov who in 1907, in relation with his study on the alternation of vowels and consonants in Pushkin's *Eugène Oneguin*, realized that the appearance of a vowel or a consonant depended strongly on whether the immediately preceding letter was a vowel or a consonant but very weakly on earlier letters.[1]

Physics provides another simple example of "recent memory". The energy transfer between the molecules of a gas can be considered a random process in continuous time. For an ideal (i.e., dilute) gas the amount of energy transferred in a given collision solely depends upon the energy of the colliding molecules at the instant of collision but it is independent of previous amounts of energy at earlier times (Oppenheim *et al.*, 1977).

We can describe these and similar situations by devising a process such that its value $X(t_n)$ at time t_n only depends on the value $X(t_{n-1})$ taken by the process at the preceding time t_{n-1} and that earlier values at $t_{n-2}, t_{n-3}, \ldots$ have no effect on $X(t_n)$. This leads to the following definition.

---

[1] Eugéne Oneguin is a novel written in verse with an atypical rhyme scheme known as "Pushkin sonnet". This surely explains Markov's interest on the structure of the text.

We say that $X(t)$ is a *Markov random process* if

$$p_n(x_n, t_n | x_{n-1}, t_{n-1}; x_{n-2}, t_{n-2}; \ldots; x_1, t_1) = p(x_n, t_n | x_{n-1}, t_{n-1}), \quad (3.2)$$

for all $n = 1, 2, 3, \ldots$ and $t_n > t_{n-1} > t_{n-2} > \cdots > t_1$. The function $p(x_n, t_n | x_{n-1}, t_{n-1})$ is called *transition probability density* of the process and plays a central role in the theory of Markovian processes.

From Eqs. (3.1) and (3.2) we write

$$p_n(x_n, t_n; \ldots; x_1, t_1) = p(x_n, t_n | x_{n-1}, t_{n-1}) p_{n-1}(x_{n-1} t_{n-1}; \ldots; x_1, t_1).$$

Iterating this equation and the repeated use of the Markov property, Eq. (3.2), yields

$$p_n(x_n, t_n; \ldots; x_1, t_1) = p(x_n, t_n | x_{n-1}, t_{n-1}) p(x_{n-1}, t_{n-1} | x_{n-2}, t_{n-2})$$
$$\cdots p(x_2, t_2 | x_1, t_1) p(x_1, t_1), \quad (3.3)$$

where $p(x_1, t_1) \equiv p_1(x_1, t_1)$ is the univariate probability density function. Equation (3.3) is an alternative definition of the Markovian property given by Eq. (3.2) and shows that to determine a Markov process we do not have to specify the infinite sequence of joint densities $p_n(x_1, t_1; \ldots, x_n, t_n)$ $(n = 1, 2, 3, \ldots)$, but only two functions: the univariate density $p(x, t)$ and the transition density $p(x, t | x', t')$ $(t > t')$. No doubt that this is a great simplification in the description of Markovian processes as opposed to non-Markovian ones for which, in principle, an infinite number of densities has to be established.

If the process can only assume a denumerable infinity of values and the units of time are discrete multiples of a single time τ, i.e., $t_n = n\tau$, the Markov processes is usually termed as a *Markov chain*. Classical examples are found within the theory of random walks (Weiss, 1994).

3.1.1 *The Chapman-Kolmogorov equation*

The Chapman-Kolmogorov equation is a nonlinear integral equation satisfied by any Markov process. It is an identity linking the values of the transition density at different instants of time.

In order to derive it we begin with the trivariate probability density function $p_3(x_3, t_3; x_2, t_2; x_1, t_1)$, where $t_3 > t_2 > t_1$. The bivariate density $p_2(x_3, t_3; x_1, t_1)$ is then given by

$$p_2(x_3, t_3; x_1, t_1) = \int_{-\infty}^{\infty} p_3(x_3, t_3; x_2, t_2; x_1, t_1) dx_2. \quad (3.4)$$

On the other hand,

$$p_2(x_3, t_3; x_1, t_1) = p(x_3, t_3 | x_1, t_1) p_1(x_1, t_1)$$

and for a Markov process we have [cf. Eq. (3.3)]

$$p_3(x_3,t_3;x_2,t_2;x_1,t_1) = p(x_3,t_3|x_2,t_2)p(x_2,t_2|x_1,t_1)p_1(x_1,t_1).$$

Substituting these results into Eq. (3.4) yields

$$p(x_3,t_3|x_1,t_1)p_1(x_1,t_1) = \left[\int_{-\infty}^{\infty} p(x_3,t_3|x_2,t_2)p(x_2,t_2|x_1,t_1)dx_2\right]p_1(x_1,t_1),$$

which after cancelling out $p_1(x_1,t_1)$ provides the Chapman-Kolmogorov equation that we write as $(t > t' > t_0)$

$$p(x,t|x_0,t_0) = \int_{-\infty}^{\infty} p(x,t|x',t')p(x',t'|x_0,t_0)dx'. \tag{3.5}$$

Suppose that $X(t)$ is a discrete Markovian process taking on a countable infinity of values $n_0, n_1, \ldots$. In this case Eq. (3.5) takes the form

$$p(n,t|n_0,t_0) = \sum_{n'} p(n,t|n',t')p(n',t'|n_0,t_0). \tag{3.6}$$

Let us also note that the nonlinear and very general character of the Chapman-Komogorov equation makes practically impossible obtaining from this equation explicit expressions of the transition density. In other words, the equation is more a defining property of any Markov process than an operative device for getting the transition density.

Multidimensional Markov processes

The concepts that we have just presented are very easily generalized to include vectorial processes. Thus a multidimensional random process,

$$\mathbf{X}(t) = (X_1(t),\ldots,X_N(t)),$$

is Markovian if

$$p_n(\mathbf{x}_n,t_n|\mathbf{x}_{n-1},t_{n-1};\mathbf{x}_{n-2},t_{n-2};\ldots;\mathbf{x}_1,t_1) = p(\mathbf{x}_n,t_n|\mathbf{x}_{n-1},t_{n-1}), \tag{3.7}$$

for all $n = 1,2,3,\ldots$ and $t_n > t_{n-1} > t_{n-2} > \cdots > t_1$.

The function $p(\mathbf{x}_n,t_n|\mathbf{x}_{n-1},t_{n-1})$ is the transition density of the multidimensional process and, as it is straightforward to see, this transition density obeys the multivariate Chapman-Kolmogorov equation

$$p(\mathbf{x},t|\mathbf{x}_0,t_0) = \int_{-\infty}^{\infty} \cdots \int_{-\infty}^{\infty} p(\mathbf{x},t|\mathbf{x}',t')p(\mathbf{x}',t'|\mathbf{x}_0,t_0)d\mathbf{x}'., \tag{3.8}$$

where $d\mathbf{x}' = dx'_1 \cdots dx'_N$ is the N-dimensional volume element.

3.2   Deterministic and Gaussian Markov processes

We now take a look at some general examples of Markov processes and
we will realize that these processes can be seen as a direct generalization
of deterministic, that is, nonrandom processes. We will also discuss that
second-order deterministic processes are non Markovian, unless we enlarge
the phase space.

Another key question which we address in this section is when a Gaussian process turns out to be Markovian. The answer is found in Doob's
theorem (see below).

3.2.1   *Deterministic processes*

Suppose $x = x(t)$ is a nonrandom dynamical process described by a first-
order differential equation,

$$\dot{x} = f(x, t), \qquad (3.9)$$

where $\dot{x}$ is the time derivative of $x(t)$. Let us denote by $x = \phi(t; x_0, t_0)$ the
solution to this equation satisfying the initial condition $x(t_0) = x_0$. Note
that the function ϕ completely determines the value of x at time t knowing
only the value x_0 at some previous time $t_0 < t$. Let us also observe that,
from the point of view of the theory of probability, the conditional density
of the deterministic process will be given by Dirac's delta function as[2]

$$p(x, t | x_0, t_0) = \delta[x - \phi(t; x_0, t_0)]. \qquad (3.10)$$

Observe that this density depends only on the value x_0 at time t_0 and
not on earlier values. In other words, $x(t)$ is a Markov process. This can
be more explicitly seen by noticing that the above density satisfies the
Chapman-Kolmogorov equation (3.5). In effect

$$\int_{-\infty}^{\infty} p(x, t | x', t') p(x', t' | x_0, t_0) dx'$$

$$= \int_{-\infty}^{\infty} \delta[x - \phi(t; x', t')] \delta[x' - \phi(t'; x_0, t_0)] dx'$$

$$= \delta\Big[x - \phi\big(t; \phi(t'; x_0, t_0), t'\big)\Big].$$

---

[2] Indeed, the density is zero unless $x(t) = \phi(t; x_0, t_0)$ which implies deterministic motion
given by the solution to Eq. (3.9). From a well known property of delta function we
have

$$\int_{-\infty}^{\infty} p(x, t | x_0, t_0) dx = \int_{-\infty}^{\infty} \delta[x - \phi(t; x_0, t_0)] dx = 1,$$

so that $p(x, t | x_0, t_0)$ is properly normalized.

But

$$\delta\Big[x - \phi\big(t; \phi(t'; x_0, t_0), t'\big)\Big] = \delta\big[x - \phi(t; x_0, t_0)\big] = p(x, t|x_0, t_0),$$

by virtue of the composition property,

$$\phi\big[t; \phi(t'; x_0, t_0), t'\big] = \phi(t; x_0, t_0)],$$

of the one-parameter group of transformations associated to the deterministic Eq. (3.9) (see, for instance, Arnold, 1978). Hence, the density (3.10) obeys the Chapman-Kolmogorov equation (3.5). We can therefore affirm that *Markov processes are a direct generalization of deterministic dynamical processes*.

Suppose now that, instead a first-order process, we have a deterministic process governed by a second-order differential equation:

$$\ddot{x} = f(x, \dot{x}, t), \tag{3.11}$$

as is the case, for example in physics, of Newton second law.

Let us now denote by $x = \phi(t; x_0, v_0, t_0)$ the solution to Eq. (3.11) satisfying the initial conditions $x(t_0) = x_0$ and $\dot{x}(t_0) = v_0$.[3] The conditional density now is

$$p(x, t|x_0, v_0, t_0) = \delta\big[x - \phi(t; x_0, v_0, t_0)\big].$$

However, this density does not represent a Markov process, for knowing v_0 is equivalent to knowing two values of $x(t)$ at *two near time intervals*. That is, x_0 at time t_0 and $x_0 - h$ at time $t_0 - \tau$ such that

$$v_0 = \lim_{\tau\to 0}\left(\frac{h}{\tau}\right).$$

Consequently the conditional density really depends on two previous values,

$$p(x, t|x_0, v_0, t_0) \approx g(x, t; x_o, t_o; x_0 - h, t_0 - \tau),$$

and not only one value, as is required by the Markov property. Therefore *second-order deterministic process are not Markovian* (see Rytov et al., 1987, for further information).

We can, nonetheless, turn the second-order process (3.11) into a Markovian process by extending the phase space. We thus introduce, in addition to x, a new independent variable $v = \dot{x}$. In other words, instead of Eq. (3.11) we consider the two-dimensional dynamical process (x, v), where

$$\dot{x} = v$$
$$\dot{v} = f(x, v, t).$$

[3]Recall that the solution of any second-order differential equation depends on two arbitrary constants, in the present case x_0 and v_0.

Let $x = \phi(t; x_0, v_0, t_0)$ and $v = \dot{\phi}(t; x_0, v_0, t_0)$ be the solution of this bidimensional system. Then the transition density

$$p(x, v, t | x_0, v_0, t_0) = \delta[x - \phi(t; x_0, v_0, t_0)] \delta[v - \dot{\phi}(t; x_0, v_0, t_0)],$$

only depends on the state (x_0, v_0) at *one previous time* t_0 and not on earlier times. The bidimensional process $(x(t), v(t))$ is, therefore, Markovian while the one-dimensional $x(t)$ is not.

3.2.2 *Gaussian Markov processes*

As we have seen in Chapter 2, Gaussian processes are ubiquitous in natural and socio-economic sciences. A question arises at once: when a Gaussian process is Markovian? For stationary processes there is a rather simple and general answer to this question given by Doob's theorem (Doob, 1942, reprinted in Wax 1954).

Theorem (Doob, 1942)

Let $X(t)$ be a stationary Gaussian process with finite variance. The process is Markovian if

$$K(\tau) = K(0)e^{-\alpha|\tau|}, \qquad (\alpha > 0),$$

where $K(\tau) = \langle X(t + \tau)X(t) \rangle - \langle X(t + \tau) \rangle \langle X(t) \rangle$.

Proof. In what follows we assume that $X(t)$ has zero mean and unit variance,[4] so that $\langle X(t) \rangle = 0$ and

$$K(t - t') = \langle X(t)X(t') \rangle, \qquad K(0) = \langle X^2(0) \rangle = 1.$$

Due to the Gaussian character of $X(t)$ we have seen in Chapter 2 that the bivariate and univariate probability densities of the process are respectively given by

$$p_2(x_2, t_2; x_1, t_1) = \frac{1}{2\pi\sqrt{1 - K^2(t_2, t_1)}}$$

$$\times \exp\left\{ -\frac{1}{1 - K^2(t_2, t_1)} \left[x_2^2 - 2K_2(t_2, t_1)x_2x_1 + x_1^2 \right] \right\},$$

and

$$p_1(x_1, t_1) = \frac{1}{\sqrt{2\pi}} e^{-x_1^2/2}.$$

---

[4]This can be assumed without loss of generality because, if the variance σ of $X(t)$ is finite, we can always work with the normalized process $Z = (X - \langle X \rangle)/\sqrt{\sigma}$ which has zero mean and unit variance.

The conditional probability density function,

$$p(x_2, t_2 | x_1, t_1) = \frac{p_2(x_2, t_2; x_1, t_1)}{p_1(x_1, t_1)},$$

then reads $(t_2 > t_1)$

$$p(x_2, t_2 - t_1 | x_1) = \frac{1}{\sqrt{2\pi[1 - K^2(t_2 - t_1)]}}$$
$$\times \exp\left\{ -\frac{[x_2 - K(t_2 - t_1)x_1]^2}{2[1 - K^2(t_2 - t_1)]} \right\}, \qquad (3.12)$$

where we have taken into account stationarity, so that

$$K(t_2, t_1) = K(t_2 - t_1).$$

Since $X(t)$ is also Markovian the transition density must obey the Chapman-Kolmogorov equation (3.5). Substituting Eq. (3.12) into the right hand side of Eq. (3.5) and integrating on the intermediate value x', we get

$$p(x, t - t_0 | x_0) = \frac{1}{\sqrt{2\pi[1 - K^2(t - t')K^2(t' - t_0)]}}$$
$$\times \exp\left\{ -\frac{[x - K(t - t')K(t' - t_0)x_0]^2}{2[1 - K^2(t - t')K^2(t' - t_0)]} \right\}.$$

Equating to Eq. (3.12) (with the replacements $x_2 \to x$, $t_2 \to t$, $x_1 \to x_0$ and $t_1 \to t_0$) we see that the correlation function obeys the functional equation

$$K(t - t_0) = K(t - t')K(t' - t_0), \qquad (t \geq t' \geq t_0),$$

which we write as

$$K(\tau + s) = K(\tau)K(s), \qquad (\tau, s \geq 0). \qquad (3.13)$$

The correlation function $K(\tau)$ has the properties: (i) it is even, $K(-\tau) = K(\tau)$; (ii) $K(0) = 1$; (iii) $|K(\tau)| \leq 1$ for all $\tau \in \mathbb{R}$. This last property is a direct consequence of the Schwarz inequality; in effect

$$|K(\tau)| = |\langle X(t + \tau)X(t)\rangle| \leq \langle X^2(t + \tau)\rangle\langle X^2(t)\rangle = K^2(0) = 1.$$

The solution to the functional equation (3.13) satisfying properties (i)-(iii) is

$$K(\tau) = e^{-\alpha|\tau|},$$

where $\alpha > 0$ is an arbitrary constant having dimensions of (time)$^{-1}$. The reciprocal, $1/\alpha$, is called the *correlation time* of the process.

□

Remark

One of most widely used random process in countless applications is the so-called *Gaussian white noise*. It is a stationary Gaussian process with zero mean and correlation function:

$$K(\tau) = \delta(\tau).$$

Despite not having the exponential form required by Doob's theorem, Gaussian white noise is also Markovian since its transition density obeys the Chapman-Kolmogorov equation, for it satisfies Eq. (3.13) because

$$\delta(\tau + s) = \delta(\tau)\delta(s)$$

($\tau \geq 0, s \geq 0$) which is a standard property of the delta function.

3.3 Kinetic equations. The Kramers-Moyal expansion

We will see in this section that the Chapman-Kolmogorov equation (3.5) –a nonlinear integral equation for the transition probability of any Markov process– is equivalent to a differential equation with an infinite number of spatial derivatives. The resulting differential equation is sometimes called *kinetic equation* and the procedure to get it *Kramers-Moyal expansion*.

We begin with the transition density of a Markov process from the initial state $X(t_0) = x_0$ to the final state $X(t + \Delta t) = x$. This transition density, $p(x, t + \Delta t | x_0, t_0)$, obeys the Chapman-Kolmogorov equation (3.5),

$$p(x, t + \Delta t | x_0, t_0) = \int_{-\infty}^{\infty} p(x, t + \Delta t | y, t) p(y, t | x_0, t_0) dy, \qquad (3.14)$$

where $t > t_0$ and $\Delta t > 0$.

We define the Fourier transform of the transition density as [5]

$$\phi(\omega, t + \Delta t | y, t) = \int_{-\infty}^{\infty} e^{i\omega(x-y)} p(x, t + \Delta t | y, t) dx. \qquad (3.15)$$

Before proceeding further we first obtain the Fourier inversion of this equation. We multiply both sides of Eq. (3.15) by $e^{-i\omega(z-y)}$ (z is an arbitrary real number) and integrate over ω, we have

$$\int_{-\infty}^{\infty} e^{-i\omega(z-y)} \phi(\omega, t + \Delta t | y, t) d\omega$$

$$= \int_{-\infty}^{\infty} d\omega e^{-i\omega(z-y)} \int_{-\infty}^{\infty} e^{i\omega(x-y)} p(x, t + \Delta t | y, t) dx$$

---

[5]We introduce the extra term $e^{-i\omega y}$ in the Fourier transform for convenience.

which, after some cancellation and the interchange of integrals, yields

$$\int_{-\infty}^{\infty} e^{-i\omega(z-y)}\phi(\omega, t+\Delta t|y, t)d\omega = \int_{-\infty}^{\infty} p(x, t+\Delta t|y, t)dx \int_{-\infty}^{\infty} e^{-i\omega(z-x)}d\omega$$

$$= 2\pi \int_{-\infty}^{\infty} p(x, t+\Delta t|y, t)\delta(z-x)dx = 2\pi p(z, t+\Delta t|y, t),$$

where we have used the following standard representation of the delta function (Vladimirov, 1984)

$$\delta(z-x) = \frac{1}{2\pi} \int_{-\infty}^{\infty} e^{\pm i\omega(x-y)}d\omega. \tag{3.16}$$

Hence, the Fourier inversion of Eq. (3.15) is

$$p(x, t+\Delta t|y, t) = \frac{1}{2\pi} \int_{-\infty}^{\infty} e^{-i\omega(x-y)}\phi(\omega, t+\Delta t|y, t)d\omega. \tag{3.17}$$

Let us return to Eq. (3.15) and expand the exponential term inside the integral with the result

$$\phi(\omega, t+\Delta t|y, t) = \sum_{n=0}^{\infty} \frac{(i\omega)^n}{n!} \int_{-\infty}^{\infty} (x-y)^n p(x, t+\Delta t|y, t)dx,$$

which, after defining the quantities

$$a_n(t+\Delta t|y, t) \equiv \int_{-\infty}^{\infty} (x-y)^n p(x, t+\Delta t|y, t)dx, \tag{3.18}$$

can be written as

$$\phi(\omega, t+\Delta t|y, t) = \sum_{n=0}^{\infty} \frac{(i\omega)^n}{n!} a_n(t+\Delta t|y, t).$$

We next substitute this expression for the function ϕ into the Fourier inversion formula (3.17) and get

$$p(x, t+\Delta t|y, t) = \frac{1}{2\pi} \sum_{n=0}^{\infty} \frac{1}{n!} a_n(t+\Delta t|y, t) \int_{-\infty}^{\infty} e^{-i\omega(x-y)}(i\omega)^n d\omega.$$

But, using Eq. (3.16) in the form

$$\delta(x-y) = \frac{1}{2\pi} \int_{-\infty}^{\infty} e^{-i\omega(x-y)}d\omega,$$

we see that

$$\frac{\partial^n}{\partial x^n}\delta(x-y) = \frac{1}{2\pi} \int_{-\infty}^{\infty} e^{-i\omega(x-y)}(-i\omega)^n d\omega,$$

that is,

$$\int_{-\infty}^{\infty} e^{-i\omega(x-y)}(i\omega)^n d\omega = 2\pi(-1)^n \frac{\partial^n}{\partial x^n}\delta(x-y).$$

Hence,

$$p(x, t+\Delta t|y, t) = \sum_{n=0}^{\infty} \frac{(-1)^n}{n!} a_n(t+\Delta t|y, t)\frac{\partial^n}{\partial x^n}\delta(x-y).$$

Substituting the last expression into the right hand side of the Chapman-Kolmogorov equation (3.14) yields

$$p(x, t+\Delta t|x_0, t_0) = \sum_{n=0}^{\infty} \frac{(-1)^n}{n!}\frac{\partial^n}{\partial x^n}\int_{-\infty}^{\infty}\delta(x-y)a_n(t+\Delta t|y, t)p(y, t|x_0, t_0)dy$$

$$= \sum_{n=0}^{\infty} \frac{(-1)^n}{n!}\frac{\partial^n}{\partial x^n}\left[a_n(t+\Delta t|x, t)p(x, t|x_0, t_0)\right],$$

which, taking into account that $a_0 = 1$,[6] reads

$$p(x, t+\Delta t|x_0, t_0) = p(x, t|x_0, t_0)+\sum_{n=1}^{\infty} \frac{(-1)^n}{n!}\frac{\partial^n}{\partial x^n}\left[a_n(t+\Delta t|x, t)p(x, t|x_0, t_0)\right].$$

We can thus write

$$\frac{p(x, t+\Delta t|x_0, t_0) - p(x, t|x_0, t_0)}{\Delta t}$$

$$= \sum_{n=1}^{\infty} \frac{(-1)^n}{n!}\frac{\partial^n}{\partial x^n}\left[\frac{a_n(t+\Delta t|x, t)}{\Delta t}p(x, t|x_0, t_0)\right]$$

and, finally,

$$\frac{\partial}{\partial t}p(x, t|x_0, t_0) = \sum_{n=1}^{\infty} \frac{(-1)^n}{n!}\frac{\partial^n}{\partial x^n}\left[A_n(x, t)p(x, t|x_0, t_0)\right], \qquad (3.19)$$

where

$$A_n(x, t) \equiv \lim_{\Delta t\to 0}\frac{a_n(t+\Delta t|x, t)}{\Delta t}. \qquad (3.20)$$

Equation (3.19) is the differential form of the Chapman-Kolmogorov equation also called (forward) *kinetic equation*.

---

[6]See Eq. (3.18) and recall the normalization of the transition density.

3.3.1   *Infinitesimal moments*

The functions $A_n(x,t)$ are defined by Eq. (3.20) but because of Eq. (3.18) they can be written as

$$A_n(x,t) = \lim_{\Delta t \to 0} \left[\frac{1}{\Delta t} \int_{-\infty}^{\infty} (y-x)^n p(y, t+\Delta t | x, t) dy \right] \qquad (3.21)$$

and are called *infinitesimal moments* of the Markov process, we will shortly see why.

Using the binomial expansion for $(y-x)^n$ we have

$$\int_{-\infty}^{\infty} (y-x)^n p(y, t+\Delta t | x, t) dy = \sum_{k=0}^{n} \binom{n}{k} (-x)^{n-k} \int_{-\infty}^{\infty} y^k p(y, t+\Delta t | x, t) dy.$$

However,

$$\int_{-\infty}^{\infty} y^k p(y, t+\Delta t | x, t) dy = \left\langle X^k(t+\Delta t) | X(t) = x \right\rangle,$$

are conditional moments of the process. Thus

$$\int_{-\infty}^{\infty} (y-x)^n p(y, t+\Delta t | x, t) dy$$

$$= \sum_{k=0}^{n} \binom{n}{k} (-x)^{n-k} \left\langle X^k(t+\Delta t) | X(t) = x \right\rangle$$

$$= \left\langle \sum_{k=0}^{n} \binom{n}{k} (-x)^{n-k} X^k(t+\Delta t) \Big| X(t) = x \right\rangle$$

$$= \left\langle \left[X(t+\Delta t) - x \right]^n \Big| X(t) = x \right\rangle$$

$$= \left\langle \left[X(t+\Delta t) - X(t) \right]^n \Big| X(t) = x \right\rangle,$$

that is,

$$\int_{-\infty}^{\infty} (y-x)^n p(y, t+\Delta t | x, t) dy = \left\langle \left(\Delta X(t) \right)^n \Big| X(t) = x \right\rangle, \qquad (3.22)$$

where

$$\Delta X(t) = X(t+\Delta t) - X(t).$$

Substituting Eq. (3.22) into Eq. (3.21) yields

$$A_n(x,t) = \lim_{\Delta t \to 0} \left[\frac{1}{\Delta t} \left\langle \left(\Delta X(t) \right)^n \Big| X(t) = x \right\rangle \right], \qquad (3.23)$$

which is equivalent to

$$A_n(x,t) \Delta t = \left\langle \left(\Delta X(t) \right)^n \Big| X(t) = x \right\rangle + O(\Delta t^2). \qquad (3.24)$$

As we will see later the first two infinitesimal moments are the most relevant. The first moment $A_1(x,t)$ is called the *drift* of the process and we usually employ the notation $f(x,t)$ to designate it. Thus

$$f(x,t) = A_1(x,t) = \lim_{\Delta t \to 0} \left\langle \frac{\Delta X}{\Delta t} \middle| X(t) = x \right\rangle. \tag{3.25}$$

The second infinitesimal moment $A_2(x,t)$ is called the *diffusion coefficient* and it is related to the conditional variance of the increment of the process. Indeed,

$$\begin{aligned}
\text{Var}\{\Delta X(t)|X(t) = x\} &\equiv \langle (\Delta X)^2 | X(t) = x \rangle - [\langle \Delta X | X(t) = x \rangle]^2 \\
&= A_2(x,t)\Delta t + O(\Delta t^2),
\end{aligned}$$

so that

$$\begin{aligned}
A_2(x,t) &= \lim_{\Delta t \to 0} \left[\frac{1}{\Delta t} \text{Var}\{\Delta X(t)|X(t) = x\} \right] \\
&= \lim_{\Delta t \to 0} \left\langle \frac{(\Delta X)^2}{\Delta t} \middle| X(t) = x \right\rangle. \tag{3.26}
\end{aligned}$$

3.3.2 *Kinetic equations for the univariate and bivariate densities*

Let us now show that the forward kinetic equation for the transition density, Eq. (3.19), is also valid for the univariate and the bivariate probability density functions, $p_1(x,t)$ and $p_2(x,t;x_0,t_0)$ respectively.

Indeed, in terms of the transition density the bivariate density is given by

$$p_2(x,t;x_0,t_0) = p(x,t|x_0,t_0)p_1(x_0,t_0).$$

If we multiply both sides of the kinetic equation (3.19) by the univariate density $p_1(x_0,t_0)$ and bearing in mind that all derivatives refer to forward variables x and t and not to x_0 and t_0, we recover the Kramers-Moyal expansion for the bivariate density:

$$\frac{\partial}{\partial t} p_2(x,t;x_0,t_0) = \sum_{n=1}^{\infty} \frac{(-1)^n}{n!} \frac{\partial^n}{\partial x^n} [A_n(x,t)p_2(x,t;x_0,t_0)]. \tag{3.27}$$

On the other hand, since

$$p_1(x,t) = \int_{-\infty}^{\infty} p_2(x,t;x_0,t_0)dx_0,$$

integrating Eq. (3.27) with respect to x_0 we see that the univariate density also obeys the Kramers-Moyal expansion:

$$\frac{\partial}{\partial t} p_1(x,t) = \sum_{n=1}^{\infty} \frac{(-1)^n}{n!} \frac{\partial^n}{\partial x^n} [A_n(x,t)p_1(x,t)]. \tag{3.28}$$

Truncating Kramers-Moyal expansions

For practical purposes as, for instance, in the numerical calculation of the transition density of a Markov process, it is very convenient to know how many terms of the Kramers-Moyal series we should retain in order to get a good approximation. The answer is given by the following theorem:

Theorem (Pawula, 1967)

In order to obtain a positive solution of the kinetic equation for the transition probability density of any Markov processes, the Kramer-Moyal expansion must be truncated to first or second order or, otherwise, maintain an infinite number of terms.

The proof of this theorem is based on the application of the generalized Schwarz inequality and we refer the reader to Risken's book (Risken, 1987) for more information.

We must, however, remark that this theorem does not affirm that Kramers-Moyal expansions truncated at some $n \geq 3$ are useless in the approximate calculation of transition densities. In fact it can be shown that better approximations for $p(x, t|x_0, t_0)$ are achieved for increasing values of n. Although for greater values than $n = 2$ the resulting functions present regions where they are negative, which is incompatible with the axioms of prbability. If these negative regions are small, the rest of the function can be a good approximation to the transition density (see Risken, 1987, for more details).

3.4   Backward kinetic equations

Kinetic equations presented up to now are "forward equations", in the sense that they relate derivatives of the transition density with respect to the forward state x and time t, while past state and time (x_0, t_0) appear as fixed parameters but not as variables. We will see next that the role of initial and final variables, (x_0, t_0) and (x, t), are interchangeable which results in "backward" kinetic equations.

As in Sec. 3.3 we begin with the Chapman-Kolmogorov equation (3.14) now written in the form:

$$p(x, t|x_0, t_0 - \Delta t_0) = \int_{-\infty}^{\infty} p(x, t|y, t_0)p(y, t_0|x_0, t_0 - \Delta t_0)dy. \qquad (3.29)$$

where $t > t_0$ and $\Delta t_0 > 0$.

Starting from the normalization condition

$$\int_{-\infty}^{\infty} p(y, t_0 | x_0, t_0 - \Delta t_0) dy = 1,$$

we write the identity

$$p(x, t | x_0, t_0) = p(x, t | x_0, t_0) \int_{-\infty}^{\infty} p(y, t_0 | x_0, t_0 - \Delta t_0) dy$$

$$= \int_{-\infty}^{\infty} p(x, t | x_0, t_0) p(y, t_0 | x_0, t_0 - \Delta t_0) dy.$$

Subtracting this equation from Eq. (3.29) yields

$$p(x, t | x_0, t_0 - \Delta t_0) - p(x, t | x_0, t_0)$$

$$= \int_{-\infty}^{\infty} [p(x, t | y, t_0) - p(x, t | x_0, t_0)] p(y, t_0 | x_0, t_0 - \Delta t_0) dy.$$

We now expand $p(x, t | y, t_0)$ in powers of y around x_0:

$$p(x, t | y, t_0) = p(x, t | x_0, t_0) + \sum_{n=1}^{\infty} \frac{(y - x_0)^n}{n!} \frac{\partial^n}{\partial x_0^n} p(x, t | x_0, t_0),$$

which, after substituting into the previous equation, yields

$$p(x, t | x_0, t_0 - \Delta t_0) - p(x, t | x_0, t_0)$$

$$= \Delta t_0 \sum_{n=1}^{\infty} \frac{1}{n!} \bar{A}_n(x_0, t_0, t_0 - \Delta t_0) \frac{\partial^n}{\partial x_0^n} p(x, t | x_0, t_0), \qquad (3.30)$$

where

$$\bar{A}_n(x_0, t_0, t_0 - \Delta t_0) \equiv \frac{1}{\Delta t_0} \int_{-\infty}^{\infty} (y - x_0)^n p(y, t_0 | x_0, t_0 - \Delta t_0) dy. \quad (3.31)$$

Before proceeding further, let us first show that, in the limit $\Delta t_0 \to 0$, the quantities $\bar{A}_n(x_0, t_0, t_0 - \Delta t_0)$ coincide with the infinitesimal moments, Eq. (3.21), evaluated at the initial state (x_0, t_0); that is

$$\lim_{\Delta t_0 \to 0} \bar{A}_n(x_0, t_0, t_0 - \Delta t_0) = A_n(x_0, t_0). \qquad (3.32)$$

Indeed, from the definition of $\bar{A}_n$ given in Eq. (3.31) and using Eq. (3.21) we see that

$$\lim_{\Delta t_0 \to 0} \bar{A}_n(x_0, t_0, t_0 - \Delta t_0)$$

$$\equiv \lim_{\Delta t_0 \to 0} \left[\frac{1}{\Delta t_0} \int_{-\infty}^{\infty} (y - x_0)^n p(y, t_0 | x_0, t_0 - \Delta t_0) dy \right]$$

$$= \lim_{\Delta t_0 \to 0} \left[\frac{1}{\Delta t_0} \int_{-\infty}^{\infty} (y - x_0)^n p(y, t_0 + \Delta t_0 | x_0, t_0) dy \right] \equiv A_n(x_0, t_0),$$

where in the passage from the first integral to the second we have taken into account that, as $\Delta t_0 \to 0$

$$\lim_{\Delta t_0 \to 0} p(y, t_0 | x_0, t_0 - \Delta t_0) = \lim_{\Delta t_0 \to 0} p(y, t_0 + \Delta t_0 - \Delta t_0 | x_0, t_0 - \Delta t_0)$$

$$= \lim_{\Delta t_0 \to 0} p(y, t_0 + \Delta t_0 | x_0, t_0).$$

We now return to the derivation of the backward Kramers-Moyal expansion. Taking the limit $\Delta t_0 \to 0$ in Eq. (3.30), using Eq. (3.32) and noticing that

$$\lim_{\Delta t_0 \to 0} \frac{p(x, t | x_0, t_0 - \Delta t_0) - p(x, t | x_0, t_0)}{\Delta t_0} = -\frac{\partial}{\partial t_0} p(x, t | x_0, t_0),$$

we finally get the *backward kinetic equation*

$$\frac{\partial}{\partial t_0} p(x, t | x_0, t_0) = -\sum_{n=1}^{\infty} \frac{1}{n!} A_n(x_0, t_0) \frac{\partial^n}{\partial x_0^n} p(x, t | x_0, t_0). \qquad (3.33)$$

Both, the forward equation (3.19) and the backward equation (3.33) satisfy the initial (or "final") condition

$$\lim_{t \to t_0} p(x, t | x_0, t_0) = \delta(x - x_0). \qquad (3.34)$$

It can be proved (see Risken, 1987, for details) that the solutions of the backward and froward equations are completely equivalent to each other. The choice of solving one equation or the other is a matter of convenience depending on the problem at hand.

We can easily see that neither the bivariate density $p_2(x, t; x_0, t_0)$ nor the univariate density $p_1(x_0, t_0)$ satisfy the backward kinetic equation (as they do in the case of the forward equation). However, the conditional distribution function,

$$F(x, t | x_0, t_0) = \int_{-\infty}^{x} p(x', t | x_0, t_0) dx',$$

does satisfy the backward equation, for integrating Eq. (3.33) with respect to x' we get

$$\frac{\partial}{\partial t_0} F(x, t | x_0, t_0) = -\sum_{n=1}^{\infty} \frac{1}{n!} A_n(x_0, t_0) \frac{\partial^n}{\partial x_0^n} F(x, t | x_0, t_0), \qquad (3.35)$$

with the "final" condition

$$\lim_{t_0 \to t} F(x, t | x_0, t_0) = \theta(x - x_0), \qquad (3.36)$$

where $\theta(x - x_0)$ is the Heavise step function. The last statement is easily proved by integrating Eq. (3.34):

$$\lim_{t_0 \to t} F(x, t | x_0, t_0) = \int_{-\infty}^{x} \delta(x' - x_0) dx'$$

$$= \int_{-\infty}^{\infty} \theta(x - x') \delta(x' - x_0) dx' = \theta(x - x_0).$$

3.5 Multidimensional kinetic equations

We now generalize the forward and backward kinetic equations for vector Markov processes $\mathbf{X}(t) = \big(X_1(t),\ldots,X_N(t)\big)$. The starting point is the multidimensional Chapman-Kolmogorov equation (3.8) that we write in the form

$$p(\mathbf{x},t+\Delta t|\mathbf{x}_0,t_0) = \int_{-\infty}^{\infty} \cdots \int_{-\infty}^{\infty} p(\mathbf{x},t+\Delta t|\mathbf{y},t)p(\mathbf{y},t|\mathbf{x}_0,t_0)d\mathbf{y}., \quad (3.37)$$

where $t > t_0$, $\Delta t > 0$ and $d\mathbf{y} = dy_1 \cdots dy_N$ is the N-dimensional volume element.

In order to derive the Kramers-Moyal expansion for the N-dimensional Markov process, we follow a different, although completely equivalent, way to that used in the one dimensional case. We begin with the following identity

$$p(\mathbf{x},t+\Delta t|\mathbf{y},t) = \int_{-\infty}^{\infty} \cdots \int_{-\infty}^{\infty} \delta^N(\mathbf{z}-\mathbf{x})p(\mathbf{z},t+\Delta t|\mathbf{y},t)d\mathbf{z}, \quad (3.38)$$

and expand the N-dimensional delta function,

$$\delta^N(\mathbf{z}-\mathbf{x}) = \delta(x_1-x_1)\cdots\delta(z_N-x_N),$$

in Taylor series around $\mathbf{z}=\mathbf{y}$,

$$\delta^N(\mathbf{z}-\mathbf{x}) = \delta^N(\mathbf{y}-\mathbf{x})$$
$$+ \sum_{n=1}^{\infty} \frac{1}{n!} \sum_{k_1=1}^{N} \cdots \sum_{k_n=1}^{N} (z_{k_1}-y_{k_1})\cdots(z_{k_n}-y_{k_n})\frac{\partial^n \delta^N(\mathbf{y}-\mathbf{x})}{\partial y_{k_1}\cdots\partial y_{k_n}}.$$

Substituting this development into Eq. (3.38) yields

$$p(\mathbf{x},t+\Delta t|\mathbf{x}_0,t_0) = \delta^N(\mathbf{y}-\mathbf{x})\int_{-\infty}^{\infty} \cdots \int_{-\infty}^{\infty} p(\mathbf{z},t+\Delta t|\mathbf{y},t)d\mathbf{z}$$
$$+ \sum_{n=1}^{\infty} \frac{1}{n!} \sum_{k_1=1}^{N} \cdots \sum_{k_n=1}^{N} a_{k_1\cdots k_n}^{(n)}(t+\Delta t|\mathbf{y},t)\frac{\partial^n \delta^N(\mathbf{y}-\mathbf{x})}{\partial y_{k_1}\cdots\partial y_{k_n}}, \quad (3.39)$$

where

$$a_{k_1\cdots k_n}^{(n)}(t+\Delta t|\mathbf{y},t) \equiv \int_{-\infty}^{\infty} \cdots \int_{-\infty}^{\infty} (z_{k_1}-y_{k_1})\cdots(z_{k_n}-y_{k_n})$$
$$\times p(\mathbf{z},t+\Delta t|\mathbf{y},t)d\mathbf{z}. \quad (3.40)$$

Taking into account the normalization of the transition density,

$$\int_{-\infty}^{\infty} \cdots \int_{-\infty}^{\infty} p(\mathbf{z},t+\Delta t|\mathbf{y},t)d\mathbf{z} = 1,$$

Equation (3.39) yields

$$p(\mathbf{x}, t + \Delta t | \mathbf{x}_0, t_0) = \delta^N(\mathbf{y} - \mathbf{x})$$

$$+ \sum_{n=1}^{\infty} \frac{1}{n!} \sum_{k_1=1}^{N} \cdots \sum_{k_n=1}^{N} a_{k_1 \cdots k_n}^{(n)}(t + \Delta t | \mathbf{y}, t) \frac{\partial^n \delta^N(\mathbf{y} - \mathbf{x})}{\partial y_{k_1} \cdots \partial y_{k_n}}, \qquad (3.41)$$

Recall that $p(\mathbf{z}, t + \Delta t | \mathbf{y}, t) \to \delta^N(\mathbf{z} - \mathbf{y})$ as $\Delta t \to 0$, so that from Eq. (3.40) we see that

$$a_{k_1 \cdots k_n}^{(n)}(t + \Delta t | \mathbf{y}, t) \to 0 \quad \text{as} \quad \Delta t \to 0,$$

which allows us to write

$$a_{k_1 \cdots k_n}^{(n)}(t + \Delta t | \mathbf{y}, t) = A_{k_1 \cdots k_n}^{(n)}(\mathbf{y}, t) \Delta t + O(\Delta t^2), \qquad (3.42)$$

where the quantities $A_{k_1 \cdots k_n}^{(n)}(\mathbf{y}, t)$ are defined by

$$A_{k_1 \cdots k_n}^{(n)}(\mathbf{y}, t) \equiv \lim_{\Delta t \to 0} \left[\frac{1}{\Delta t} a_{k_1 \cdots k_n}^{(n)}(t + \Delta t | \mathbf{y}, t) \right]. \qquad (3.43)$$

Introducing Eq. (3.42) into Eq. (3.41) we have

$$p(\mathbf{x}, t + \Delta t | \mathbf{x}_0, t_0) = \delta^N(\mathbf{y} - \mathbf{x})$$

$$+ \Delta t \left[\sum_{n=1}^{\infty} \frac{1}{n!} \sum_{k_1=1}^{N} \cdots \sum_{k_n=1}^{N} A_{k_1 \cdots k_n}^{(n)}(\mathbf{y}, t) \frac{\partial^n \delta^N(\mathbf{y} - \mathbf{x})}{\partial y_{k_1} \cdots \partial y_{k_n}} \right] + O(\Delta t^2),$$

Substituting this expression into the Chapman-Kolmogorov equation (3.37) yields

$$\frac{p(\mathbf{x}, t + \Delta t | \mathbf{x}_0, t_0) - p(\mathbf{x}, t | \mathbf{x}_0, t_0)}{\Delta t}$$

$$= \sum_{n=1}^{\infty} \frac{1}{n!} \sum_{k_1=1}^{N} \cdots \sum_{k_n=1}^{N} p(\mathbf{y}, t | \mathbf{x}_0, t_0) A_{k_1 \cdots k_n}^{(n)}(\mathbf{y}, t) \frac{\partial^n \delta^N(\mathbf{y} - \mathbf{x})}{\partial y_{k_1} \cdots \partial y_{k_n}} + O(\Delta t).$$

Taking the limit $\Delta t \to 0$ and implementing the standard property of the delta function,

$$\int f(y) \frac{\partial^n \delta(y - x)}{\partial y^n} dy = (-1)^n \frac{d^n f(x)}{dx^n},$$

we obtain the *multidimensional forward kinetic equation:*

$$\frac{\partial}{\partial t} p(\mathbf{x}, t | \mathbf{x}_0, t_0) \qquad (3.44)$$

$$= \sum_{n=1}^{\infty} \frac{(-1)^n}{n!} \sum_{k_1=1}^{N} \cdots \sum_{k_n=1}^{N} \frac{\partial^n}{\partial x_{k_1} \cdots \partial x_{k_n}} \left[A_{k_1 \cdots k_n}^{(n)}(\mathbf{x}, t) p(\mathbf{x}, t | \mathbf{x}_0, t_0) \right].$$

Following an analogous procedure we can find the backward equation (see Risken, 1987, for more information)

$$\frac{\partial}{\partial t_0} p(\mathbf{x}, t | \mathbf{x}_0, t_0) \tag{3.45}$$

$$= -\sum_{n=1}^{\infty} \frac{1}{n!} \sum_{k_1=1}^{N} \cdots \sum_{k_n=1}^{N} A_{k_1 \cdots k_n}^{(n)}(\mathbf{x}_0, t_0) \frac{\partial^n}{\partial x_{k_1}^0 \cdots \partial x_{k_n}^0} p(\mathbf{x}, t | \mathbf{x}_0, t_0),$$

where we have indicated the components of the initial state as $\mathbf{x}_0 = (x_1^0, \ldots, x_N^0)$.

The quantities $A_{k_1 \cdots k_n}^{(n)}$ appearing in both expansions (forward and backward) are called the *multidimensional infinitesimal moments* of order n of the vector Markov process $\mathbf{X}(t)$. They are defined by Eq. (3.43) which, combined with Eq. (3.40), can be written as

$$A_{k_1 \cdots k_n}^{(n)}(\mathbf{x}, t) \tag{3.46}$$

$$= \lim_{\Delta t \to 0} \left[\frac{1}{\Delta t} \int_{-\infty}^{\infty} \cdots \int_{-\infty}^{\infty} (z_{k_1} - x_{k_1}) \cdots (z_{k_n} - x_{k_n}) p(\mathbf{z}, t + \Delta t | \mathbf{x}, t) d\mathbf{z} \right].$$

Let us observe that the first infinitesimal moment –which corresponds to $n = 1$ and called the drift of the process– is the N-dimensional vector

$$A_k(\mathbf{x}, t) = \lim_{\Delta t \to 0} \left[\frac{1}{\Delta t} \int_{-\infty}^{\infty} \cdots \int_{-\infty}^{\infty} (z_k - x_k) p(\mathbf{z}, t + \Delta t | \mathbf{x}, t) d\mathbf{z} \right], \tag{3.47}$$

$(k = 1, 2, \ldots, N)$. The second infinitesimal moment $(n = 2)$ is the $N \times N$ matrix

$$A_{jk}(\mathbf{x}, t) \tag{3.48}$$

$$= \lim_{\Delta t \to 0} \left[\frac{1}{\Delta t} \int_{-\infty}^{\infty} \cdots \int_{-\infty}^{\infty} (z_j - x_j)(z_k - x_k) p(\mathbf{z}, t + \Delta t | \mathbf{x}, t) d\mathbf{z} \right],$$

$(j, k = 1, 2, \ldots, N)$. In these expressions we have dropped the superscript representing the order of the infinitesimal moments for the ease of notation. Higher-order moments with $n \geq 3$ are represented by higher-order matrices, $A_{jkl}, A_{jklr}, \ldots$ Although they usually have no role in most applications.

Finally, backward and forward equations both must satisfy the initial (or final) condition:

$$\lim_{t \to t_0} p(\mathbf{x}, t | \mathbf{x}_0, t_0) = \delta^N(\mathbf{x} - \mathbf{x}_0). \tag{3.49}$$

3.6 Homogeneous and stationary Markov processes

In this section we analyze when Markov processes are stationary. We begin with one-dimensional processes and at the end of the section we briefly generalize the analysis to include multidimensional processes.

A Markov process is (time) *homogeneous* when the transition density depends on time differences:

$$p(x,t|y,t') = p(x,t-t'|y,0), \qquad (t \geq t'). \tag{3.50}$$

The most important characteristic of homogeneous processes is that their infinitesimal moments are independent of time. Indeed, from Eq. (3.21) and taking into account that now

$$p(y,t+\Delta t|x,t) = p(y,\Delta t|x,0)$$

depends only on the time difference Δt, we see that

$$A_n(x,t) = \lim_{\Delta t \to 0} \left[\frac{1}{\Delta t} \int_{-\infty}^{\infty} (y-x)^n p(y,\Delta t|x,0)dy \right] \equiv A_n(x), \tag{3.51}$$

and the infinitesimal moments are time independent.

Let us next analyze the following question: is a homogeneous process also stationary? Recall from Chapter 2 that a random process is (strictly) stationary if all its finite-dimensional distributions are invariant under time translations, that is to say

$$p_n(x_1,t_1;x_2,t_2;\ldots;x_n,t_n) = p_n(x_1,t_1+h;x_2,t_2+h;\ldots;x_n,t_n),$$

$n=1,3,3,\ldots$ and $h \in \mathbb{R}$ arbitrary. Choosing $h=-t_1$ stationarity means that

$$p_n(x_1,t_1;x_2,t_2;\ldots;x_n,t_n) = p_n(x_1,0;x_2,t_2-t_1;\ldots;x_n,t_n-t_1).$$

For Markov processes we have the following theorem

Theorem

A Markov process is stationary when:

(1) The transition density is invariant under time translations,

$$p(x,t|y,t') = p(x,t-t'|y,0).$$

(2) The univariate density is independent of time,

$$p_1(x,t) = p_1(x,0),$$

where $p_1(x,0) \equiv p_s(x)$ is called the stationary density of the process.

Proof. We know that any Markov process is defined by two functions, the transition and the univariate density's, since any finite dimensional density can be written in terms of these two functions as [cf. Eq. (3.3)]

$$p_n(x_1, t_1; x_2, t_2; \ldots; x_n, t_n) = \prod_{j=2}^{n} p(x_j, t_j | x_{j-1}, t_{j-1}) p_1(x_1, t_1),$$

$(n = 2, 3, \ldots)$. Assuming conditions (1) and (2) we have

$$p(x_j, t_j | x_{j-1}, t_{j-1}) = p(x_j, t_j - t_{j-1} | x_{j-1}, 0)$$
$$= p(x_j, t_j - t_1 - (t_{j-1} - t_1) | x_{j-1}, 0) = p(x_j, t_j - t_1 | x_{j-1}, t_{j-1} - t_1),$$

and

$$p_1(x_1, t_1) = p_1(x_1, 0) = p_1(x_1, t_1 - t_1).$$

Therefore,

$$p_n(x_1, t_1; x_2, t_2; \ldots; x_n, t_n) = \prod_{j=2}^{n} p(x_j, t_j - t_1 | x_{j-1}, t_{j-1} - t_1) p_1(x_1, t_1 - t_1)$$
$$= p_n(x_1, 0; x_2, t_2 - t_1; \ldots; x_n, t_n - t_1),$$

and the process is stationary.

□

We can now answer the question posed above and affirm that *any homogeneous Markov process is stationary when the univariate density,*

$$p_1(x, t) = p_s(x),$$

is independent of time. Indeed

$$p_1(x, t) = \int_{-\infty}^{\infty} p_2(x, t; y, t') dy = \int_{-\infty}^{\infty} p(x, t - t' | y, 0) p_1(y, t') dy,$$

and since $p_1(x, t) = p_s(x)$ we see that the homogeneous process will be stationary if there exists the so-called *invariant density* which is the non trivial (i.e., nonzero) solution of the integral equation

$$p_s(x) = \int_{-\infty}^{\infty} p(x, \tau | y) p_s(y) dy, \qquad (3.52)$$

where $p(x, \tau | y)$ ($\tau > 0$) is the transition density of the process.

Let us observe that if the transition density $p(x, \tau | y)$ goes, as $\tau \to \infty$, to a finite and non vanishing limit, which is independent of the initial value y, that is

$$\lim_{\tau \to 0} p(x, \tau | y) = \phi(x) \neq 0,$$

then, taking the limit $\tau \to \infty$ in Eq. (3.52) and recalling normalization, we get

$$p_s(x) = \phi(x) \int_{-\infty}^{\infty} p_s(y)dy = \phi(x).$$

We, therefore, see that the *stationary density* $p_s(x)$ is the long-time limit of the transition density

$$p_s(x) = \lim_{\tau \to \infty} p(x, \tau | y) \tag{3.53}$$

as far as the limit exists, it is independent of y and non vanishing.

Let us incidentally note that, since $p(x, t | x_0, t_0) = p(x, t - t_0 | x_0)$, the stationary limit $t - t_0 \to \infty$ can be either achieved by letting $t \to \infty$ (i.e., in the distant future) or $t_0 \to -\infty$ (i.e., the remote past).

On the other hand for any stationary process –in fact, for any homogeneous process– the choice of the initial time t_0 is irrelevant and we may take $t_0 = 0$ at our convenience. Indeed, since the process is invariant under time translations we can always define a new time scale $t \to t - t_0$ in which $t_0 = 0$.

We finally remark that the results of this section are straightforwardly extended to multidimensional Markov processes. Thus $\mathbf{X}(t)$ is time homogeneous if the transition density depends on time differences

$$p(\mathbf{x}, t | \mathbf{y}, t') = p(\mathbf{x}, t - t' | \mathbf{y}).$$

The multidimensional infinitesimal moments defined in Eq. (3.46) are now independent of time

$$A^{(n)}_{k_1 \cdots k_n}(\mathbf{x}) \tag{3.54}$$

$$= \lim_{\Delta t \to 0} \left[\frac{1}{\Delta t} \int_{-\infty}^{\infty} \cdots \int_{-\infty}^{\infty} (z_{k_1} - x_{k_1}) \cdots (z_{k_n} - x_{k_n}) p(\mathbf{z}, \Delta t | \mathbf{x}) d\mathbf{z} \right].$$

The multidimensional process is, therefore, stationary when, in addition of being time homogeneous, the univariate density is time independent

$$p_1(\mathbf{x}, t) = p_s(\mathbf{x}).$$

In such a case, if $p(\mathbf{x}, t - t_0 | \mathbf{x}_0)$ tends, as $t - t_0 \to \infty$, to a finite and non-vanishing limit independent of the initial state $\mathbf{x}_0$, the stationary density is given by

$$p_s(\mathbf{x}) = \lim_{t - t_0 \to \infty} p(\mathbf{x}, t - t_0 | \mathbf{x}_0).$$

Chapter 4

Diffusion processes

Diffusion processes constitute one of the most important classes of Markov processes because of their intensive use in the representation of countless phenomena in natural and social sciences. They are essentially Markov processes with continuous trajectories; that is to say, Markov processes whose sample paths present no jumps.

4.1 Definition and general properties

Diffusion processes can be instrumentally defined as Markov processes having different from zero only the first and second infinitesimal moments. Focusing first on one dimensional processes, this means that

$$A_n(x,t) = 0, \qquad n = 3,4,5,\ldots, \tag{4.1}$$

where the A_n's are the infinitesimal moments defined as [cf. Eqs. (3.21) and (3.23) of Chapter 3]

$$A_n(x,t) = \lim_{\Delta t \to 0} \left[\frac{1}{\Delta t} \left\langle (\Delta X)^n | X(t) = x \right\rangle \right]$$

$$= \lim_{\Delta t \to 0} \left[\frac{1}{\Delta t} \int_{-\infty}^{\infty} (y - x)^n p(y, t + \Delta t | x, t) dy \right], \tag{4.2}$$

where $\Delta X = X(t + \Delta t) - X(t)$.

Before proceeding further let us observe that since now

$$A_3 = A_4 = \cdots = 0$$

the (forward) Kramers-Moyal expansion (3.19) reduces to a second-order partial differential equation,

$$\frac{\partial p(x,t|x_0,t_0)}{\partial t} = -\frac{\partial}{\partial x} \left[A_1(x,t) p(x,t|x_0,t_0) \right]$$

$$+ \frac{1}{2} \frac{\partial^2}{\partial x^2} \left[A_2(x,t) p(x,t|x_0,t_0) \right], \tag{4.3}$$

89

called the *Fokker-Planck equation*. In an analogous way we see that the backward kinetic equation (3.33) reduces to the backward Fokker-Planck equation, also called *Kolmogorov equation*:

$$\frac{\partial}{\partial t_0} p(x,t|x_0,t_0) = -A_1(x_0,t_0)\frac{\partial}{\partial x_0} p(x,t|x_0,t_0)$$
$$-\frac{1}{2}A_2(x_0,t_0)\frac{\partial^2}{\partial x_0^2} p(x,t|x_0,t_0). \qquad (4.4)$$

4.1.1   Continuity of sample paths

From Eqs. (4.1) and (4.2) we see that any diffusion processes satisfies

$$\lim_{\Delta t \to 0}\left[\frac{1}{\Delta t}\left\langle (\Delta X)^n | X(t) = x \right\rangle\right] = 0, \qquad (n = 3,4,5,\dots). \qquad (4.5)$$

This condition means that significant changes in the process during a time interval Δt (that is, large changes in ΔX on average) go to zero faster than $\Delta t \to 0$. In other words, during small intervals of time the probability of great changes in the process is very small. This means that $X(t)$ as a function of t (i.e., the trajectories, realizations, or sample paths of the process) are continuous functions, obviously in the sense of probability, as explained in Chapter 1.

The probability that the trajectory of the process has a discontinuity at time t will be given by the limit

$$\lim_{\Delta t \to 0} P\{|X(t+\Delta t) - X(t)| > \epsilon | X(t) = x\},$$

for any $\epsilon > 0$. On the other hand, from Chebyshev's theorem, which states that given $n > 0$ and arbitrary the following inequality holds (see Chapter 1 and Gnedenko 1976)

$$P\{|\Delta X(t)| > \epsilon | X(t) = x\} \le \frac{1}{\epsilon^n}\left\langle |\Delta X|^n | X(t) = x \right\rangle, \quad \forall \epsilon > 0,$$

we see that

$$\lim_{\Delta t \to 0}\left[\frac{1}{\Delta t}P\{|\Delta X(t)| > \epsilon | X(t) = x\}\right] \le \frac{1}{\epsilon^n}\lim_{\Delta t \to 0}\left[\frac{1}{\Delta t}\left\langle |\Delta X|^n | X(t) = x \right\rangle\right],$$

($\forall \epsilon > 0$). For a diffusion process Eq. (4.5) holds for $n \ge 3$ and, since probability cannot be negative, we conclude that

$$\lim_{\Delta t \to 0}\left[\frac{1}{\Delta t}P\{|X(t+\Delta t) - X(t)| > \epsilon | X(t) = x\}\right] = 0, \qquad \forall \epsilon > 0, \quad (4.6)$$

which means that the probability for $X(t+\Delta t)$ to be arbitrarily different of $X(t)$ vanishes faster than Δt. In other words, the trajectory is continuous in probability. Let us note that in terms of the transition density condition (4.6) can be written as

$$\lim_{\Delta t \to 0} \left[\frac{1}{\Delta t} \int_{|z-x|>\epsilon} p(z, t + \Delta t | x, t) dz \right] = 0, \qquad (4.7)$$

for any $\epsilon > 0$ and uniformly in x, t and Δt. This equation is frequently termed as the *Lindeberg condition*.

All of the above sustains the alternative definition: *Diffusion processes are Markov processes with continuous trajectories*. We remark again that continuity of the sample paths is in stochastic sense. It might happen that some trajectories are discontinuous, but the set of such discontinuous realizations has zero measure.

Before proceeding further, let us clarify a common misconception. Thus the fact that a given random process takes on continuous values does not mean it has continuous trajectories. Think, for instance, of a gas whose molecules are moving with random velocities $\mathbf{v}_i(t)$. Clearly these velocities can take values within a continuous range (some subset of $\mathbb{R}^3$). However, in an ideal gas, molecules undergo elastic collisions meaning that, in colliding two molecules, their velocities instantaneously change from one velocity to another. In other words, the time functions $\mathbf{v}_i(t)$ are discontinuous. Positions $\mathbf{r}_i(t)$ are also discontinuous since they execute little jumps after each collision. However, compared with the mean free path of any ideal gas, these discontinuities are infinitesimal. As a result, and contrary to the velocity, positions are very well approximated by a random process with continuous sample paths.

4.1.2 *Characterization of trajectories*

Let us further describe diffusion processes for which only the first two infinitesimal moments can be different from zero. In what follows we will use a more standard notation and write $f(x, t)$, called *drift*, and $D(x, t)$, called *diffusion coefficient*, instead of A_1 and A_2 respectively.

Using Eqs. (3.25) and (3.26) of Chapter 3 we see that

$$\langle \Delta X(t) | X(t) = x \rangle = f(x, t) \Delta t + O(\Delta t^2), \qquad (4.8)$$

and

$$\text{Var}\{\Delta X(t) | X(t) = x\} = D(x, t) \Delta t + O(\Delta t^2). \qquad (4.9)$$

Therefore, and from a heuristic point of view, we can approximate the increment of the diffusion process, $\Delta X(t) = X(t+\Delta t) - X(t)$ with $X(t) = x$ known, as

$$\Delta X(t) = f(x,t)\Delta t + Z\sqrt{D(x,t)\Delta t} + O(\Delta t^2), \qquad (4.10)$$

where Z is a random variable with $\langle Z \rangle = 0$ and $\text{Var}\{Z\} = 1$. Indeed, from this representation we easily see that Eqs. (4.8) and (4.9) hold. Moreover, taking into account that for diffusion processes higher-order moments of $\Delta X(t)$ are zero, one can see that higher-order moments of the random variable Z must agree with those of the Gaussian distribution with zero mean and unit variance (Ghiman and Skorohod, 1972). Hence Z is a Gaussian random variable.

Equation (4.10) is frequently used for obtaining numerical simulations of the trajectories of diffusion processes. Note that it is equivalent to

$$\frac{\Delta X(t)}{\Delta t} = f(x,t) + \frac{1}{\sqrt{\Delta t}}Z\sqrt{D(x,t)} + O(\Delta t). \qquad (4.11)$$

If we interpret $X(t)$ as the position of a particle moving under the effect of random disturbances represented by Z, then Eqs. (4.10) and (4.11) show that the drift represents the average velocity,

$$f(x,t) \simeq \left\langle \frac{\Delta X(t)}{\Delta t} \right\rangle,$$

while the diffusion coefficient $D(x,t)$ gives a measure of of the fluctuations of $\Delta X(t)$ around the average value. We also see from these expressions that for any diffusion process:

(1) *Trajectories are continuous functions of time*:

$$\lim_{\Delta t \to 0} \Delta X(t) = 0.$$

(2) *Trajectories are not differentiable*:

$$\lim_{\Delta t \to 0} \frac{\Delta X(t)}{\Delta t} = \infty.$$

4.1.3   *Deterministic processes*

Recall from Eq. (4.9) that if the diffusion coefficient is identically zero, then the variance of the increment of the process is negligibly small (smaller than Δt). This means that $\Delta X(t)$ is nearly equal to its mean value, so that the process is virtually deterministic.

We will see that this is exactly the case and show that if $D(x,t) = 0$ for all x and t, the diffusion process is deterministic and no longer random.

In effect, since now all infinitesimal moments are zero except the first one,

$$A_2(x,t) = A_3(x,t) = \cdots = 0,$$

the Kramers-Moyal expansion (3.19) for the transition probability density $p(x,t|x_0,t_0)$ becomes a first-order differential equation:

$$\frac{\partial}{\partial t} p(x,t|x_0,t_0) = -\frac{\partial}{\partial x}\left[f(x,t)p(x,t|x_0,t_0)\right], \qquad (4.12)$$

called *Liouville equation*. Let us now prove that the solution to Eq. (4.12) with the initial condition

$$\lim_{t \to t_0} p(x,t|x_0,t_0) = \delta(x - x_0), \qquad (4.13)$$

is given by

$$p(x,t|x_0,t_0) = \delta[x - x(t;x_0,t_0)], \qquad (4.14)$$

where $x(t;x_0,t_0)$ is the solution to the deterministic equation

$$\frac{dx}{dt} = f(x,t), \qquad (4.15)$$

with initial condition $x(t_0) = x_0$.

it is obvious that (4.14) satisfies the initial condition (4.13). since $x(t_0;x_0,t_0) = x_0$. We next prove that (4.14) identically satisfies Liouville equation (4.12). In effect, let us call for simplicity $x(t)$ instead of $x(t;x_0,t_0)$ as the solution of the deterministic equation (4.15). Substituting then Eq. (4.14) into the right hand side of Eq. (4.12), and bearing in mind that

$$f(x)\delta(x - a) = f(a)\delta(x - a),$$

we write

$$-\frac{\partial}{\partial x}\left[f(x,t)p(x,t|x_0,t_0)\right] = -\frac{\partial}{\partial x}\left[f(x,t)\delta(x - x(t))\right]$$

$$= -\frac{\partial}{\partial x}\left[f(x(t),t)\delta(x - x(t))\right] = -f(x(t),t)\frac{\partial}{\partial x}\delta(x - x(t))$$

$$= f(x(t),t)\frac{\partial}{\partial x(t)}\delta(x - x(t)) = \frac{dx(t)}{dt}\frac{\partial}{\partial x(t)}\delta(x - x(t)),$$

where we have used the fact $(\partial/\partial x)\delta(x - y) = -(\partial/\partial y)\delta(x - y)$ and Eq. (4.15). But

$$\frac{dx(t)}{dt}\frac{\partial}{\partial x(t)} = \frac{\partial}{\partial t},$$

hence

$$\frac{\partial}{\partial t}\delta[x - x(t)] = -\frac{\partial}{\partial x}\left[f(x,t)\delta(x - x(t))\right]$$

which proves that (4.14) solves the Liouville equation (4.12).

4.2   Multidimensional diffusions

The generalization of the above development to include vectorial process is straightforward. Thus, multidimensional diffusion processes are Markov processes such that

$$A^{(n)}_{k_1 \cdots k_n}(\mathbf{x}, t) = 0, \quad \text{for } n \geq 3 \text{ and } k_j = 1, \ldots, N,$$

where $A^{(n)}_{k_1 \cdots k_n}(\mathbf{x}, t)$ are the N-dimensional infinitesimal moments of order n defined in Eq. (3.46). As in the one-dimensional case, this condition implies that the probability of notable changes in the process during a time interval Δt tends to zero very quickly as $\Delta t \to 0$. In other words, all sample paths of the process are continuous functions of time (in probability). The Lindeberg condition which assures this stochastic continuity is now given by [cf. Eq. (4.7)]

$$\lim_{\Delta t \to 0} \left[\frac{1}{\Delta t} \int \cdots \int_{|\mathbf{z}-\mathbf{x}|>\epsilon} p(\mathbf{z}, t + \Delta t | \mathbf{x}, t) d\mathbf{z} \right] = 0.$$

The only non-vanishing infinitesimal moments correspond to $n = 1$ and $n = 2$. The first moment is a N dimensional vector called drift while the second moment is a $N \times N$ square matrix called diffusion matrix. They are usually represented by f_k and D_{jk} instead of the previous notation $A^{(1)}_k$ and $A^{(2)}_{jk}$. They are given by Eqs. (3.47) and (3.48) respectively which we write in the following alternative form (see the one-dimensional analog given by Eq. (4.2))

$$f_k(\mathbf{x}, t) = \lim_{\Delta t \to 0} \left\{ \frac{1}{\Delta t} \Big\langle \big[X_k(t + \Delta t) - X_k(t) \big] \Big| \mathbf{X}(t) = \mathbf{x} \Big\rangle \right\}, \qquad (4.16)$$

and

$$D_{jk}(\mathbf{x}, t) = \lim_{\Delta t \to 0} \left[\frac{1}{\Delta t} \Big\langle \big[X_j(t + \Delta t) - X_j(t) \big] \right.$$

$$\left. \times \big[X_k(t + \Delta t) - X_k(t) \big] \Big| \mathbf{X}(t) = \mathbf{x} \Big\rangle \right], \qquad (4.17)$$

$(j, k = 1, \ldots, N)$.

We finally observe that the forward Kramers-Moyal expansion, Eq. (3.44), for the transition probability $p(\mathbf{x}, t | \mathbf{x}_0, t_0)$ reduces to the multivariate (forward) Fokker-Planck equation

$$\frac{\partial}{\partial t}p(\mathbf{x}, t|\mathbf{x}_0, t_0) = -\sum_{k=1}^{N} \frac{\partial}{\partial x_k}\left[f_k(\mathbf{x}, t)p(\mathbf{x}, t|\mathbf{x}_0, t_0)\right]$$

$$+\frac{1}{2}\sum_{j,k=1}^{N} \frac{\partial^2}{\partial x_j x_k}\left[D_{jk}(\mathbf{x}, t)p(\mathbf{x}, t|\mathbf{x}_0, t_0)\right], \quad (4.18)$$

with initial condition

$$\lim_{t \to t_0} p(\mathbf{x}, t|\mathbf{x}_0, t_0) = \delta(\mathbf{x} - \mathbf{x}_0). \tag{4.19}$$

In an analogous way the backward Kramers-Moyal expansion, Eq. (3.45), reduces to the multivariate (backward) Fokker-Planck equation

$$\frac{\partial p}{\partial t_0} = -\sum_{j=1}^{N} f_j(\mathbf{x}_0, t_0)\frac{\partial}{\partial x_{0j}}p(\mathbf{x}, t|\mathbf{x}_0, t_0)$$

$$-\frac{1}{2}\sum_{j,k=1}^{N} D_{jk}(\mathbf{x}_0, t_0)\frac{\partial^2}{\partial x_{0j}\partial x_{0k}}p(\mathbf{x}, t|\mathbf{x}_0, t_0), \tag{4.20}$$

with final condition given by Eq. (4.19).

4.3 Purely discontinuous processes. The master equation *

We have seen that Markov processes have continuous trajectories when all higher-order infinitesimal moments, $A_3 = A_4 = \cdots = 0$, are identically zero. This indicates that continuous trajectories are connected with the non-vanishing of the first two infinitesimal moments while the rest of moments are associated with discontinuous motion.

We briefly pause our account of diffusion processes for a short description on Markov processes with discontinuities, we refer the reader to the literature for more information (see, for instance Gardiner, 1986).

A Markov process is discontinuous if some (or all) infinitesimal moments of order higher than 2 are different from zero,

$$A_n(x, t) \neq 0, \qquad n = 3, 4, 5, \ldots.$$

If in addition

$$A_1(x, t) = A_2(x, t) = 0,$$

we have a *purely discontinuous process*. In this case the variations of $X(t)$ only occur in instantaneous jumps. That is, the process keeps its value during some time intervals (which may be random) and after each interval the process performs an instantaneous and random jump to another value. A classical example is the random telegraph signal in which the process switches from ± 1 to ∓ 1 at random Poisson times. Another example is furnished by the velocities of the molecules of an ideal gas which between collisions keep constant values but jump to another value when a collision takes place.

There are mixed situations to this picture when A_1 and A_2 are also different from zero. In such a case the system evolves as a diffusion process but having discontinuities. The case where $A_1 \neq 0$ but $A_2 = 0$ corresponds to a deterministic trajectory driven by the drift $A_1(x, t)$ but having randomly distributed jumps (Gardiner, 1986).

Returning to purely discontinuous processes, we observe that their main characteristic is that during a small time interval the process will remain almost surely in the original state while the probability for the transition to a new value is small. We can quantify this characteristic by assuming that during some time interval $(t, t + \Delta t)$ the probability that the $X(t)$ remains unchanged and equal to x is

$$1 - a(t, x)\Delta t + O(\Delta t^2),$$

while the probability that a jump has occurred is

$$a(t, x)\Delta t + O(\Delta t^2),$$

where $0 < a(t, x) < 1$. Let us incidentally note that these assumptions indicate that the probability that two or more changes have occurred during Δt is $O(\Delta t^2)$ and, hence, highly improbable if Δt is small.[1]

In order to proceed further we need to define an additional conditional density governing the magnitude of the possible discontinuities. We thus denote by $h(z|t, x)$ the conditional probability density that $X(t)$ reaches the value z provided that a jump occurred at time t from the value $X(t-0) = x$.

In terms of these two auxiliary functions, the transition density $p(z, t + \Delta t|x, t)$ of the process is

$$
\begin{aligned}
p(z, t + \Delta t|x, t) &= a(t, x)h(z|t, x)\Delta t \\
&\quad + \left[1 - a(t, x)\Delta t\right]\delta(z - x) + O(\Delta t^2). \quad (4.21)
\end{aligned}
$$

In effect, the first term on the right hand side indicates that a jump has occurred during the time interval $(t, t + \Delta t)$ from x to z, while the second

---

[1]These conditions imply that the set of random times $t_1, t_2, \cdots$, where the discontinuities of $X(t)$ occur, is a Poisson set of events (see Chapter 2).

term implies that no jump occurred so that the process has remained at the original position x.

Observe also that $a(t, x)$ and $h(z|t, x)$ do not depend on the values that the process had taken at any previous time before the discontinuity. Therefore, the process is Markovian and the transition density must obey the Chapman-Kolmogorov equation:

$$p(z, t + \Delta t|x, t) = \int_{-\infty}^{\infty} p(x, t + \Delta t|y, t)p(y, t|x_0, t_0)dy.$$

Substituting for (4.21) we have

$$p(x, t + \Delta t|x, t) = \Delta t \int_{-\infty}^{\infty} a(t, y)h(x|t, y)p(y, t|x_0, t_0)dy$$

$$+ \int_{-\infty}^{\infty} [1 - a(t, y)\Delta t]\delta(x - y)p(y, t|x_0, t_0)dy + O(\Delta t^2),$$

which, applying the delta function and rearranging terms, yields

$$\frac{1}{\Delta t}[p(x, t + \Delta t|x, t) - p(x, t|x_0, t_0)] =$$

$$-a(t, x)p(x, t|x_0, t_0) + \int_{-\infty}^{\infty} a(t, y)h(x|t, y)p(y, t|x_0, t_0)dy + O(\Delta t),$$

and as $\Delta t \to 0$ we finally get

$$\frac{\partial}{\partial t}p(x, t|x_0, t_0) = -a(t, x)p(x, t|x_0, t_0)$$

$$+ \int_{-\infty}^{\infty} a(t, y)h(x|t, y)p(y, t|x_0, t_0)dy, \quad (4.22)$$

which is an integro-differential equation for the transition density called the *Kolmogorov-Feller equation*.

Let us write this equation in an alternative form. To this end we define

$$W(x|t, y) \equiv a(t, y)h(x|t, y), \quad (4.23)$$

where $W(x|t, y)$ is the joint probability density for having one jump at time t form value y to x. Moreover, integrating Eq. (4.23) with respect to x and bearing in mind the normalization of $h(x|t, y)$, we see that

$$\int_{-\infty}^{\infty} W(x|t, y)dx = a(t, y)\int_{-\infty}^{\infty} h(x|t, y)dx = a(t, y).$$

Substituting this result into the Kolmogorov-Feller equation (4.22) and applying definition (4.23), we obtain the so-called *master equation*:

$$\frac{\partial}{\partial t}p(x, t|x_0, t_0) = \int_{-\infty}^{\infty} [W(x|y, t)p(y, t|x_0, t_0)$$

$$-W(y|x, t)p(x, t|x_0, t_0)]dy, \quad (4.24)$$

The master equation shows more clearly than Kolmogorov-Feller equation (4.22) that the temporal variation of the transition density is a balance between two fluxes of probability accounting for jumps in one direction and in the other.

4.4   The one-dimensional Fokker-Planck equation

We now return to diffusion processes and study the most relevant features of the equation governing them, the Fokker-Planck equation (FPE). We will examine both the forward and backward versions of the FPE. We center first on the one-dimensional case and leave multidimensional equations for the next chapter.

4.4.1   *General properties*

We have seen in Sect. 4.1 that for diffusion processes the forward Kramers-Moyal expansion reduces to the Fokker-Planck equation (FPE):

$$\frac{\partial p(x,t|x_0,t_0)}{\partial t} = -\frac{\partial}{\partial x}\big[f(x,t)p(x,t|x_0,t_0)\big] + \frac{1}{2}\frac{\partial^2}{\partial x^2}\big[D(x,t)p(x,t|x_0,t_0)\big],$$

$$(4.25)$$

where in writing Eq. (4.25) we have used the standard notations, $f(x,t)$ (the drift) for the first infinitesimal moment $A_1(x,t)$, and $D(x,t)$ (the diffusion coefficient) for the second infinitesimal moment $A_2(x,t)$.

In a similar way, the backward Fokker-Planck equation, Eq. (4.4), also known as Kolmogorov equation, reads

$$\frac{\partial p(x,t|x_0,t_0)}{\partial t_0} = -f(x_0,t_0)\frac{\partial p(x,t|x_0,t_0)}{\partial x_0} - \frac{1}{2}D(x_0,t_0)\frac{\partial^2 p(x,t|x_0,t_0)}{\partial x_0^2},$$

$$(4.26)$$

Both equations have to be solved under the initial (final) condition

$$\lim_{t \to t_0} p(x,t|x_0,t_0) = \delta(x - x_0). \qquad (4.27)$$

As we have discussed in the previous chapter in relation to forward and backward kinetic equations, the two diffusion equations (4.25) and (4.26) are equivalent to each other, in the sense that solutions to Eq. (4.25) with the initial condition (4.27), and given boundary conditions, are also solutions to Eqs. (4.26)-(4.27) with the same boundary conditions and conversely. The basic difference between both equations resides in the set of variables that is considered as fixed. In the forward equation x_0 and t_0 are fixed and we look for solutions at $t > t_0$, while in the backward equation x and t are fixed and we look for solutions at $t_0 < t$.

The forward equation can directly provide the average values of the process (which are usually the observable magnitudes) at the observation time t. This is the reason why forward equations are more widely used in applications. On the other hand, backward equations may be more convenient from a mathematical point of view because the functions f and D need not be differentiable. They are also very helpful in the study of first-passage and escape problems as we will see in Part III of this book.

The univariate probability density $p(x, t)$, which in terms of the transition density is given by

$$p(x, t) = \int_{-\infty}^{\infty} p(x, t | x_0, t_0) p(x_0, t_0) dx_0,$$

satisfies the forward Fokker-Planck equation as well (but not the backward equation) [2]

$$\frac{\partial p(x, t)}{\partial t} = -\frac{\partial}{\partial x} \left[f(x, t) p(x, t) \right] + \frac{1}{2} \frac{\partial^2}{\partial x^2} \left[D(x, t) p(x, t) \right], \qquad (4.28)$$

with initial condition

$$\lim_{t \to t_0} p(x, t) = p(x_0, t_0), \qquad (4.29)$$

where $p(x_0, t_0)$ is a given initial probability density function.

Let us also observe that all admissible solutions to the forward or backward Fokker-Planck equations must satisfy the normalization condition

$$\int_{-\infty}^{\infty} p(x, t | x_0, t_0) dx = 1. \qquad (4.30)$$

4.4.2 *Probability current. Some boundary conditions*

The FPE can be interpreted as an expression for the conservation law of probability. Indeed, the forward FPE (4.25) can be written as a continuity equation:

$$\frac{\partial}{\partial t} p(x, t | x_0, t_0) + \frac{\partial}{\partial x} J(x, t | x_0, t_0) = 0, \qquad (4.31)$$

where

$$J(x, t | x_0, t_0) \equiv f(x, t) p(x, t | x_0, t_0) - \frac{1}{2} \frac{\partial}{\partial x} \left[D(x, t) p(x, t | x_0, t_0) \right], \qquad (4.32)$$

is the *probability current*.

---

[2]We have seen in the previous chapter that the univariate and bivariate densities, $p(x, t)$ and $p(x, t; x_0, t_0)$ respectively, satisfy the forward kinetic equation but not the backward equation. Obviously the same applies to diffusion processes.

The FPE, in the form given by continuity equation (4.31), shows that any time variation of the transition density, given by its time derivative, is exactly balanced by the spatial variation of the probability flux. We can thus look at the FPE as the equation for the conservation of probability.

Let us now integrate Eq. (4.31) over the entire real line

$$\frac{\partial}{\partial t} \int_{-\infty}^{\infty} p(x,t|x_0,t_0)dx + \int_{-\infty}^{\infty} \frac{\partial}{\partial x} J(x,t|x_0,t_0)dx = 0.$$

Taking into account normalization, Eq. (4.30), which implies the vanishing of the time derivative, the integration of the second summand yields

$$J(\infty,t|x_0,t_0) = J(-\infty,t|x_0,t_0), \qquad (4.33)$$

which is usually called "periodic boundary condition" at infinity, meaning that any probability current disappearing at one end materializes at the other end. Another more restrictive boundary condition is

$$J(\infty,t|x_0,t_0) = J(-\infty,t|x_0,t_0) = 0, \qquad (4.34)$$

meaning that there is no probability current at the boundary.

In many applications and theoretical settings, such as diffusion in confined geometries as well as in escape and first-passage problems, the diffusion process is not defined on the entire real line but inside a finite interval (a,b), the *diffusion interval*.[3] In this case, since the process cannot take values outside (a,b), we will have

$$p(x,t|x_0,t_0) = 0, \qquad \text{for all } x \notin (a,b),$$

and normalization now is

$$\int_a^b p(x,t|x_0,t_0)dx = 1.$$

4.4.3   Time homogeneous processes

Let us suppose next that the diffusion process is time homogeneous. As we have seen in Sect. 3.6 of Chapter 3, this means that the infinitesimal moments $A_1(x) = f(x)$ and $A_2(x) = D(x)$ do not depend of time and the transition density,

$$p(x,t|x_0,t_0) = p(x,t-t_0|x_0),$$

only depends on time differences. The FPE (4.25) reads

$$\frac{\partial p}{\partial t} = -\frac{\partial}{\partial x}\left[f(x)p(x,t-t_0|x_0)\right] + \frac{1}{2}\frac{\partial^2}{\partial x^2}\left[D(x)p(x,t-t_0|x_0)\right], \qquad (4.35)$$

---

[3]The semi-infinite intervals $(-\infty,b)$ and (a,∞) are also considered, specially in first passage problems as we will see in Part III.

with initial condition

$$p(x, 0|x_0) = \delta(x - x_0). \tag{4.36}$$

In an analogous way, the backward FPE (4.26) can be written as

$$\frac{\partial p}{\partial t_0} = -f(x_0)\frac{\partial p(x, t - t_0|x_0)}{\partial x_0} - \frac{1}{2}D(x_0)\frac{\partial^2 p(x, t - t_0|x_0)}{\partial x_0^2}, \tag{4.37}$$

with the final condition, as $t_0 \to t$, given by Eq. (4.36). Since the transition density is invariant under time translations, we can set a new time scale as $t' = t - t_0$. Hence $\partial/\partial t_0 = -\partial/\partial t'$, and Eq. (4.37) reads (we drop the prime in the time variable)

$$\frac{\partial p}{\partial t} = f(x_0)\frac{\partial p(x, t|x_0)}{\partial x_0} + \frac{1}{2}D(x_0)\frac{\partial^2 p(x, t|x_0)}{\partial x_0^2}. \tag{4.38}$$

A similar manipulation can be performed on the forward FPE, although, in this case, the new time scale $t \to t - t_0$ leaves Eq. (4.35) invariant:

$$\frac{\partial p}{\partial t} = -\frac{\partial}{\partial x}[f(x)p(x, t|x_0)] + \frac{1}{2}\frac{\partial^2}{\partial x^2}[D(x)p(x, t|x_0)]. \tag{4.39}$$

For both equations, backward and forward, the initial condition is given by Eq. (4.36).

4.4.4 *Stationary processes*

We know that any diffusion process is stationary when, in addition of being time homogeneous, the univariate density is independent of time:

$$p(x, t) = p_s(x).$$

In such a case, if the transition density tends, as $t - t_0 \to \infty$, to a finite and non-vanishing function of x independent of the initial state x_0, the stationary density turns out to be [4]

$$p_s(x) = \lim_{t-t_0 \to \infty} p(x, t - t_0|x_0). \tag{4.40}$$

Taking the limit $t_0 \to -\infty$ (t arbitrary) in the FPE (4.35), interchanging that limit with the time derivative and bearing in mind Eq. (4.40), we see that, since $\partial p_s(x)/\partial t = 0$, the stationary density $p_s(x)$ must be a finite and non-vanishing solution to the following ordinary differential equation

$$-\frac{d}{dx}[f(x)p_s(x)] + \frac{1}{2}\frac{d^2}{dx^2}[D(x)p_s(x)] = 0,$$

---

[4]Recall from Chapter 3 (Sect. 3.6) that the limit $t - t_0 \to \infty$ can be achieved either by assuming t arbitrary and $t_0 \to -\infty$, this is, assuming the process started off in the distant past; or when t_0 is arbitrary and $t \to \infty$, that is, looking at the process in the remote future.

or, equivalently,

$$\frac{d}{dx}\left\{f(x)p_s(x) - \frac{1}{2}\frac{d}{dx}[f(x)p_s(x)]\right\} = \frac{d}{dx}J_s(x) = 0, \qquad (4.41)$$

where

$$J_s(x) \equiv f(x)p_s(x) - \frac{1}{2}\frac{d}{dx}[f(x)p_s(x)], \qquad (4.42)$$

is the *stationary current*. Note that Eq. (4.41) tells us that the stationary current is constant,

$$J_s(x) = J_0,$$

where J_0 is an arbitrary constant. That is,

$$f(x)p_s(x) - \frac{1}{2}\frac{d}{dx}[D(x)p_s(x)] = J_0.$$

The direct integration of this equation yields the stationary density

$$p_s(x) = \frac{1}{D(x)}\left[N - 2J_0\int^x \exp\left\{2\int^y \frac{f(z)}{D(z)}dz\right\}dy\right]$$

$$\times \exp\left\{2\int^x \frac{f(y)}{D(y)}dy\right\}, \qquad (4.43)$$

where N and J_0 are arbitrary constants to be determined by normalization and boundary conditions.

If we suppose that there is no flux of probability at the boundary [see Eq. (4.34)], then $J_s(x) = J_0 = 0$ and the stationary density (4.43) reduces to

$$p_s(x) = \frac{N}{D(x)}\exp\left\{2\int^x \frac{f(y)}{D(y)}dy\right\}, \qquad (4.44)$$

where N is determined by the normalization of $p_s(x)$. This is the so-called *potential solution*.

4.5   Some examples

We end this chapter by presenting two widely used diffusion processes: the Wiener process, also called Brownian motion, and the Ornstein-Uhlenbeck process.

4.5.1 *The Wiener process*

Brownian motion –the wandering of macroscopic particles moving inside a fluid– is the simplest and most fundamental diffusion process having no drift $f(x,t) = 0$ and constant diffusion coefficient $D(x,t) = D$. When $D = 1$, the motion is usually called Wiener process and denoted by $W(t)$. Therefore, $W(t)$ is a time-homogeneous diffusion process whose FPE for the transition density reads

$$\frac{\partial}{\partial t}p(x,t|x_0) = \frac{1}{2}\frac{\partial^2}{\partial x^2}p(x,t|x_0), \qquad (4.45)$$

where, due to temporal homogeneity and without loss of generality, we have set $t_0 = 0$.[5] The initial condition attached to Eq. (4.45) is

$$p(x,0|x_0) = \delta(x - x_0). \qquad (4.46)$$

We easily see from Eqs. (4.45) – (4.46) that the Fourier transform of the transition density,

$$\tilde{p}(\omega,t|x_0) = \int_{-\infty}^{\infty} e^{i\omega x}p(x,t|x_0)dx,$$

satisfies the following first-order ordinary differential equation

$$\frac{d\tilde{p}}{dt} = -\frac{1}{2}\omega^2\tilde{p},$$

with initial condition

$$\tilde{p}(\omega,0|x_0) = e^{i\omega x_0}.$$

Finding the solution to this problem is straightforward and we have

$$\tilde{p}(\omega,t|x_0) = e^{i\omega x_0 - \omega^2 t/2},$$

which, after Fourier inverting, yields

$$p(x,t|x_0) = \frac{1}{\sqrt{2\pi t}}\exp\left\{-\frac{(x - x_0)^2}{2t}\right\}. \qquad (4.47)$$

We thus see that the Wiener process $W(t)$ is Gaussian with mean and variance given by

$$\langle W(t)\rangle = x_0, \qquad \langle[W(t) - x_0]^2\rangle = t. \qquad (4.48)$$

Note that the process is centered at the initial value x_0 but with a dispersion around the mean value which increases linearly with time. The latter

---

[5]Since the process is also homogeneous in space, as can be seen from the invariance of FPE (4.45) under the spatial translation $x \to x - x_0$, we could also take $x_0 = 0$ without loss of generality.

is a key characteristic of the Wiener process and, by extension, of any diffusion process. The initial distribution initially peaked at x_0 is, therefore, spreading out on the entire real line at a rate governed by the diffusion coefficient.

Let us also observe from Eq. (4.47) that, despite being time homogeneous, the Wiener process is not stationary because the transition density vanishes as time increases

$$\lim_{t \to \infty} p(x, t | x_0) = 0.$$

The main properties of the Wiener process are summarized in the following result

Theorem (Properties of the Wiener process)

Let $W(t)$ be a Wiener process with increments defined as $\Delta W(t) = W(t + \Delta t) - W(t)$. Then

(1) $\Delta W(t)$ is also a Wiener process.
(2) The increments are independent.
(3) The correlation function of $W(t)$ is $\langle W(t_1) W(t_2) \rangle = \min(t_1, t_2)$.
(4) The trajectories of $W(t)$ are continuous but not differentiable.
(5) The Wiener process is a martingale.[6]

Proof. (1) From Eq. (4.47) we clearly see that the Wiener process, besides being time homogeneous, is also state homogeneous because the transition density only depends on the difference $x - x_0$. Thus the transition density from $W(t_1) = x_1$ to $W(t_2) = x_2$ is

$$p(x_2, t_2 | x_1, t_1) = \frac{1}{\sqrt{2\pi(t_2 - t_1)}} \exp\left\{ -\frac{(x_2 - x_1)^2}{2(t_2 - t_1)} \right\} \qquad (4.50)$$

$(t_1 < t_2)$. From which we see that

$$p(x_2, t_2 | x_1, t_1) = p(x_2 - x_1, t_2 - t_1).$$

Suppose that $x_1 = x$, $t_1 = t$, $x_2 = x + \Delta x$ and $t_2 = t + \Delta t$, then Eq. (4.50) shows that the transition density of the increment of the Wiener process,

$$\Delta W(t) = W(t + \Delta t) - W(t),$$

---

[6]A random process $X(t)$ is said to be a *martingale* if

$$\left\langle X(t) \middle| X(t_0) = x_0 \right\rangle = x_0, \qquad (4.49)$$

for any $t \geq t_0$.

is

$$p(\Delta x, \Delta t) = \frac{1}{\sqrt{2\pi\Delta t}} \exp\left\{-\frac{(\Delta x)^2}{2\Delta t}\right\}. \tag{4.51}$$

In other words, $\Delta W(t)$ is a Wiener process.

(2) Since the Wiener process is a Markov process, the joint density can be written as

$$p(x_0, t_0; x_1, t_1; \ldots; x_n, t_n) = \prod_{j=1}^{n} p(x_j, t_j | x_{j-1}, t_{j-1}) p(x_0, t_0),$$

which, recalling Eqs. (4.50) and (4.51), yields

$$p(x_0, t_0; x_1, t_1; \ldots; x_n, t_n)$$

$$= \prod_{j=1}^{n} [2\pi(t_j - t_{j-1})]^{-1/2} \exp\{-(x_j - x_{j-1})^2/2(t_j - t_{j-1})\} p(x_0, t_0)$$

$$= \prod_{j=1}^{n} (2\pi\Delta t_j)^{-1/2} \exp\{-(\Delta x_j)^2/2\Delta t_j\} p(x_0, t_0)$$

$$= \prod_{j=1}^{n} p(\Delta x_j, \Delta t_j) p(x_0, t_0) = p(\Delta x_n, \Delta t_n; \Delta x_{n-1}, \Delta t_{n-1}; \ldots; x_0, t_0).$$

That is,

$$p(\Delta x_n, \Delta t_n; \ldots; \Delta x_1, \Delta t_1; x_0, t_0)$$
$$= p(\Delta x_n, \Delta t_n) \ldots p(\Delta x_1, \Delta t_1) p(x_0, t_0), \tag{4.52}$$

and the increments are independent. Note that this equation implies

$$\langle \Delta X(t_n) \cdots \Delta X(t_1) X(t_0)\rangle = \langle \Delta X(t_n)\rangle \cdots \langle \Delta X(t_1)\rangle \langle X(t_0)\rangle. \tag{4.53}$$

(3) We will use the above result to compute the correlation function of the process. In what follows we will assume, without loss of generality that $x_0 = 0$ which implies $\langle W(t)\rangle = 0$ [cf. Eq. (4.48)]. We first suppose $t_2 > t_1$ and the correlation function is thus given by

$$\begin{aligned}\langle W(t_1)W(t_2)\rangle &= \langle W(t_1)W(t_2) - W(t_1)^2 + W(t_1)^2\rangle \\ &= \langle W(t_1)[W(t_2) - W(t_1)] + W(t_1)^2\rangle \\ &= \langle W(t_1)\rangle\langle W(t_2) - W(t_1)\rangle + \langle W(t_1)^2\rangle \\ &= \langle W(t_1)^2\rangle = t_1,\end{aligned}$$

$(t_2 > t_1)$, where we have taken into account the independence of increments, as expressed in Eq. (4.53), and also that $\langle W(t)\rangle = 0$. We can easily convince ourselves that if $t_2 < t_1$ a similar reasoning leads to $\langle W(t_1)W(t_2)\rangle = t_2$. Therefore,

$$\langle W(t_1)W(t_2)\rangle = \min(t_1, t_2). \tag{4.54}$$

If $x_0 \neq 0$ and $t_0 \neq 0$ the result is:

$$\langle W(t_1)W(t_2)\rangle = \min(t_1 - t_0, t_2 - t_0) + x_0^2.$$

(4) The continuity of trajectories directly follows from the fact that the Wiener process is a diffusion process.

In order to prove the no differentiability of trajectories we will evaluate the probability that the rate $\Delta W(t)/\Delta t$ grows without bounds as $\Delta t \to 0$ and show that this probability approaches 1. We have ($\eta > 0$)

$$P\left\{\left|\frac{\Delta W(t)}{\Delta t}\right| > \eta\right\} = P\{-\eta\Delta t > \Delta W(t) > \eta\Delta t\}$$

$$= \int_{-\infty}^{-\eta\Delta t} p(y, \Delta t|0, 0)dy + \int_{\eta\Delta t}^{\infty} p(y, \Delta t|0, 0)dy$$

$$= \frac{2}{\sqrt{2\pi\Delta t}} \int_{\eta\Delta t}^{\infty} e^{-y^2/2\Delta t}dy,$$

where we assumed $\Delta t > 0$ and used Eq. (4.51). Therefore,

$$P\left\{\left|\frac{dW(t)}{dt}\right| > \eta\right\} = \lim_{\Delta t \to 0} P\left\{\left|\frac{\Delta W(t)}{\Delta t}\right| > \eta\right\}$$

$$= \lim_{\Delta t \to 0} \frac{2}{\sqrt{2\pi\Delta t}} \int_{0}^{\infty} e^{-y^2/2\Delta t}dy = 1.$$

Since $\eta > 0$ is arbitrary, the probability that the derivative $W(t)$ takes arbitrarily large values is 1. In other words, the Wiener process is not differentiable with probability one.

(5) In order to prove that $W(t)$ is a martingale, let us evaluate the conditional average $\langle W(t)|W(t_0) = x_0\rangle$ [cf. Eq. (4.49)]. Using Eq. (4.50) we write

$$\langle W(t)|W(t_0) = x_0\rangle = \frac{1}{\sqrt{2\pi(t - t_0)}} \int_{-\infty}^{\infty} xe^{-(x-x_0)^2/2(t-t_0)}dx$$

$$= \frac{1}{\sqrt{2\pi(t - t_0)}} \int_{-\infty}^{\infty} (x - x_0)e^{-(x-x_0)^2/2(t-t_0)}dx$$

$$+ x_0 \frac{1}{\sqrt{2\pi(t - t_0)}} \int_{-\infty}^{\infty} e^{-(x-x_0)^2/2(t-t_0)}dx,$$

but

$$\frac{1}{\sqrt{2\pi(t-t_0)}} \int_{-\infty}^{\infty} (x-x_0)e^{-(x-x_0)^2/2(t-t_0)}dx = 0,$$

since the integrand is an odd function and, by normalization,

$$\frac{1}{\sqrt{2\pi(t-t_0)}} \int_{-\infty}^{\infty} e^{-(x-x_0)^2/2(t-t_0)}dx = 1.$$

Therefore,

$$\left\langle W(t) \middle| W(t_0) = x_0 \right\rangle = x_0$$

and the Wiener process is a martingale.

$\square$

Observation

The no differentiability of the Wiener process can be seen from a heuristic point of view as follows. Recall that in Sect. 4.1 [cf. Eq. (4.10)] we represented the increment of any diffusion processes $X(t)$ with drift $f(x,t)$ and diffusion coefficient $D(x,t)$ as

$$\Delta X(t) = f(x,t)\Delta t + Z\sqrt{D(x,t)\Delta t} + O(\Delta t^2), \qquad (4.55)$$

where $\Delta X(t) = X(t+\Delta t) - x$ and Z is a Gaussian variable with zero mean and unit variance. For the Wiener process $f(x,t) = 0$ and $D(x,t) = 1$ and this representation reduces to

$$\Delta W(t) = Z\sqrt{\Delta t} + O(\Delta t^2). \qquad (4.56)$$

This equation provides a simple and useful algorithm for the numerical simulation of the Wiener process. We also see that

$$\Delta W(t) \to 0 \qquad \text{as} \qquad \Delta t \to 0,$$

showing the continuity of the trajectories.

Let us finally note that the ratio $\Delta W(t)/\Delta t$ reads

$$\frac{\Delta W(t)}{\Delta t} = Z\frac{1}{\sqrt{\Delta t}} + O(\Delta t), \qquad (4.57)$$

clearly showing the no differentiability of the Wiener process (more on this point below).

4.5.2   *Gaussian white noise*

We have briefly met the so-called white noise in Chapter 3 (cf. Sect. 3.2). Strictly speaking, Gaussian white noise is not a diffusion process because its sample paths are not continuous. However, it is closely related to the Wiener process and this is the reason why we present here its main properties.

Let us recall that Gaussian white noise is a Gaussian random process, usually denoted by $\xi(t)$, with zero mean and delta correlation[7]

$$\langle \xi(t) \rangle = 0, \qquad \langle \xi(t)\xi(t') \rangle = \delta(t - t'). \tag{4.58}$$

Since $\delta(t-t') = 0$ for $t \neq t'$, we see that $\xi(t)$ is completely uncorrelated (i.e., independent) of $\xi(t')$ except with $t = t'$. Obviously this is an idealization, since in nature the value of any process $X(t)$ affects any subsequent value $X(t + \tau)$ $(\tau > 0)$. However, if we are observing the process at time scales in which τ is very small, we can approximate the process by a white noise.

It is clear that the trajectories of a process with independent values at every instant are highly irregular and one can prove that *the trajectories of any white noise are discontinuous with probability one* (Doob, 1966). The fact that we treat such a discontinuous process within the framework of diffusion processes which are, by definition, continuous, is because *Gaussian white noise can be considered as the time derivative of the Wiener process.* In effect, suppose $\xi(t)$ is Gaussian white noise and let $X(t)$ be a Gaussian process with mean and correlation function defined as

$$\langle X(t) \rangle = \int_0^t \langle \xi(t') \rangle dt', \qquad \langle X(t_1)X(t_2) \rangle = \int_0^{t_1} dt_1' \int_0^{t_2} \langle \xi(t_1')\xi(t_2') \rangle dt_2'.$$

From Eq. (4.58) we see that $X(t)$ has zero mean,

$$\langle X(t) \rangle = 0,$$

and correlation function

$$\langle X(t_1)X(t_2) \rangle = \int_0^{t_1} dt_1' \int_0^{t_2} \delta(t_1' - t_2')dt_2'$$

$$= \int_0^\infty \Theta(t_1 - t_1')dt_1' \int_0^\infty \Theta(t_2 - t_2')\delta(t_1' - t_2')dt_2'$$

$$= \int_o^\infty \Theta(t_1 - t_1')\Theta(t_2 - t_1')dt_1' = \min(t_1, t_2).$$

---

[7]In fact, any process satisfying (4.58) is called white noise regardless its statistics. We can thus have Poisson white noise, Gaussian white noise, etc.

That is,

$$\langle X(t_1)X(t_2)\rangle = \min(t_1, t_2),$$

and $X(t) = W(t)$ is the Wiener process. This somewhat justifies the above statement that Gaussian white noise is the derivative of the Wiener process, written symbolically as

$$dW(t) = \xi(t)dt, \tag{4.59}$$

despite the fact that the latter is not differentiable. One can, nonetheless, carry out a rigorous treatment of white noise within the context of generalized function and we refer the interested reader to more specialized literature (see, for instance, Gelfand and Shilov, 2016b).

4.5.3 *The Ornstein-Uhlenbeck process*

After the Wiener process, which describes the free Brownian motion, perhaps one of the most utilized random processes is the Ornstein-Uhlenbeck (OU) process which incorporates a linear force (or quadratic potential as in the harmonic oscillator) in the evolution of the process.

The OU process is a diffusion process with linear drift and constant diffusion coefficient[8]

$$f(x,t) = -\alpha x, \qquad D(x,t) = D,$$

where $\alpha > 0$. We , therefore, see that the OU process is time-homogeneous and the FPE for the transition density reads

$$\frac{\partial}{\partial t}p(x,t|x_0) = \frac{\partial}{\partial x}\left[\alpha x p(x,t|x_0)\right] + \frac{1}{2}D\frac{\partial^2}{\partial t^2}p(x,t|x_0), \tag{4.60}$$

with the initial condition

$$p(x,0|x_0) = \delta(x - x_0). \tag{4.61}$$

We will proceed to solve this initial-value problem by Fourier methods. The Fourier transform of the transition density,

$$\tilde{p}(\omega,t|x_0) = \int_{-\infty}^{\infty} e^{i\omega x}p(x,t|x_0)dx,$$

satisfies a partial differential equation of first order

$$\frac{\partial\tilde{p}}{\partial t} + \alpha\omega\frac{\partial\tilde{p}}{\partial\omega} = -\frac{1}{2}D\omega^2\tilde{p}, \tag{4.62}$$

---

[8]A more general form of the OU process is provided by the linear drift $f(x) = -\alpha x + \beta$. We can easily include the latter form in the former by the simple change of variables $x' = x - \beta/\alpha$.

with initial condition

$$\tilde{p}(\omega, 0|x_0) = e^{i\omega x_0}. \qquad (4.63)$$

In writing Eq. (4.62) we have taken into account the following properties of the Fourier transform (Erdelyi, 1954)

$$\mathcal{F}\left\{\frac{\partial}{\partial x}(xp)\right\} = -\omega\frac{\partial\tilde{p}}{\partial\omega}, \qquad \mathcal{F}\left\{\frac{\partial^2 p}{\partial x^2}\right\} = -\omega^2\tilde{p}. \qquad (4.64)$$

The initial-value problem (4.62)-(4.63) can be solved by standard methods of solution for partial differential equation of first order –such as the methods of characteristics (see, for instance, Courant and Hilbert, 1991)–. In the present case it is, however, faster and easier to proceed in the following way. Since the OU process is still linear it is plausible to assume that the process is Gaussian and seek a solution of the form:

$$\tilde{p}(\omega, t|x_0) = \exp\{-A(t)\omega^2 + B(t)\omega\}, \qquad (4.65)$$

where $A(t)$ and $B(t)$ are unknown functions. Substituting this ansatz into Eq. (4.62) we have

$$-(\dot{A} + 2\alpha A - D/2)\omega^2 + (\dot{B} + \alpha B)\omega = 0,$$

and since this is an identity valid for all values of ω, we obtain

$$\dot{A} + 2\alpha A = \frac{1}{2}D, \qquad \dot{B} + \alpha B = 0,$$

which are two uncoupled ordinary differential equations of first order. The initial conditions attached to them read [cf. Eqs. (4.63) and (4.65)]

$$A(0) = 0, \qquad B(0) = ix_0.$$

The solution is

$$A(t) = \frac{D}{4\alpha}\left(1 - e^{-\alpha t}\right), \qquad B(t) = ix_0 e^{-\alpha t}.$$

Finally,

$$\tilde{p}(\omega, t|x_0) = \exp\left\{-\frac{D}{4\alpha}\left(1 - e^{-\alpha t}\right) + i\omega x_0 e^{-\alpha t}\right\}, \qquad (4.66)$$

which corresponds to a Gaussian process with mean and variance respectively given by

$$\langle X(t)\rangle = x_0 e^{-\alpha t}, \qquad \langle X^2(t)\rangle = \frac{D}{2\alpha}\left(1 - e^{-2\alpha t}\right). \qquad (4.67)$$

In consequence the transition density is given by

$$p(x, t|x_0) = \left(\frac{\alpha/\pi D}{1 - e^{-2\alpha t}}\right)^{1/2}\exp\left\{-\frac{(\alpha/D)\left(x - x_0 e^{-\alpha t}\right)^2}{1 - e^{-2\alpha t}}\right\}. \qquad (4.68)$$

Note that there exists a stationary distribution given by the limit $p(x, t|x_0) \to p_s(x)$ as $t \to \infty$. That is

$$p_s(x) = \left(\frac{\alpha}{\pi D}\right)^{1/2} e^{-\alpha x^2/D}. \qquad (4.69)$$

Let us finally remark that this stationary distribution coincides with the potential solution given in Eq. (4.44).

4.6 Anomalous diffusion and fractional motion*

Suppose $X(t)$ represents the Brownian motion discussed in the Sect. 4.5.1. Besides the fact that $X(t)$ is a Gaussian process, another major characteristic is that its variance grows linearly with time (cf. Eq. (4.48))

$$\langle \Delta X(t)^2 \rangle = Dt, \qquad (4.70)$$

where $\Delta X(t) = X(t) - x_0$ and D is the diffusion constant.[9]

Before proceeding further let us note that the linear growth in time of the variance is not exclusive of Brownian motion but of any diffusion process. Indeed, as we have seen at the beginning of this chapter, the (conditional) variance of the increment, $\Delta X = X(t + \Delta t) - X(t)$, given that $X(t) = x$ –that is, the mean square deviation– grows linearly with time (cf. Eq. (4.9)):

$$\text{Var}\{\Delta X(t)|X(t) = x\} = D(x,t)\Delta t + O(\Delta t^2), \qquad (4.71)$$

where $D(x,t)$ is the diffusion coefficient.

The linear growth of the variance showed by diffusion processes is, however, not fully universal. In many practical settings, mostly arising in extremely disordered media and fractal structures (Bouchaud and Georges, 1990; Ben-Avraham and Havlin, 2000), it appears the so-called "anomalous diffusion" –also termed "anomalous transport"– for which its most distinctive characteristic is that the mean square deviation follows the asymptotic law (Bouchaud and Georges, 1990; Metzler and Klafter, 2000)

$$\langle \Delta X(t)^2 \rangle \sim t^\alpha, \qquad (4.72)$$

($t \to \infty$), where $\alpha > 0$ is any positive real number. The range $0 < \alpha < 1$ describes *subdiffusion*, $\alpha = 1$ corresponds to (normal) diffusive transport such as in Eqs. (4.70) and (4.71), and $\alpha > 1$ describes *superdiffusion*.

For more than two decades the anomalous behavior has been the object of intense research with countless applications in many areas of physics, chemistry and natural and socio-economic sciences. There is an immense literature on the subject and for complete reports we may cite Balescu (2005; 2007), Bouchaud and Georges (1990), Eliazar (2013), Klafter and Sokolov (2005), Metzler and Klafter (2000; 2004), West *et al.* (2003), West (2014; 2016) and Zaslavsky (2002), among many others.

The concept of anomalous diffusion first emerged from the theory of random processes, specifically from continuous time random walks (see Chapter 12) and it was first applied to diffusion of charge carriers in organic semiconductors by Scher and Montroll in the 1970's (Weiss, 1994).

---

[9]In Sect. 4.5.1 we have assumed $D = 1$.

In the force-free case where no drift is present and if space is homogeneous and isotropic, the anomalous diffusion process is described by the so-called *fractional Brownian motion* for which the ordinary diffusion equation (4.45) for the transition density is replaced by a *fractional diffusion equation* which, in the one dimensional case, can be written as

$$\frac{\partial^\alpha p}{\partial t^\alpha} = \frac{1}{2} D \frac{\partial^{2\gamma} p}{\partial x^{2\gamma}}, \qquad (4.73)$$

where $\alpha > 0$ and $\gamma > 0$ can be any positive real numbers, although the most usual case found in practice is when $0 < \alpha \leq 2$ and $0 < \gamma \leq 1$. The symbols $\partial^\alpha / \partial t^\alpha$ and $\partial^{2\gamma} / \partial x^{2\gamma}$ stand respectively for the fractional Caputo and the Riesz-Feller fractional derivatives to be defined below.

For a general process in which $f(x,t) \neq 0$ and/or $D = D(x,t)$ –as, for instance, the case of particles diffusing under the influence of an external field of force– the fractional diffusion equation (4.73) is replaced by the fractional Fokker-Planck equation,

$$\frac{\partial^\alpha p}{\partial t^\alpha} = -\frac{\partial^\gamma}{\partial x^\gamma}[f(x,t)p] + \frac{1}{2}\frac{\partial^{2\gamma}}{\partial x^{2\gamma}}[D(x,t)p]. \qquad (4.74)$$

We will not treat this case in this introduction, and refer the interested reader to the literature for further information (see, for example, Metzler and Klafter, 2000, 2004).

The mathematical properties of the solutions to the fractional diffusion equation have been thoroughly studied and very clearly exposed by Mainardi and collaborators (Mainardi, 1996; Mainardi *et al.*, 2001; Gorenflo *et al.*, 2007). One of these properties is the scaling relation (Bouchaud and Georges, 1990; Mainardi *et al.*, 2001; Balescu, 2007)

$$p(x,t) = t^{-\alpha/2\gamma} g\left(\frac{x}{t^{\alpha/2\gamma}}\right). \qquad (4.75)$$

Using this scaling relation we see that

$$\langle X^2(t) \rangle \equiv \int_{-\infty}^{\infty} x^2 p(x,t) dx = t^{-\alpha/2\gamma} \int_{-\infty}^{\infty} x^2 g\left(\frac{x}{t^{\alpha/2\gamma}}\right) dx,$$

which after the change of variable $z = x/t^{\alpha/2\gamma}$ yields

$$\langle X^2(t) \rangle = K t^{\alpha/\gamma}, \qquad (4.76)$$

where

$$K = \int_{-\infty}^{\infty} z^2 g(z) dz.$$

From Eq. (4.76) we see that subdiffusion appears when $\alpha < \gamma$ and superdiffusion when $\alpha > \gamma$. Although very appealing, Eq. (4.76) has a limited range because K turns out to be infinite when $0 < \gamma < 1$ and the mean square displacement loses its sense except when $\gamma = 1$ (Balescu, 2007).

When $\gamma = 1$ but α is not an integer we have the so-called "time-fractional diffusion", the case $0 < \alpha < 1$ corresponding to subdiffusion while $1 < \alpha \leq 2$ to superdiffusion. When $\alpha = 1$ but γ is not an integer, the fractional diffusion equation (4.73) describes a Lévy process; this case is always associated to superdiffusion and it is termed "space-fractional diffusion" (Metzler and Klafter, 2004; Balescu, 2007).

Let us recall that the formulation of the fractional Brownian motion was first addressed within the continuous time random walk formalism (Weiss, 1994). In consequence derivations of the fractional diffusion equation (4.73) are mainly based on this formalism as we will explain in our brief report on the continuous time random walk we present in Chapter 12. In any case there also exist alternative approaches based on master equations or (fractional) Chapman-Kolmogorov expansions (see, for instance, Metzler and Klafter, 2000, for further information).

4.6.1 *Space-time fractional diffusion equation*

We will now obtain the solution of the space-time fractional diffusion equation,

$$\frac{\partial^{\alpha} p}{\partial t^{\alpha}} = \frac{1}{2} D \frac{\partial^{2\gamma} p}{\partial x^{2\gamma}}, \tag{4.77}$$

where $0 < \alpha \leq 2$ and $0 < \gamma \leq 1$. The initial conditions that accompany this fractional equation depend on the range of values the fractional index $\alpha > 0$ may take. When $\alpha \leq 1$ there is only one initial condition which is the usual one [10]

$$p(x,0) = \delta(x). \tag{4.78}$$

If, however, $1 < \alpha \leq 2$ there is an additional initial condition on the time derivative of $p(x,t)$ evaluated at $t = 0$. In what follows, and when this is the case, we will assume that

$$\left. \frac{\partial p(x,t)}{\partial t} \right|_{t=0} = 0. \tag{4.79}$$

Hence, when $1 < \alpha \leq 2$ the initial conditions are

$$p(x,0) = \delta(x), \qquad \left. \frac{\partial p(x,t)}{\partial t} \right|_{t=0} = 0.$$

---

[10]We assume that the process $X(t)$ is initially at the origin.

In order to get the solution to that initial value problem we need first to introduce some mathematical formalism concerning fractional derivatives.

The Caputo fractional derivative of order $\alpha > 0$ of a function $\phi(t)$ is defined by the functional (Balescu, 2007; Mainardi, 1996; Mainardi *et al.*, 2001; Gorenflo *et al.*, 2007; Podbury, 1999)

$$\frac{\partial^\alpha \phi(t)}{\partial t^\alpha} = \begin{cases} \dfrac{1}{\Gamma(n-\alpha)} \displaystyle\int_0^t \dfrac{\phi^{(n)}(t')dt'}{(t-t')^{1+\alpha-n}}, & n-1 < \alpha < n, \\ \phi^{(n)}(t), & \alpha = n, \end{cases} \qquad (4.80)$$

$(n = 1, 2, 3, \dots)$.

Using this definition we can readily obtain the Laplace transform of the Caputo derivative,

$$\mathcal{L}\left\{\frac{\partial^\alpha \phi(t)}{\partial t^\alpha}\right\} = \int_0^\infty e^{-st} \frac{\partial^\alpha \phi(t)}{\partial t^\alpha} dt.$$

Indeed, Laplace transforming Eq. (4.80) and using the convolution theorem, we have

$$\mathcal{L}\left\{\frac{\partial^\alpha \phi(t)}{\partial t^\alpha}\right\} = \frac{1}{\Gamma(n-\alpha)} \mathcal{L}\left\{\phi^{(n)}(t)\right\} \mathcal{L}\left\{t^{n-\alpha-1}\right\}.$$

With the explicit forms (Roberts and Kaufman, 1966)

$$\mathcal{L}\left\{\phi^{(n)}(t)\right\} = s^n \hat{\phi}(s) - \sum_{k=0}^{n-1} s^{n-1-k} \phi^{(k)}(0),$$

where $\hat{\phi}(s) = \mathcal{L}\{\phi(t)\}$, and

$$\mathcal{L}\left\{t^{n-\alpha-1}\right\} = \Gamma(n-\alpha) s^{\alpha-n},$$

the Laplace transform of the Caputo derivative is found to be

$$\mathcal{L}\left\{\frac{\partial^\alpha \phi(t)}{\partial t^\alpha}\right\} = s^\alpha \hat{\phi}(s) - s^{\alpha-1} \phi(0) - \sum_{k=1}^{n-1} s^{\alpha-1-k} \phi^{(k)}(0), \qquad (4.81)$$

$(n-1 < \alpha \le n, \ n = 1, 2, 3, \dots)$.

Note that when $0 < \alpha \le 1$ we have $n = 1$ and this transform reduces to

$$\mathcal{L}\left\{\frac{\partial^\alpha \phi(t)}{\partial t^\alpha}\right\} = s^\alpha \hat{\phi}(s) - s^{\alpha-1} \phi(0). \qquad (4.82)$$

Let us also note that if $n-1 < \alpha \le n$ $(n = 2, 3, 4, \dots)$ but $\phi^{(k)}(0) = 0$ $(k = 1, 2, \dots, n-1)$, the expression for the Laplace transform of the Caputo derivative is also given by Eq. (4.82).

The second kind of fractional derivative we need to define is the Riesz-Feller fractional derivative of order β ($0 < \beta \leq 2$) of a function $g(x)$ such that $g(x) \to 0$ as $x \to \pm\infty$. There are several equivalent ways to define it (Podbury, 1999), although one of the simplest and most operative definitions is through Fourier analysis. We thus define this operator as (Balescu, 2007)

$$\frac{\partial^\beta g(x)}{\partial |x|^\beta} = \mathcal{F}^{-1}\left\{-|\omega|^\beta \tilde{g}(\omega)\right\} \qquad (4.83)$$

($0 < \beta \leq 2$), where $\mathcal{F}^{-1}\{\cdot\}$ stands for the inverse Fourier transform, and

$$\tilde{g}(\omega) \equiv \int_{-\infty}^{\infty} e^{i\omega x} g(x) dx$$

is the direct transform.

Before proceeding further let us observe that Eq. (4.77) for the one-dimensional fractional Brownian motion can be easily generalized to account for anomalous diffusions in higher dimensions. In the isotropic and force-free case, the N-dimensional fractional diffusion equation is

$$\frac{\partial^\alpha p}{\partial t^\alpha} = \frac{1}{2}D\boldsymbol{\nabla}^{2\gamma}p \qquad (4.84)$$

($0 < \alpha \leq 2$, $0 < \gamma \leq 1$), where $\partial^\alpha/\partial t^\alpha$ is the fractional Caputo derivative defined above, and $\boldsymbol{\nabla}^{2\gamma}$ is the Riesz-Feller fractional Laplacian defined as the N-dimensional inverse Fourier transform

$$\boldsymbol{\nabla}^\beta g(\mathbf{x}) = \mathcal{F}^{-1}\left\{-|\boldsymbol{\omega}|^\beta \tilde{g}(\boldsymbol{\omega})\right\}, \qquad (4.85)$$

($0 < \beta \leq 2$), where $|\boldsymbol{\omega}| = (\omega_1^2 + \cdots + \omega_N^2)^{1/2}$ and

$$\tilde{g}(\boldsymbol{\omega}) = \int_{\mathbb{R}^N} e^{i\boldsymbol{\omega}\cdot\mathbf{x}} g(\mathbf{x}) d^N\mathbf{x}$$

($\boldsymbol{\omega} \cdot \mathbf{x} = \omega_1 x_1 + \cdots + \omega_N x_n$) is the direct Fourier transform.

Let us return to the one dimensional case and solve the fractional diffusion equation (4.77) under the initial conditions (4.78)–(4.79). The analytical solution to the problem cannot unfortunately be obtained for the original density $p(x,t)$ but only for the characteristic function which, recall, is the Fourier transform of the probability density,

$$\tilde{p}(\omega, t) = \int_{-\infty}^{\infty} e^{i\omega x} p(x,t) dx.$$

Taking the Fourier transform of Eqs. (4.77) and (4.78)–(4.79) and recalling Eq. (4.83) yields

$$\frac{\partial^\alpha \tilde{p}}{\partial t^\alpha} = -\frac{1}{2} D \omega^{2\gamma} \tilde{p}. \qquad (4.86)$$

The initial condition is

$$\tilde{p}(\omega, 0) = 1,$$

when $0 < \alpha \le 1$, or

$$\tilde{p}(\omega, 0) = 1 \quad \text{and} \quad \left.\frac{\partial \tilde{p}(\omega, t)}{\partial t}\right|_{t=0} = 0,$$

when $1 < \alpha \le 2$.

Laplace transforming Eq. (4.86) and using Eq. (4.82), we have

$$s^\alpha \hat{\tilde{p}} - s^{\alpha-1} = -\frac{1}{2} D \omega^{2\gamma} \hat{\tilde{p}},$$

from which we obtain the exact solution

$$\hat{\tilde{p}}(\omega, s) = \frac{s^{\alpha-1}}{s^\alpha + D\omega^{2\gamma}/2}, \qquad (4.87)$$

where

$$\hat{\tilde{p}}(\omega, s) = \int_0^\infty e^{-st} \tilde{p}(\omega, t) dt,$$

is the joint Fourier-Laplace transform of the probability density function $p(x, t)$.

Obtaining the density $p(x, t)$ in real time and space by inverting the joint transform (4.87) seems to be out of reach. It is, nonetheless, possible obtaining the exact expression for the characteristic function $\tilde{p}(\omega, t)$. We thus proceed to Laplace inverting Eq. (4.87). The first step is given by the expansion

$$\hat{\tilde{p}}(\omega, s) = \frac{s^{\alpha-1}}{s^\alpha + D\omega^{2\gamma}/2} = \frac{1/s}{1 + (D\omega^{2\gamma}/2s^\alpha)} = \frac{1}{s} \sum_{n=0}^\infty \left(-D\omega^{2\gamma}/2s^\alpha\right)^n,$$

that is,

$$\hat{\tilde{p}}(\omega, s) = \sum_{n=0}^\infty \frac{(-D\omega^{2\gamma}/2)^n}{s^{1+n\alpha}}. \qquad (4.88)$$

But (Roberts and Kaufman, 1966)

$$\mathcal{L}^{-1}\left\{\frac{1}{s^\beta}\right\} = \frac{t^{\beta-1}}{\Gamma(\beta)}, \qquad (\beta > 0), \qquad (4.89)$$

and Laplace inverting Eq. (4.88) we get

$$\tilde{p}(\omega,t) = \sum_{n=0}^{\infty} \frac{(-D\omega^{2\gamma}t^{\alpha}/2)^n}{\Gamma(1+n\alpha)},$$

which, in terms of the Mittag-Leffler function defined by the power series (Erdelyi, 1953)

$$E_{\alpha}(z) = \sum_{n=0}^{\infty} \frac{z^n}{\Gamma(1+n\alpha)}, \qquad (4.90)$$

can be written as

$$\tilde{p}(\omega,t) = E_{\alpha}\big(-D\omega^{2\gamma}t^{\alpha}/2\big). \qquad (4.91)$$

Let us finally see that when $\alpha = 1$ (no fractional time but fractional space) this result agrees with the Lévy distribution. Indeed, taking into account that $\Gamma(1+n) = n!$ we see from Eq. (4.90) that

$$E_1(z) = \sum_{n=0}^{\infty} \frac{z^n}{n!} = e^z,$$

and Eq. (4.91) yields

$$\tilde{p}(\omega,t) = e^{-D\omega^{2\gamma}t/2}, \qquad (4.92)$$

which corresponds to the Lévy distribution with zero mean.

4.6.2 *Time-fractional diffusion equation*

The time-fractional diffusion equation is a particular case of the space-time fractional diffusion equation (4.77) when $\gamma = 1$:

$$\frac{\partial^{\alpha} p}{\partial t^{\alpha}} = \frac{1}{2} D \frac{\partial^2 p}{\partial x^2}, \qquad (4.93)$$

$(0 < \alpha \leq 2)$.[11] In the Fourier-Laplace space the solution to Eq. (4.93) with the initial conditions (4.78)–(4.79) is given by (cf. Eq. (4.87) with $\gamma = 1$)

$$\hat{\tilde{p}}(\omega,s) = \frac{s^{\alpha-1}}{s^{\alpha} + D\omega^2/2}. \qquad (4.94)$$

Contrary to the space-fractional case where $\gamma < 1$, the Fourier inversion of this expression is now possible. In effect, recalling the Fourier inversion [Erdelyi (1954)]

$$\mathcal{F}\left\{\frac{a}{b+c^2\omega^2}\right\} = \frac{a}{2\sqrt{bc}}e^{-|x|b^{1/2}/c},$$

---

[11]Note that when $\alpha = 1$, Eq. (4.93) reduces to the diffusion equation, while if $\alpha = 2$ we obtain the wave equation.

we get ($a = s^{\alpha-1}$, $b = s^{\alpha}$ and $c^2 = D/2$)

$$\hat{p}(x,s) = \frac{1}{\sqrt{2D}s^{1-\alpha/2}} e^{-|x|s^{\alpha/2}\sqrt{2/D}}. \tag{4.95}$$

The inverse Laplace inversion of $\hat{p}(x,s)$ is defined by

$$p(x,t) = \frac{1}{2\pi i} \int_{\text{Br}} e^{st}\hat{p}(x,s)ds,$$

where Br is the Bromwich contour (Roberts and Kaufman, 1966). Substituting for (4.95) and defining $\sigma = st$ as the new integration variable we have

$$p(x,t) = \frac{1}{\sqrt{2Dt^{\alpha}}} \frac{1}{2\pi i} \int_{\text{Br}} e^{\sigma - |x|\sqrt{2/Dt^{\alpha}}\sigma^{\alpha/2}} \frac{d\sigma}{\sigma^{1-\alpha/2}}.$$

In terms of Mainardi's function defined as the following contour integral in the complex plane (Mainardi, 1996)

$$M_{\beta}(z) = \frac{1}{2\pi i} \int_{\text{Br}} e^{\sigma - z\sigma^{\beta}} \frac{d\sigma}{\sigma^{1-\beta}} \tag{4.96}$$

($0 < \beta < 1$), we finally write

$$p(x,t) = \frac{1}{\sqrt{2Dt^{\alpha}}} M_{\alpha/2}\left(|x|\sqrt{2/Dt^{\alpha}}\right) \tag{4.97}$$

($0 < \alpha < 2$).

It can be proved that Mainardi's function $M_{\beta}(z)$ is an entire function of z for $0 < \beta < 1$ that is a special case of the Wright function (Erdelyi, 1953), which is, in turn, closely related to the rather cumbersome Fox function, the latter frequently used in the anomalous diffusion literature (Metzler and Klafter, 2000). It can also be proved that $M_{\beta}(z)$ has the power expansion (see Mainardi, 1996, for details)

$$M_{\beta}(z) = \sum_{n=0}^{\infty} \frac{(-1)^n z^n}{n!\Gamma(-\beta n + 1 - \beta)}, \qquad 0 < \beta < 1. \tag{4.98}$$

By combining this expansion with Eq. (4.97) we can readily obtain the following asymptotic approximation which is useful for large values of time

$$p(x,t) = \frac{1}{\sqrt{2D}\Gamma(1-\alpha/2)} t^{-\alpha/2} + O(t^{-\alpha}). \tag{4.99}$$

The case of the ordinary Brownian motion ($\alpha = 1$) is evidently included in Eq. (4.97). Indeed, since (Mainardi, 1996)

$$M_{1/2}(z) = \frac{1}{\sqrt{\pi}} e^{-z^2/4},$$

we obtain from Eq. (4.97) the usual Gaussian density

$$p(x,t) = \frac{1}{\sqrt{2\pi Dt}} e^{-x^2/2Dt}.$$

We will finally evaluate the mean square displacement, a basic characteristic of anomalous diffusion, for the time-fractional Brownian motion. Let us recall that moments $\langle X^n(t) \rangle$ are related to derivatives of the characteristic function $\tilde{p}(\omega, t)$ with respect to ω evaluated at $\omega = 0$. Hence, in terms of the Laplace transform of the characteristic function $\hat{\tilde{p}}(\omega, s)$, we may write

$$\mathcal{L}\{\langle X^n(t) \rangle\} = i^{-n} \left. \frac{\partial \hat{\tilde{p}}(\omega, s)}{\partial \omega^n} \right|_{\omega=0}.$$

Using Eq. (4.94) it is a simple exercise for the reader to show that the first moment is zero while the Laplace transform of the second moment reads

$$\mathcal{L}\{\langle X^2(t) \rangle\} = \frac{D}{s^{1+\alpha}},$$

which by Laplace inversion (cf. Eq. (4.89)) results in anomalous diffusion

$$\langle X^2(t) \rangle = \frac{D}{\Gamma(1+\alpha)} t^\alpha, \qquad (4.100)$$

with sunbdiffusion when $0 < \alpha < 1$ and superdiffusion when $1 < \alpha < 2$.

We finish here this brief review on fractional diffusion but in Chapters 11 and 12 we will present more information on this topic in relation to first passage problems and continuous time random walks.

Chapter 5

Fokker-Planck equations in several dimensions

5.1 The multidimensional Fokker-Planck equation

Let $\mathbf{X}(t) = \big(X_1(t), \ldots, X_N(t)\big)$ be a N-dimensional diffusion process with vector drift

$$\mathbf{f}(\mathbf{x}, t) = \big(f_1(\mathbf{x}, t), \ldots, f_N(\mathbf{x}, t)\big),$$

and diffusion matrix

$$\mathbf{D}(\mathbf{x}, t) = \big(D_{jk}(\mathbf{x}, t)\big), \qquad (j, k = 1, 2, \ldots, N),$$

where $\mathbf{x} = (x_1, \ldots, x_N) \in \mathbb{R}^N$. As we have seen in Chapter 4 the transition density of the process, $p(\mathbf{x}, t|\mathbf{x}_0, t_0)$, satisfies the forward FPE [cf. Eq. (4.18)–(4.19)]

$$\frac{\partial p}{\partial t} = -\sum_{j=1}^{N} \frac{\partial}{\partial x_j} \big[f_j(\mathbf{x}, t) p(\mathbf{x}, t|\mathbf{x}_0, t_0) \big]$$

$$+ \frac{1}{2} \sum_{j,k=1}^{N} \frac{\partial^2}{\partial x_j \partial x_k} \big[D_{jk}(\mathbf{x}, t) p(\mathbf{x}, t|\mathbf{x}_0, t_0) \big], \qquad (5.1)$$

with initial condition

$$p(\mathbf{x}, t_0|\mathbf{x}_0, t_0) = \delta(\mathbf{x} - \mathbf{x}_0). \qquad (5.2)$$

We now present an alternative vector notation for the FPE that is quite convenient in many cases. Indeed, the use of the nabla operator

$$\boldsymbol{\nabla} = \left(\frac{\partial}{\partial x_1}, \ldots, \frac{\partial}{\partial x_N} \right),$$

allows us to write

$$\sum_{j=1}^{N} \frac{\partial}{\partial x_j}(f_j p) = \boldsymbol{\nabla} \cdot (\mathbf{f} p), \qquad \sum_{j,k=1}^{N} \frac{\partial^2}{\partial x_j \partial x_k}(D_{jk} p) = \boldsymbol{\nabla}^2 (\mathbf{D} p),$$

where $\nabla^2 = \nabla \cdot \nabla$ is the Laplacian. Equation (5.1) can thus be written in the following more compact form

$$\frac{\partial p}{\partial t} = -\nabla \cdot (\mathbf{f}p) + \frac{1}{2}\nabla^2(\mathbf{D}p). \tag{5.3}$$

As in the one dimensional case, we define the *probability current*, now a N-dimensional vector $\mathbf{J} = (J_1, \ldots, J_N)$ with components

$$J_k(\mathbf{x}, t|\mathbf{x}_0, t_0) = f_k(\mathbf{x}, t)p(\mathbf{x}, t|\mathbf{x}_0, t_0)$$
$$-\frac{1}{2}\sum_{j=1}^{N} \frac{\partial}{\partial x_j}\big[D_{jk}(\mathbf{x}, t)p(\mathbf{x}, t|\mathbf{x}_0, t_0)\big], \tag{5.4}$$

or, equivalently,

$$\mathbf{J}(\mathbf{x}, t|\mathbf{x}_0, t_0) = \mathbf{f}(\mathbf{x}, t)p(\mathbf{x}, t|\mathbf{x}_0, t_0) - \frac{1}{2}\nabla\big[\mathbf{D}(\mathbf{x}, t)p(\mathbf{x}, t|\mathbf{x}_0, t_0)\big]. \tag{5.5}$$

The vector current allows us to write the FPE (5.1) in the form of a continuity equation,

$$\frac{\partial p}{\partial t} + \sum_{k=1}^{N} \frac{\partial}{\partial x_k} J_k(\mathbf{x}, t|\mathbf{x}_0, t_0) = 0, \tag{5.6}$$

or [cf. Eq. (5.3)]

$$\frac{\partial p}{\partial t} + \nabla \cdot \mathbf{J} = 0, \tag{5.7}$$

where $\nabla \cdot \mathbf{J}$ is the divergence of the vector current.

5.2 Boundary conditions for the forward Fokker-Planck equation

We will now engage in a detailed analysis on the boundary conditions attached to the multivariate FPE. In Chapter 4 (Sect. 4.4) we have made, for the one dimensional FPE, a preliminary and rather incomplete approach to this issue (which is central in first-passage and escape problems, as we will see in Part III). We will now look at the question in a more thorough way.

Suppose that the diffusion process $\mathbf{X}(t)$ is defined inside some region $\mathrm{R} \subset \mathbb{R}^N$ with boundary S.[1] We will analyze what boundary conditions are appropriate to describe the diffusive process. We start with the forward FPE and postpone for the next section the discussion on the backward equation.

---

[1] Region R may be the entire real space $\mathbb{R}^N$ or some subset of it.

In terms of the transition density, $p(\mathbf{x}, t|\mathbf{x}_0, t_0)$, the probability that the diffusion process stays inside some N-dimensional region R, given that it is initially in this region, $\mathbf{X}(t_0) = \mathbf{x}_0 \in$ R, is given by the volume integral

$$P(\mathrm{R}, t|\mathbf{x}_0, t_0) = \int_{\mathrm{R}} p(\mathbf{x}, t|\mathbf{x}_0, t_0) d^N \mathbf{x},$$

where $d^N \mathbf{x} = dx_1 \cdots dx_N$ is the volume element. Integrating Eq. (5.7) over region R we get

$$\frac{\partial P(\mathrm{R}, t)}{\partial t} + \int_{\mathrm{R}} \boldsymbol{\nabla} \cdot \mathbf{J} \, d^N \mathbf{x} = 0, \tag{5.8}$$

where $\mathbf{J} = \mathbf{J}(\mathbf{x}, t|\mathbf{x}_0, t_0)$ is the probability current and for the ease of notation we have omitted the initial variables $\mathbf{x}_0$ and t_0 in the argument of the probability P.

Let us now suppose that the region R is bounded by a surface S, then Gauss theorem tells us that

$$\int_{\mathrm{R}} \boldsymbol{\nabla} \cdot \mathbf{J} \, d^N \mathbf{x} = \int_{\mathrm{S}} \mathbf{J} \cdot d\mathbf{S},$$

where $d\mathbf{S} = \mathbf{n} dS$, dS is the surface element and $\mathbf{n}$ is the (unitary) outward pointing normal to S. The use of Gauss theorem in Eq. (5.8) yields

$$\frac{\partial P(\mathrm{R}, t)}{\partial t} = - \int_{\mathrm{S}} \mathbf{J} \cdot d\mathbf{S}. \tag{5.9}$$

This balance equation clearly indicates that the rate of change of probability inside any region R is given by the surface integral of the flux $\mathbf{J}$ over the boundary of that region. If the surface integral of the flux is positive then $\partial P(\mathrm{R}, t)/\partial t < 0$ which implies a loss of probability. If the integral is negative, $\partial P(\mathrm{R}, t)/\partial t > 0$ and there is a gain of probability inside R.

We can now discuss the possible boundary conditions for higher dimensional diffusions. They are basically two:

(i) *Absorbing barrier*

Let us suppose that when the process reaches the boundary S of the region R, it disappears. In other words, the barrier absorbs the process. Since there is no process at S anymore, the probability of finding the process at the absorbing boundary must be zero. This implies that the transition density must also be zero at the surface

$$p(\mathbf{x}, t|\mathbf{x}_0, t_0) = 0, \qquad (\mathbf{x} \in \mathrm{S}). \tag{5.10}$$

(ii) *Reflecting barrier*

Let us now suppose the process cannot leave some region R (think, for instance, of a diffusing particle which is confined to move inside R). This means that the probability of finding the process inside R must be constant at any time. In other words, there is no gain or loss of probability inside R, that is, $\partial P(\mathrm{R}, t)/\partial t = 0$ and Eq. (5.9) implies a zero net flow of probability across the border S of the region. In such a case we, therefore, require

$$\mathbf{n} \cdot \mathbf{J}(\mathbf{x}, t) = 0, \qquad (\mathbf{x} \in \mathrm{S}). \tag{5.11}$$

That is to say, the orthogonal component of the probability flux across the boundary of the region must be zero. Note that, since the process cannot cross S, it must then be reflected there and hence the name of "reflecting barrier".

In some practical applications one can meet more complicated situations like, for instance, when the boundary S of the region R is divided into two parts, $\mathrm{S_a}$ and $\mathrm{S_r}$, such that the process is absorbed when reaches $\mathrm{S_a}$ and reflected when it hits $\mathrm{S_r}$. In such a case boundary conditions are

$$p(\mathbf{x}, t | \mathbf{x}_0, t_0) = 0, \qquad (\mathbf{x} \in \mathrm{S_a})., \tag{5.12}$$

and

$$\mathbf{n} \cdot \mathbf{J}(\mathbf{x}, t) = 0, \qquad (\mathbf{x} \in \mathrm{S_r}). \tag{5.13}$$

5.3   The backward Fokker-Planck equation and its boundary conditions *

Let us recall that, besides the forward Fokker-Planck equation (5.1), the transition density of any diffusion process also satisfies the backward Fokker-Planck equation. We have seen in Chapter 4 (Sect. 4.2) that the multidimensional backward FPE is

$$\frac{\partial p}{\partial t_0} = -\sum_{j=1}^{N} f_j(\mathbf{x}_0, t_0) \frac{\partial}{\partial x_{0j}} p(\mathbf{x}, t | \mathbf{x}_0, t_0)$$

$$-\frac{1}{2} \sum_{j,k=1}^{N} D_{jk}(\mathbf{x}_0, t_0) \frac{\partial^2}{\partial x_{0j} \partial x_{0k}} p(\mathbf{x}, t | \mathbf{x}_0, t_0), \tag{5.14}$$

with final condition

$$p(\mathbf{x}, t | \mathbf{x}_0, t) = \delta(\mathbf{x} - \mathbf{x}_0). \tag{5.15}$$

Obviously solving the backward problem (5.14)-(5.15) is equivalent to solving the forward problem (5.1)-(5.2).

Suppose that the process is confined in an N-dimensional region R with boundary S and we want to obtain the appropriate boundary conditions for the backward problem.

Since the diffusion process is Markovian, the density obeys the Chapman-Kolmogorov equation,

$$p(\mathbf{x}, t | \mathbf{x}_0, t_0) = \int_R p(\mathbf{x}, t | \mathbf{x}', t') p(\mathbf{x}', t' | \mathbf{x}_0, t_0) d^N \mathbf{x}', \qquad (5.16)$$

where $t_0 < t' < t$, i.e., t' is any intermediate time between initial and final times. However, and due to the Markovian character of the diffusion process, the transition density $p(\mathbf{x}, t | \mathbf{x}_0, t_0)$ does not depend on any intermediate time between t_0 and t. Hence

$$\frac{\partial}{\partial t'} p(\mathbf{x}, t | \mathbf{x}_0, t_0) = 0,$$

and taking the derivative with respect to t' on both sides of Eq. (5.16) we have

$$\int_R \left[p(\mathbf{x}', t' | \mathbf{x}_0, t_0) \frac{\partial}{\partial t'} p(\mathbf{x}, t | \mathbf{x}', t') + p(\mathbf{x}, t | \mathbf{x}', t') \frac{\partial}{\partial t'} p(\mathbf{x}', t' \mathbf{x}_0, t_0) \right] d^N \mathbf{x}' = 0.$$
$$(5.17)$$

Our subsequent development becomes easier if we simplify notation by writing

$$p' = p(\mathbf{x}', t' | \mathbf{x}_0, t_0), \quad \text{and} \quad p = p(\mathbf{x}, t | \mathbf{x}', t'). \qquad (5.18)$$

Thus Eq. (5.17) reads

$$\int_R \left(p' \frac{\partial p}{\partial t'} + p \frac{\partial p'}{\partial t'} \right) d^N \mathbf{x}' = 0. \qquad (5.19)$$

We next use the forward FPE for the time derivative $\partial p' / \partial t'$ and the backward equation for $\partial p / \partial t'$. The use of several manipulations, which we detail at the end of this section, finally result in the expression

$$\int_S \sum_{j=1}^N dS_j \left\{ \left[f_j p' - \frac{1}{2} \sum_{k=1}^N \frac{\partial}{\partial x'_k} (D_{jk} p') \right] p \right\}$$
$$+ \frac{1}{2} \int_S \sum_{j=1}^N dS_j \left(\sum_{k=1}^N D_{jk} \frac{\partial p}{\partial x'_k} \right) p' = 0, \qquad (5.20)$$

where dS_j $(j = 1, \ldots, n)$ are the components of the infinitesimal surface vector $d\mathbf{S} = \mathbf{n} dS$, where $\mathbf{n}$ is the unitary outward normal to S and dS is the surface element.

From Eq. (5.20) we will obtain the boundary conditions for the backward FPE.

(i) *Absorbing barrier*

Suppose that S is an absorbing boundary, the forward boundary condition (5.10) requires that $p(\mathbf{x}', t'|\mathbf{x}_0, t_0) = 0$ when $\mathbf{x}' \in$ S. That is $p' = 0$ on the boundary and from Eq. (5.20) we have

$$\int_S p \sum_{j,k=1}^{N} \frac{\partial}{\partial x'_k} \left(D_{jk} p' \right) dS_j = 0,$$

which necessarily implies that $p = 0$ on the boundary. Since $p = p(\mathbf{x}, t|\mathbf{x}', t')$, the absorbing boundary condition for the backward FPE is

$$p(\mathbf{x}, t|\mathbf{x}', t') = 0, \quad \text{for} \quad \mathbf{x}' \in \text{S}. \qquad (5.21)$$

Note that this condition means that the probability for the diffusion process to re-enter R from an absorbing boundary is zero.

(ii) *Refecting barrier*

Since [cf. Eq. (5.4)]

$$J'_j = f_j p' - \frac{1}{2} \sum_{k=1}^{N} \frac{\partial}{\partial x'_k} \left(D_{jk} p' \right),$$

$(j = 1, \ldots, N)$, is the flux of probability, $\mathbf{J}' = \mathbf{J}(\mathbf{x}', t'|\mathbf{x}_0, t_0)$, from the point $\mathbf{x}_0$ at t_0 to $\mathbf{x}'$ at $t' > t_0$, we can write Eq. (5.20) as

$$\int_S p \sum_{j=1}^{N} J'_j dS_j + \frac{1}{2} \int_S \sum_{j=1}^{N} dS_j \left(\sum_{k=1}^{N} D_{jk} \frac{\partial p}{\partial x'_k} \right) p' = 0. \qquad (5.22)$$

If S is a reflecting boundary, then the forward boundary condition (5.11) demands that

$$\mathbf{J}' = \mathbf{J}(\mathbf{x}', t'|\mathbf{x}_0, t_0) = 0 \quad \text{for} \quad \mathbf{x}' \in \text{S},$$

and Eq. (5.22) reads

$$\int_S p' \sum_{j,k=1}^{N} D_{jk} \frac{\partial p}{\partial x'_k} dS_j = 0. \qquad (5.23)$$

Since $dS_j = n_j dS$, where n_j is the jth component of the unitary normal vector pointing outside the surface S and dS is the infinitesimal surface element, Eq. (5.23) implies that for an arbitrary $p' = p(\mathbf{x}', t' | \mathbf{x}_0, t_0)$ we necessarily have

$$\sum_{j,k=1}^{N} n_j D_{jk}(\mathbf{x}', t') \frac{\partial}{\partial x'_k} p(\mathbf{x}, t | \mathbf{x}', t') = 0, \qquad (\mathbf{x}' \in S), \qquad (5.24)$$

which constitutes the backward reflecting boundary condition. let us finally note that in one dimension condition (5.24) reduces to

$$\left. \frac{\partial}{\partial x'} p(x, t | x', t') \right|_{x'=a} = 0, \qquad (5.25)$$

where a is a reflecting point.

Like the case of the forward boundary conditions when the boundary S is divided in two surfaces, absorbing and reflecting: $S = S_a \cup S_r$, the backward boundary conditions are [cf. Eqs. (5.21) and (5.24)]

$$p(\mathbf{x}, t | \mathbf{x}', t') = 0, \quad \text{for} \quad \mathbf{x}' \in S_a, \qquad (5.26)$$

and

$$\sum_{j,k=1}^{N} n_j D_{jk}(\mathbf{x}', t') \frac{\partial}{\partial x'_k} p(\mathbf{x}, t | \mathbf{x}', t') = 0, \quad \text{for} \quad \mathbf{x}' \in S_r. \qquad (5.27)$$

Derivation of Eq. (5.20)

We detail the derivation of Eq. (5.20). What follows is rather technical and we advise the uninterested reader to skip this part.

We start from Eq. (5.19):

$$\int_R \left(p' \frac{\partial p}{\partial t'} + p \frac{\partial p'}{\partial t'} \right) d^N \mathbf{x}' = 0, \qquad (5.28)$$

where p' and p are defined in Eq. (5.18):

$$p' = p(\mathbf{x}', t' | \mathbf{x}_0, t_0), \quad \text{and} \quad p = p(\mathbf{x}, t | \mathbf{x}', t').$$

The time derivative $\partial p / \partial t' = (\partial / \partial t') p(\mathbf{x}, t | \mathbf{x}', t')$ can be written in terms of the backward FPE as [cf Eq. (5.14) with $t_0 = t'$ and $\mathbf{x}_0 = \mathbf{x}'$]

$$\frac{\partial p}{\partial t'} = -\sum_{j=1}^{N} f_j \frac{\partial p}{\partial x'_j} - \frac{1}{2} \sum_{j,k=1}^{N} D_{jk} \frac{\partial^2 p}{\partial x'_j \partial x'_k}, \qquad (5.29)$$

and the derivative $\partial p'/\partial t' = (\partial/\partial t')p(\mathbf{x}',t'|\mathbf{x}_0,t_0)$ by the forward FPE as [cf Eq. (5.1) with $t = t'$ and $\mathbf{x} = \mathbf{x}'$]

$$\frac{\partial p'}{\partial t'} = -\sum_{j=1}^{N}\frac{\partial}{\partial x_j'}(f_j p') + \frac{1}{2}\sum_{j,k=1}^{N}\frac{\partial^2}{\partial x_j'\partial x_k'}(D_{jk}p'), \qquad (5.30)$$

where in both equations $f_j = f_j(\mathbf{x}',t')$ and $D_{jk} = D_{jk}(\mathbf{x}',t')$.

Substituting Eqs. (5.29) and (5.30) into Eq. (5.28) and rearranging terms yield

$$\int_{\mathrm{R}}\left\{\sum_{j=1}^{N}\left[p'f_j\frac{\partial p}{\partial x_j'} + p\frac{\partial}{\partial x_j'}(f_j p')\right]\right.$$

$$\left. +\frac{1}{2}\sum_{j,k=1}^{N}\left[p'D_{jk}\frac{\partial^2 p}{\partial x_j'\partial x_k'} - p\frac{\partial^2}{\partial x_j'\partial x_k'}(D_{jk}p')\right]\right\}d^N\mathbf{x}' = 0. \qquad (5.31)$$

Note that

$$p'f_j\frac{\partial p}{\partial x_j'} + p\frac{\partial}{\partial x_j'}(f_j p') = \frac{\partial}{\partial x_j'}(f_j p p'). \qquad (5.32)$$

Moreover

$$\frac{\partial}{\partial x_j'}\left[p'D_{jk}\frac{\partial p}{\partial x_k'} - p\frac{\partial}{\partial x_k'}(D_{jk}p')\right]$$

$$= \frac{\partial p}{\partial x_k'}\frac{\partial}{\partial x_j'}(D_{jk}p') + p'D_{jk}\frac{\partial^2 p}{\partial x_j'\partial x_k'}$$

$$- \frac{\partial p}{\partial x_j'}\frac{\partial}{\partial x_k'}(D_{jk}p') - p\frac{\partial^2}{\partial x_j'\partial x_k'}(D_{jk}p'). \qquad (5.33)$$

Taking into account that $D_{jk} = D_{kj}$ is a symmetrical matrix we see that

$$\sum_{j,k=1}^{N}\frac{\partial p}{\partial x_j'}\frac{\partial}{\partial x_k'}(D_{jk}p') = \sum_{j,k=1}^{N}\frac{\partial p}{\partial x_k'}\frac{\partial}{\partial x_j'}(D_{jk}p'),$$

and from Eq. (5.33) we get

$$\sum_{j,k=1}^{N}\left[p'D_{jk}\frac{\partial^2 p}{\partial x_j'\partial x_k'} - p\frac{\partial^2}{\partial x_j'\partial x_k'}(D_{jk}p')\right]$$

$$= \sum_{j=1}^{N}\frac{\partial}{\partial x_j'}\sum_{k=1}^{N}\left[p'D_{jk}\frac{\partial p}{\partial x_k'} - p\frac{\partial}{\partial x_k'}(D_{jk}p')\right]. \qquad (5.34)$$

Substituting Eqs. (5.32) and (5.34) into Eq. (5.31) yields

$$\int_R \sum_{j=1}^N \frac{\partial}{\partial x_j'} \left\{ f_j pp' + \frac{1}{2} \sum_{k=1}^N \left[p' D_{jk} \frac{\partial p}{\partial x_k'} - p \frac{\partial}{\partial x_k'} (D_{jk} p') \right] \right\} d^N \mathbf{x}' = 0. \qquad (5.35)$$

Defining the vector field **A** of components

$$A_j = f_j pp' + \frac{1}{2} \sum_{k=1}^N \left[p' D_{jk} \frac{\partial p}{\partial x_k'} - p \frac{\partial}{\partial x_k'} (D_{jk} p') \right],$$

we write Eq. (5.35) as

$$0 = \int_R \boldsymbol{\nabla} \cdot \mathbf{A} d^N \mathbf{x} = \int_S \mathbf{A} \cdot d\mathbf{S},$$

where the last equality comes from Gauss theorem. S is the boundary of the (closed) region R and $d\mathbf{S} = (dS_1, \ldots, dS_N)$ is the infinitesimal surface vector orthogonal to S and pointing outward. In components this equation reads

$$\int_S \sum_{j=1}^N dS_j \left\{ \left[f_j p' - \frac{1}{2} \sum_{k=1}^N \frac{\partial}{\partial x_k'} (D_{jk} p') \right] p \right\}$$
$$+ \frac{1}{2} \int_S \sum_{j=1}^N dS_j \left(\sum_{k=1}^N D_{jk} \frac{\partial p}{\partial x_k'} \right) p' = 0, \qquad (5.36)$$

which is Eq. (5.20).

□

5.4   Stationary distributions and potential conditions

Like in the one-dimensional case, a multidimensional diffusion process is time homogeneous if the transition density depends only on time differences which in turn implies that the drift and the diffusion matrix do not explicitly depend of time. In such a case the FPE (5.1) reads

$$\frac{\partial p}{\partial t} = -\sum_{j=1}^N \frac{\partial}{\partial x_j} \left[f_j(\mathbf{x}) p(\mathbf{x}, t - t_0 | \mathbf{x}_0) \right]$$
$$+ \frac{1}{2} \sum_{j,k=1}^N \frac{\partial^2}{\partial x_j \partial x_k} \left[D_{jk}(\mathbf{x}) p(\mathbf{x}, t - t_0 | \mathbf{x}_0) \right], \qquad (5.37)$$

with the initial condition

$$p(\mathbf{x}, 0 | \mathbf{x}_0) = \delta(\mathbf{x} - \mathbf{x}_0). \qquad (5.38)$$

In an analogous way, the backward FPE (5.14) reads

$$\frac{\partial p}{\partial t_0} = -\sum_{j=1}^{N} f_j(\mathbf{x}_0) \frac{\partial}{\partial x_{0j}} p(\mathbf{x}, t - t_0 | \mathbf{x}_0)$$

$$-\frac{1}{2}\sum_{j,k=1}^{N} D_{jk}(\mathbf{x}_0) \frac{\partial^2}{\partial x_{0j} \partial x_{0k}} p(\mathbf{x}, t - t_0 | \mathbf{x}_0), \qquad (5.39)$$

with condition given by Eq. (5.38) when $t_0 = t$. If, as we did in the one dimensional case, we set the new time scale $t' = t - t_0$, then $\partial/\partial t_0 = -\partial/\partial t'$ and Eq. (5.39) can be written as (we drop the prime in the new time variable)

$$\frac{\partial p}{\partial t} = \sum_{j=1}^{N} f_j(\mathbf{x}_0) \frac{\partial}{\partial x_{0j}} p(\mathbf{x}, t | \mathbf{x}_0)$$

$$+\frac{1}{2}\sum_{j,k=1}^{N} D_{jk}(\mathbf{x}_0) \frac{\partial^2}{\partial x_{0j} \partial x_{0k}} p(\mathbf{x}, t | \mathbf{x}_0). \qquad (5.40)$$

On the other hand, this change of time scale, $t \to t - t_0$, leaves the forward FPE (5.37) unchanged:

$$\frac{\partial p}{\partial t} = -\sum_{j=1}^{N} \frac{\partial}{\partial x_j} \left[f_j(\mathbf{x}) p(\mathbf{x}, t | \mathbf{x}_0) \right]$$

$$+\frac{1}{2}\sum_{j,k=1}^{N} \frac{\partial^2}{\partial x_j \partial x_k} \left[D_{jk}(\mathbf{x}) p(\mathbf{x}, t | \mathbf{x}_0) \right]. \qquad (5.41)$$

The initial condition attached to both equations is given by Eq. (5.38).

As we have seen in previous chapters, the time-homogeneous process will be stationary when the univariate PDF, $p(\mathbf{x}, t) = p_s(\mathbf{x})$, does not depend on time. In this case if the transition density tends, as $t - t_0 \to \infty$, to a finite and non-zero limit independent of $\mathbf{x}_0$. The stationary density $p_s(\mathbf{x})$ is given by

$$p_s(\mathbf{x}) = \lim_{t - t_0 \to \infty} p(\mathbf{x}, t - t_0 | \mathbf{x}_0). \qquad (5.42)$$

Assuming t fixed and $t_0 \to -\infty$ [2] in the multidimensional FPE (5.37) we see that $p_s(\mathbf{x})$ must be a finite and non-vanishing solution to

$$-\sum_{j=1}^{N} \frac{\partial}{\partial x_j} \left[f_j(\mathbf{x}) p_s(\mathbf{x}) \right] + \frac{1}{2}\sum_{j,k=1}^{N} \frac{\partial^2}{\partial x_j \partial x_k} \left[D_{jk}(\mathbf{x}) p_s(\mathbf{x}) \right] = 0. \qquad (5.43)$$

---

[2] That is, assuming that the process began in the remote past

In vector form this equation can be written as

$$\nabla \cdot \mathbf{J}_s(\mathbf{x}) = 0, \tag{5.44}$$

where $\mathbf{J}_s(\mathbf{x})$ is the stationary probability current defined by

$$J_s(\mathbf{x}) = \mathbf{f}(\mathbf{x})p_s(\mathbf{x}) - \frac{1}{2}\nabla\big[\mathbf{D}(\mathbf{x})p_s(\mathbf{x})\big]. \tag{5.45}$$

Obviously $J_s(\mathbf{x}) = \text{constant}$ is a solution of Eq. (5.44) but this solution, contrary to the one-dimensional case, is not unique. Indeed, suppose $J_s(\mathbf{x}) = \nabla \times \mathbf{A}(\mathbf{x})$ is the rotational of some arbitrary vector field $\mathbf{A}(\mathbf{x})$, then

$$\nabla \cdot \mathbf{J}_s(\mathbf{x}) = \nabla \times \nabla \mathbf{A}_s(\mathbf{x}) = 0$$

and Eq. (5.44) is identically satisfied.

In any case, we call *potential solution* to the stationary density such that the associated flux $\mathbf{J}_s = 0$ vanishes. Thus, the potential solution, $p_s(\mathbf{x})$, satisfies

$$\mathbf{f}p_s - \frac{1}{2}\nabla(\mathbf{D}p_s) = 0, \tag{5.46}$$

a vector equation which is, in fact, a system of N partial differential equations of first order for p_s. Let us see under which conditions Eq. (5.46) can be solved.

Recalling that $\nabla(\mathbf{D}p_s)$ is a vector with components [compare Eq. (5.4) with Eq. (5.5)]

$$\Big(\nabla(\mathbf{D}p_s)\Big)_k = \sum_{j=1}^{N} \frac{\partial}{\partial x_j}(D_{jk}p_s),$$

we see that Eq. (5.46) can be written as

$$\mathbf{f}p_s - \frac{1}{2}\mathbf{D}(\nabla p_s) - \frac{1}{2}(\nabla\mathbf{D})p_s = 0,$$

or, equivalently,

$$\frac{1}{2}\mathbf{D}\frac{\nabla p_s}{p_s} = \mathbf{f}p_s - \frac{1}{2}(\nabla\mathbf{D}),$$

and, since

$$\frac{\nabla p_s}{p_s} = \nabla \ln p_s,$$

we have

$$\nabla \ln p_s = \mathbf{D}^{-1}\big(2\mathbf{f} - \nabla\mathbf{D}\big), \tag{5.47}$$

where $\mathbf{D}^{-1}$ is the inverse diffusion matrix.

Unfortunately, Eq. (5.47) cannot be true for any choice of $\mathbf{f}$ and $\mathbf{D}$. Indeed, the vector field

$$\mathbf{F}(\mathbf{x}) = \mathbf{D}^{-1}(2\mathbf{f} - \boldsymbol{\nabla}\mathbf{D}),$$

on the right hand side of Eq. (5.47) cannot be arbitrary since, being the gradient of a scalar field [in this case the gradieent of $\ln p_s(\mathbf{x})$], the rotational of $\mathbf{F}$ must be zero, $\boldsymbol{\nabla} \times \mathbf{F}(\mathbf{x}) = 0$. In other words,

$$\boldsymbol{\nabla} \times \left[\mathbf{D}^{-1}(2\mathbf{f} - \boldsymbol{\nabla}\mathbf{D})\right] = 0, \tag{5.48}$$

which, in components, is equivalent to

$$\frac{\partial}{\partial x_j}\left[\mathbf{D}^{-1}(2\mathbf{f} - \boldsymbol{\nabla}\mathbf{D})\right]_k = \frac{\partial}{\partial x_k}\left[\mathbf{D}^{-1}(2\mathbf{f} - \boldsymbol{\nabla}\mathbf{D})\right]_j \tag{5.49}$$

$(j, k = 1, \ldots, N)$.

Equation (5.48) or equivalently, (5.49) constitute the so-called *potential conditions* that must satisfy the drift and the diffusion matrix in order to have a stationary solution in the form given by Eq. (5.47). When these conditions are met the integration of (5.47) is straightforward and yields

$$\ln p_s(\mathbf{x}) = \int \mathbf{D}^{-1}(\mathbf{x})\left[2\mathbf{f}(\mathbf{x}) - \boldsymbol{\nabla}\mathbf{D}(\mathbf{x})\right] \cdot d\mathbf{x} + \text{constant}.$$

Therefore, *potential solution*[3] is

$$p_s(\mathbf{x}) = N\exp\left\{\int \mathbf{D}^{-1}(\mathbf{x})\left[2\mathbf{f}(\mathbf{x}) - \boldsymbol{\nabla}\mathbf{D}(\mathbf{x})\right] \cdot d\mathbf{x}\right\}, \tag{5.50}$$

where N is the normalization constant.

5.5   Detailed balance *

The existence of a stationary distribution corresponding to a vanishing probability current, that is, the potential solution, is closely related to the so-called *detailed balance* which, roughly speaking, means that in stationary Markovian processes any possible transition is balanced by the reversed transition.

---

[3]The adjective "potential", either applied to conditions (5.48)–(5.49) or to the solution (5.50), comes from the fact that we have derived the vector field $\mathbf{F}(\mathbf{x})$ from the gradient of a scalar field, $\ln p_s(\mathbf{x})$, which is termed a potential, $-V(\mathbf{x})$, in the language of physics. So that $p_s(\mathbf{x}) = \exp[-V(\mathbf{x})]$.

5.5.1 *Physical motivation*

The concept of detailed balance comes from physics and is connected to the reversibility of the equations of motion. From classical mechanics we know that if the Hamiltonian of a system is an even function of the velocities, the equations of motion are invariant under time inversion[4]

$$t \to -t; \qquad \mathbf{r} \to \mathbf{r}; \qquad \mathbf{v} \to -\mathbf{v}.$$

Consider a mechanical system of point particles with positions $\mathbf{r}$ and velocities $\mathbf{v}$ in a stationary –i.e., equilibrium– state (think, for instance, of a gas in confined geometries). The typical transition corresponds to a particle that at some time t has position $\mathbf{r}$ and velocity $\mathbf{v}$, while at a later time $t' > t$ has position $\mathbf{r}'$ and velocity $\mathbf{v}'$. We write this transition as

$$(\mathbf{r}, \mathbf{v}, t) \longrightarrow (\mathbf{r}', \mathbf{v}', t'), \qquad (t < t').$$

The probability for this transition to happen is determined by the joint density $p(\mathbf{r}', \mathbf{v}', t'; \mathbf{r}, \mathbf{v}, t)$ (which is, in turn, related to the number of particles per unit volume of the phase-space undergoing such a transition).

On the other hand, under time inversion, the direct transition turns into the reverse

$$(\mathbf{r}', -\mathbf{v}', -t') \longrightarrow (\mathbf{r}, -\mathbf{v}, -t), \qquad (-t' < -t),$$

and the probability density of such a transition is given by $p(\mathbf{r}, -\mathbf{v}, -t; \mathbf{r}', -\mathbf{v}', -t')$.

The *principle of detailed balance demands that, in stationary systems, direct and reverse transitions are equally likely.* In other words:

$$p(\mathbf{r}', \mathbf{v}', t'; \mathbf{r}, \mathbf{v}, t) = p(\mathbf{r}, -\mathbf{v}, -t; \mathbf{r}', -\mathbf{v}', -t'). \qquad (5.51)$$

There are several equivalent forms of detailed balance that are sometimes more convenient than (5.51). Thus, recalling that $t' > t$ we set $t' = t + \tau \ (\tau > 0)$ and write

$$p(\mathbf{r}', \mathbf{v}', t + \tau; \mathbf{r}, \mathbf{v}, t) = p(\mathbf{r}, -\mathbf{v}, -t; \mathbf{r}', -\mathbf{v}', -t - \tau). \qquad (5.52)$$

However, and due to time homogeneity, the joint density is invariant under arbitrary time translations. In particular, for the density appearing on the right hand side of Eq. (5.52), we have

$$p(\mathbf{r}, -\mathbf{v}, -t; \mathbf{r}', -\mathbf{v}', -t - \tau) = p(\mathbf{r}, -\mathbf{v}, -t + h; \mathbf{r}', -\mathbf{v}', -t - \tau + h),$$

---

[4]In the presence of a magnetic field $\mathbf{B}$ of vector potential $\mathbf{A}$, the Hamiltonian of the system will contain terms of the form $(m\mathbf{v} - q\mathbf{A})^2$ (q is the electrical charge) which are not even in the velocity. In such a case –and in order to have invariance under time inversion– one has to add the condition $\mathbf{B} \to -\mathbf{B}$. Moreover, if there is an overall rotation of angular velocity $\mathbf{\Omega}$ we also have to add the condition $\mathbf{\Omega} \to -\mathbf{\Omega}$.

and since h is arbitrary we choose $h = 2t + \tau$, hence

$$p(\mathbf{r}, -\mathbf{v}, -t; \mathbf{r}', -\mathbf{v}', -t - \tau) = p(\mathbf{r}, -\mathbf{v}, t + \tau; \mathbf{r}', -\mathbf{v}', t).$$

Substituting this into the right hand side of Eq. (5.52) we see that the condition for detailed balance can be written as

$$p(\mathbf{r}', \mathbf{v}', t + \tau; \mathbf{r}, \mathbf{v}, t) = p(\mathbf{r}, -\mathbf{v}, t + \tau; \mathbf{r}', -\mathbf{v}', t). \tag{5.53}$$

which, using again temporal homogeneity, is equivalent to

$$p(\mathbf{r}', \mathbf{v}', \tau; \mathbf{r}, \mathbf{v}, 0) = p(\mathbf{r}, -\mathbf{v}, \tau; \mathbf{r}', -\mathbf{v}', 0). \tag{5.54}$$

Let us note that in terms of the transition density we can write

$$p(\mathbf{r}', \mathbf{v}', t + \tau | \mathbf{r}, \mathbf{v}, t) p(\mathbf{r}, \mathbf{v}, t) = p(\mathbf{r}, -\mathbf{v}, t + \tau | \mathbf{r}', -\mathbf{v}', t) p(\mathbf{r}', -\mathbf{v}', t).$$

But, due to stationarity,

$$p(\mathbf{r}', \mathbf{v}', t + \tau | \mathbf{r}, \mathbf{v}, t) = p(\mathbf{r}', \mathbf{v}', \tau | \mathbf{r}, \mathbf{v}, 0) \equiv p(\mathbf{r}', \mathbf{v}', \tau | \mathbf{r}, \mathbf{v}),$$

and

$$p(\mathbf{r}, \mathbf{v}, t) = p_s(\mathbf{r}, \mathbf{v}),$$

which is the stationary density of the process. Detailed balance thus demands

$$p(\mathbf{r}', \mathbf{v}', \tau | \mathbf{r}, \mathbf{v}) p_s(\mathbf{r}, \mathbf{v}) = p(\mathbf{r}, -\mathbf{v}, \tau | \mathbf{r}', -\mathbf{v}') p_s(\mathbf{r}', -\mathbf{v}'). \tag{5.55}$$

5.5.2   *Definition and some properties*

We next present the formal definition of detailed balance for a multidimensional Markov process. Let $\mathbf{x} = (x_1, \ldots, x_N)$ be a N-dimensional real vector which under time reversal, $t \to -t$, transforms as

$$x_j \to \epsilon_j x_j, \qquad \epsilon_j = \pm 1. \tag{5.56}$$

If $\epsilon_j = 1$ the variable x_j is said to be *even* under time reversal, while $\epsilon_j = -1$ corresponds to an *odd variable*. In the preceding example, position is even and velocity is odd.

We say that a stationary Markov process obeys detailed balance if [see Eq. (5.53)]

$$p(\mathbf{x}', t + \tau; \mathbf{x}, t) = p(\boldsymbol{\epsilon}\mathbf{x}, t + \tau; \boldsymbol{\epsilon}\mathbf{x}', t), \tag{5.57}$$

where $\boldsymbol{\epsilon}$ is a diagonal matrix with components[5]

$$(\boldsymbol{\epsilon})_{jk} = \epsilon_j \delta_{jk}, \tag{5.58}$$

---

[5]Unless we state the contrary, we do not assume Einstein summation convention on repeated indices.

so that

$$\epsilon\mathbf{x} = (\epsilon_1 x_1, \ldots, \epsilon_N x_N).$$

With this notation and looking at Eq. (5.55), we write

$$p(\mathbf{x}', \tau | \mathbf{x}) p_s(\mathbf{x}) = p(\epsilon\mathbf{x}, \tau | \epsilon\mathbf{x}') p_s(\epsilon\mathbf{x}').$$

Note that for $\tau = 0$ this condition reads

$$\delta(\mathbf{x}' - \mathbf{x}) p_s(\mathbf{x}) = \delta(\epsilon\mathbf{x} - \epsilon\mathbf{x}') p_s(\epsilon\mathbf{x}').$$

But $\delta(\epsilon\mathbf{x} - \epsilon\mathbf{x}') p_s(\epsilon\mathbf{x}') = \delta(\epsilon\mathbf{x} - \epsilon\mathbf{x}') p_s(\epsilon\mathbf{x})$ and

$$\delta(\epsilon\mathbf{x} - \epsilon\mathbf{x}') = \delta(\mathbf{x} - \mathbf{x}'), \tag{5.59}$$

since the delta function is invariant under changes of sign in its argument. Hence,

$$p_s(\epsilon\mathbf{x}) = p_s(\mathbf{x}). \tag{5.60}$$

We, therefore, rewrite the condition for detail balance as

$$p(\mathbf{x}', \tau | \mathbf{x}) p_s(\mathbf{x}) = p(\epsilon\mathbf{x}, \tau | \epsilon\mathbf{x}') p_s(\mathbf{x}'). \tag{5.61}$$

Some properties

(i) An important consequence of Eq. (5.60) is

$$\langle X_k \rangle_{\text{st}} = \epsilon_k \langle X_k \rangle_{\text{st}}, \tag{5.62}$$

where

$$\langle X_k \rangle_{\text{st}} = \int_{-\infty}^{\infty} \cdots \int_{-\infty}^{\infty} x_k p_s(x_1, \ldots, x_N) dx_1 \cdots dx_N,$$

is the stationary mean value. Observe that Eq. (5.62) implies that *the stationary mean value of any odd variable is zero*. Indeed, if $X_k(t)$ is odd then $\epsilon_k = -1$ and from (5.62) we see that $2\langle X_k \rangle_s = 0$.

(ii) The (stationary) autocorrelation function is the matrix

$$C_{jk}(\tau) = \langle X_j(\tau) X_k(0) \rangle.$$

Let us show that[6]

$$C_{jk}(\tau) = \epsilon_j \epsilon_k C_{kj}(\tau). \tag{5.63}$$

---

[6]In matrix notation we can write $\mathbf{C}(\tau) = \epsilon\mathbf{C}^T(\tau)\epsilon$, where $\mathbf{C}^T(\tau)$ is the transpose autocorrelation matrix.

Proof. In terms of the bivariate density $p(x_j, \tau; x_k, 0)$ we have

$$C_{jk}(\tau) = \int_{-\infty}^{\infty} dx_j \int_{-\infty}^{\infty} x_j x_k p(x_j, \tau; x_k, 0) dx_k,$$

which after the change of variables $\mathbf{x} = \epsilon \mathbf{y}$ reads

$$C_{jk}(\tau) = \int_{-\infty/\epsilon_j}^{\infty/\epsilon_j} \epsilon_j dy_j \int_{-\infty/\epsilon_k}^{\infty/\epsilon_k} (\epsilon_j y_j)(\epsilon_k y_k) p(\epsilon_j y_j, \tau; \epsilon_k y_k, 0) \epsilon_k dy_k.$$

Note that this change of variables is, at most, a change of sign. If x_j is an even variable the change is the identity, $y_j = x_j$, while if x_j is odd it amounts to $y_j = -x_j$. Hence,

$$\int_{-\infty/\epsilon_j}^{\infty/\epsilon_j} \epsilon_j dy_j = \int_{-\infty}^{\infty} dy_j, \qquad (5.64)$$

and similarly for x_k. We thus have

$$C_{jk}(\tau) = \int_{-\infty}^{\infty} dy_j \int_{-\infty}^{\infty} (\epsilon_j y_j)(\epsilon_k y_k) p(\epsilon_j y_j, \tau; \epsilon_k y_k, 0) dy_k.$$

Detailed balance demands that [cf. Eq. (5.57)]

$$p(\epsilon_j y_j, \tau; \epsilon_k y_k, 0) = p(y_k, \tau; y_j, 0).$$

Therefore,

$$C_{jk}(\tau) = \epsilon_j \epsilon_k \int_{-\infty}^{\infty} dy_j \int_{-\infty}^{\infty} y_j y_k p(y_k, \tau; y_j, 0) dy_k = \epsilon_j \epsilon_k \langle X_k(\tau) X_j(0) \rangle,$$

that is,

$$C_{jk}(\tau) = \epsilon_j \epsilon_k C_{kj}(\tau),$$

which is Eq. (5.63).

$\square$

(iii) Let us now see the form taken by detailed balance in terms of the characteristic function. We know that the characteristic function is the Fourier transform of the PDF. Therefore, the joint characteristic function $\tilde{p}(\boldsymbol{\omega}, t + \tau; \boldsymbol{\omega}', t)$, and the joint density are related by

$$p(\mathbf{x}, t + \tau; \mathbf{x}', t) \qquad (5.65)$$
$$= \frac{1}{(2\pi)^N} \int d\boldsymbol{\omega} \int d\boldsymbol{\omega}' \exp[-i\boldsymbol{\omega} \cdot \mathbf{x} - i\boldsymbol{\omega}' \cdot \mathbf{x}'] \tilde{p}(\boldsymbol{\omega}, t + \tau; \boldsymbol{\omega}', t),$$

where the integrals extend over the entire $\mathbb{R}^N$. We thus have

$$p(\epsilon \mathbf{x}', t + \tau; \epsilon \mathbf{x}, t) \qquad (5.66)$$
$$= \frac{1}{(2\pi)^N} \int d\boldsymbol{\omega} \int d\boldsymbol{\omega}' \exp[-i\boldsymbol{\omega} \cdot \epsilon \mathbf{x} - i\boldsymbol{\omega}' \cdot \epsilon \mathbf{x}'] \tilde{p}(\boldsymbol{\omega}', t + \tau; \boldsymbol{\omega}, t),$$

but

$$\boldsymbol{\omega} \cdot \boldsymbol{\epsilon}\mathbf{x} = \sum_{j=1}^{N} \omega_j \epsilon_j x_j = \epsilon\boldsymbol{\omega} \cdot \mathbf{x}$$

and defining the vector $\boldsymbol{\alpha}$ such that

$$\boldsymbol{\omega} = \epsilon\boldsymbol{\alpha},$$

we have

$$\boldsymbol{\omega} \cdot \boldsymbol{\epsilon}\mathbf{x} = \epsilon\boldsymbol{\omega} \cdot \mathbf{x} = \epsilon^2 \boldsymbol{\alpha} \cdot \mathbf{x} = \boldsymbol{\alpha} \cdot \mathbf{x},$$

because ϵ^2 is the identity matrix. On the other hand [see Eq. (5.64)],

$$\int_{\mathbb{R}^N} d\boldsymbol{\omega} = \prod_{j=1}^{N} \left[\int_{-\infty}^{\infty} d\omega_j \right] = \prod_{j=1}^{N} \left[\int_{-\infty/\epsilon_j}^{\infty/\epsilon_j} \epsilon_j d\alpha_j \right]$$

$$= \prod_{j=1}^{N} \left[\int_{-\infty}^{\infty} d\alpha_j \right] = \int_{\mathbb{R}^N} d\boldsymbol{\alpha}.$$

Thus, making the changes of variable $\boldsymbol{\omega} = \epsilon\boldsymbol{\alpha}$ and $\boldsymbol{\omega}' = \epsilon\boldsymbol{\alpha}'$ in Eq. (5.66), we write

$$p(\boldsymbol{\epsilon}\mathbf{x}', t + \tau; \boldsymbol{\epsilon}\mathbf{x}, t) \qquad (5.67)$$

$$= \frac{1}{(2\pi)^N} \int d\boldsymbol{\alpha} \int d\boldsymbol{\alpha}' \exp\left[-i\boldsymbol{\alpha} \cdot \mathbf{x} - i\boldsymbol{\alpha}' \cdot \mathbf{x}' \right] \tilde{p}(\epsilon\boldsymbol{\alpha}', t + \tau; \epsilon\boldsymbol{\alpha}, t).$$

The condition for detailed balance tells us that $p(\mathbf{x}, t + \tau; \mathbf{x}', t) = p(\boldsymbol{\epsilon}\mathbf{x}', t + \tau; \boldsymbol{\epsilon}\mathbf{x}, t)$. Thus, comparing Eqs. (5.65) and (5.67), we obtain

$$\tilde{p}(\boldsymbol{\omega}, t + \tau; \boldsymbol{\omega}', t) = \tilde{p}(\epsilon\boldsymbol{\omega}', t + \tau; \epsilon\boldsymbol{\omega}, t) \qquad (5.68)$$

as the condition for detailed balance of the characteristic function.

5.6 Detailed balance in diffusion processes *

Up to now we have developed the concept of detailed balance for stationary Markov processes, we now add that these processes have continuous sample paths. We will thus see what specific requirements are needed for a diffusion process to meet detailed balance. We will finally see that this is closely related with the existence of the potential solution, Eq. (5.50), for the stationary distribution.

For stationary diffusion processes. the conditions of detailed balance are given by the following result:

Theorem

Suppose a stationary multidimensional diffusion process with stationary density $p_s(\mathbf{x})$. If drift, $\mathbf{f}(\mathbf{x})$, and diffusion matrix, $\mathbf{D}(\mathbf{x}) = \big(D_{jk}(\mathbf{x})\big)$ $(j, k = 1, \ldots, N)$, verify the conditions:

$$\big[\mathbf{f}(\mathbf{x}) + \epsilon \mathbf{f}(\epsilon \mathbf{x})\big] p_s(\mathbf{x}) = \boldsymbol{\nabla}\big[\mathbf{D}(\mathbf{x}) p_s(\mathbf{x})\big] \qquad (5.69)$$

and

$$\epsilon_j \epsilon_k D_{jk}(\epsilon \mathbf{x}) = D_{jk}(\mathbf{x}), \qquad (5.70)$$

the process fulfills detailed balance.

Proof. We will prove that if conditions (5.69) and (5.70) hold then

$$p(\mathbf{x}, t|\mathbf{x}') p_s(\mathbf{x}') = p(\epsilon \mathbf{x}', t|\epsilon \mathbf{x}) p_s(\mathbf{x}) \qquad (5.71)$$

and detailed balance is satisfied [cf. Eq. (5.61)].

We define the quantity

$$q(\mathbf{x}, t|\mathbf{x}') \equiv p(\epsilon \mathbf{x}', t|\epsilon \mathbf{x}) \frac{p_s(\mathbf{x})}{p_s(\mathbf{x}')}, \qquad (5.72)$$

and we will find the conditions under which q coincides with the transition density, i.e., $q(\mathbf{x}, t|\mathbf{x}') = p(\mathbf{x}, t|\mathbf{x}')$, so that detailed balance (5.71) holds.

We apply to both sides of Eq. (5.72) the (forward) Fokker-Planck operator,

$$L_{FP}(\mathbf{x}) = -\sum_{j=1}^{N} \frac{\partial}{\partial x_j} f_j(\mathbf{x}) + \frac{1}{2} \sum_{j,k=1}^{N} \frac{\partial^2}{\partial x_j \partial x_k} D_{jk}(\mathbf{x}),$$

that is

$$p_s(x') L_{FP}(\mathbf{x}) q(\mathbf{x}, t|\mathbf{x}') = L_{FP}(\mathbf{x}) \big[p(\epsilon \mathbf{x}', t|\epsilon \mathbf{x}) p_s(\mathbf{x})\big]. \qquad (5.73)$$

In order to ease notation, we call

$$p'_s = p_s(\mathbf{x}'), \quad p_s = p_s(\mathbf{x}), \quad p'(\epsilon \mathbf{x}) = p(\epsilon \mathbf{x}', t|\epsilon \mathbf{x}), \qquad (5.74)$$

and the right hand side of Eq. (5.73) reads

$$L_{FP}(\mathbf{x}) \big[p'(\epsilon \mathbf{x}) p_s(\mathbf{x})\big]$$

$$= -\sum_{j=1}^{N} \frac{\partial}{\partial x_j} \big[f_j(\mathbf{x}) p'(\epsilon \mathbf{x}) p_s(\mathbf{x})\big] + \frac{1}{2} \sum_{j,k=1}^{N} \frac{\partial^2}{\partial x_j \partial x_k} \big[D_{jk}(\mathbf{x}) p'(\epsilon \mathbf{x}) p_s(\mathbf{x})\big],$$

so that

$$L_{FP}\left[p'(\boldsymbol{\epsilon}\mathbf{x})p_s(\mathbf{x})\right]$$

$$= -\sum_{j=1}^{N}\left[p'(\boldsymbol{\epsilon}\mathbf{x})\frac{\partial}{\partial x_j}(f_j p_s) + f_j p_s\frac{\partial}{\partial x_j}p'(\boldsymbol{\epsilon}\mathbf{x})\right]$$

$$+\frac{1}{2}\sum_{j,k=1}^{N}\left[p'(\boldsymbol{\epsilon}\mathbf{x})\frac{\partial^2}{\partial x_j \partial x_k}(D_{jk}p_s) + 2\frac{\partial}{\partial x_j}(D_{jk}p_s)\frac{\partial}{\partial x_k}p'(\boldsymbol{\epsilon}\mathbf{x})\right.$$

$$\left.+D_{jk}p_s\frac{\partial^2}{\partial x_j \partial x_k}p'(\boldsymbol{\epsilon}\mathbf{x})\right].$$

Rearranging terms we get

$$L_{FP}(\mathbf{x})\left[p'(\boldsymbol{\epsilon}\mathbf{x})p_s(\mathbf{x})\right]$$

$$= \left[-\sum_{j=1}^{N}\frac{\partial}{\partial x_j}(f_j p_s) + \frac{1}{2}\sum_{j,k=1}^{N}\frac{\partial^2}{\partial x_j \partial x_k}(D_{jk}p_s)\right]p'(\boldsymbol{\epsilon}\mathbf{x})$$

$$-\sum_{j=1}^{N}f_j p_s\frac{\partial}{\partial x_j}p'(\boldsymbol{\epsilon}\mathbf{x}) + \sum_{j,k=1}^{N}\frac{\partial}{\partial x_j}(D_{jk}p_s)\frac{\partial}{\partial x_k}p'(\boldsymbol{\epsilon}\mathbf{x})$$

$$+\frac{1}{2}\sum_{j,k=1}^{N}D_{jk}p_s\frac{\partial^2}{\partial x_j \partial x_k}p'(\boldsymbol{\epsilon}\mathbf{x}). \tag{5.75}$$

But $p_s = p_s(\mathbf{x})$ obeys the forward FPE:

$$-\sum_{j=1}^{N}\frac{\partial}{\partial x_j}(f_j p_s) + \frac{1}{2}\sum_{j,k=1}^{N}\frac{\partial^2}{\partial x_j \partial x_k}(D_{jk}p_s) = 0.$$

Moreover, due to the symmetry of the diffusion matrix, $D_{jk} = D_{kj}$, and taking into account the sum on dummy indices j and k, we have

$$\sum_{j,k=1}^{N}\frac{\partial}{\partial x_j}(D_{jk}p_s)\frac{\partial}{\partial x_k}p'(\boldsymbol{\epsilon}\mathbf{x}) = \sum_{j,k=1}^{N}\frac{\partial}{\partial x_k}(D_{jk}p_s)\frac{\partial}{\partial x_j}p'(\boldsymbol{\epsilon}\mathbf{x}).$$

Substituting these two equations into Eq. (5.75) and returning to unabridged notation [cf. Eq. (5.74)], we write

$$
L_{FP}(\mathbf{x})\left[p(\epsilon\mathbf{x}',t|\epsilon\mathbf{x})p_s(\mathbf{x})\right]
$$
$$
= \sum_{j=1}^{N}\left[-f_j(\mathbf{x})p_s(\mathbf{x}) + \sum_{k=1}^{N}\frac{\partial}{\partial x_k}\left(D_{jk}(\mathbf{x})p_s(\mathbf{x})\right)\right]\frac{\partial}{\partial x_j}p(\epsilon\mathbf{x}',t|\epsilon\mathbf{x})
$$
$$
+\frac{1}{2}\left[\sum_{j,k=1}^{N}D_{jk}(\mathbf{x})\frac{\partial^2}{\partial x_j \partial x_k}p(\epsilon\mathbf{x}',t|\epsilon\mathbf{x})\right]p_s(\mathbf{x}). \tag{5.76}
$$

We now suppose that the drift and the diffusion matrix satisfy, under time reversal, conditions (5.69) and (5.70). Then, writing the condition given in Eq. (5.69) as

$$
-f_j(\mathbf{x})p_s(\mathbf{x}) + \sum_{k=1}^{N}\frac{\partial}{\partial x_k}D_{jk}(\mathbf{x})p_s(\mathbf{x}) = \epsilon_j f_j(\epsilon\mathbf{x})p_s(\mathbf{x}),
$$

and substituting it along with condition (5.70) into Eq. (5.76), we get

$$
L_{FP}(\mathbf{x})\left[p(\epsilon\mathbf{x}',t|\epsilon\mathbf{x})p_s(\mathbf{x})\right]= \left[\sum_{j=1}^{N}\epsilon_j f_j(\epsilon\mathbf{x})\frac{\partial}{\partial x_j}p(\epsilon\mathbf{x}',t|\epsilon\mathbf{x})\right.
$$
$$
+\frac{1}{2}\sum_{j,k=1}^{N}\epsilon_j\epsilon_k D_{jk}(\epsilon\mathbf{x})\frac{\partial^2}{\partial x_j \partial x_k}p(\epsilon\mathbf{x}',t|\epsilon\mathbf{x})\left.\right]p_s(\mathbf{x}). \tag{5.77}
$$

We now set

$$
\mathbf{y} = \epsilon\mathbf{x}, \qquad \mathbf{y}' = \epsilon\mathbf{x}',
$$

which, taking into account that $\epsilon^2 = \mathbf{I}$ (the identity matrix), implies

$$
\mathbf{x} = \epsilon\mathbf{y}, \qquad \mathbf{x}' = \epsilon\mathbf{y}'.
$$

Therefore,

$$
L_{FP}\left[p(\mathbf{y}',t|\mathbf{y})p_s(\mathbf{x})\right]= \left[\sum_{j=1}^{N}f_j(\mathbf{y})\frac{\partial}{\partial y_j}p(\mathbf{y}',t|\mathbf{y})\right.
$$
$$
+\frac{1}{2}\sum_{j,k=1}^{N}D_{jk}(\mathbf{y})\frac{\partial^2}{\partial y_j \partial y_k}p(\mathbf{y}',t|\mathbf{y})\left.\right]p_s(\mathbf{x}). \tag{5.78}
$$

Note that the term in square brackets is related to the backward Fokker-Planck operator [cf. Eq. (5.14)]

$$L_{FP}^{\dagger}(\mathbf{y})p(\mathbf{y}',t|\mathbf{y}) = -\sum_{j=1}^{N} f_j(\mathbf{y})\frac{\partial}{\partial y_j}p(\mathbf{y}',t|\mathbf{y})$$
$$-\frac{1}{2}\sum_{j,k=1}^{N} D_{jk}(\mathbf{y})\frac{\partial^2}{\partial y_j\partial y_k}p(\mathbf{y}',t|\mathbf{y}),$$

so that

$$L_{FP}(\mathbf{x})\left[p(\epsilon\mathbf{x}',t|\epsilon\mathbf{x})p_s(\mathbf{x})\right] = -\left[L_{FP}^{\dagger}(\mathbf{x})p(\epsilon\mathbf{x}',t|\epsilon\mathbf{x})\right]p_s(\mathbf{x})$$
$$= \frac{\partial}{\partial t}p(\epsilon\mathbf{x}',t|\epsilon\mathbf{x}),$$

where we have taken into account that the process is time homogeneous and $p(\epsilon\mathbf{x}',t|\epsilon\mathbf{x}) = p(\epsilon\mathbf{x}',0|\epsilon\mathbf{x},-t)^7$. Substituting this into the right hand side of Eq. (5.73), we get

$$p_s(x')L_{FP}(\mathbf{x})q(\mathbf{x},t|\mathbf{x}') = \left[\frac{\partial}{\partial t}p(\epsilon\mathbf{x}',t|\epsilon\mathbf{x})\right]p_s(\mathbf{x}).$$

Equivalently [see Eq. (5.72)]

$$L_{FP}(\mathbf{x})q(\mathbf{x},t|\mathbf{x}') = \left[\frac{\partial}{\partial t}p(\epsilon\mathbf{x}',t|\epsilon\mathbf{x})\frac{p_s(\mathbf{x})}{p_s(\mathbf{x}')}\right]$$
$$= \frac{\partial}{\partial t}q(\mathbf{x},t|\mathbf{x}').$$

Therefore, the quantity q obeys the forward FPE. On the other hand, from the definition of $q(\mathbf{x},t|\mathbf{x}')$ we see that the initial condition for the FPE is the same than that of $p(\mathbf{x},t|\mathbf{x}')$ [cf. Eq. (5.59)]

$$q(\mathbf{x},0|\mathbf{x}') = p(\epsilon\mathbf{x}',0|\epsilon\mathbf{x}) = \delta(\mathbf{x}-\mathbf{x}').$$

However, solutions of the FPE with the same initial condition are unique. We, therefore, conclude that, as long as Eqs. (5.69) and (5.70) hold, $q(\mathbf{x},t|\mathbf{x}') = p(\mathbf{x},t|\mathbf{x}')$ and detailed balance is satisfied.

□

Observations

(i) Equations (5.69) and (5.70) are also *necessary conditions* for detailed balance. Thus, assuming that detailed balance is satisfied [cf. Eq. (5.71)] one can see –after using the definitions of drift and diffusion matrix as the

---

[7]See the backward FPE (5.14) with the change $\partial/\partial t_0 \to -\partial/\partial t$.

first and second infinitesimal moments respectively– that Eqs. (5.69) and (5.70) hold (see Gardiner, 1986, for more details).

(ii) For only even variables, $\epsilon \mathbf{x} = \mathbf{x}$, the condition on the diffusion matrix given in Eq. (5.70) is trivial and gives no further information. On the other hand, condition (5.69) reduces to

$$\mathbf{f}(\mathbf{x})p_s(\mathbf{x}) = \frac{1}{2}\boldsymbol{\nabla}\left[\mathbf{D}(\mathbf{x})p_s(\mathbf{x})\right]. \tag{5.79}$$

However, this condition equals Eq. (5.46) which implies the vanishing of the stationary probability current J_s. Therefore, Eq. (5.79) is equivalent to the potential condition. In other words, *detailed balance for even variables implies the existence of the potential solution (5.50) for the stationary PDF*.

5.6.1   Kramers equation for Brownian motion

In 1905 Einstein presented a physical explanation for the Brownian motion. Assuming that it is a continuous Markov process, Einstein showed that the transition density obeys the diffusion equation. Some time later, in 1908, Langevin presented an alternative explanation based on Newton laws of motion. The basic assumption is that the total force acting on a Brownian particle of mass m moving inside a fluid, and bounded by a field with potential energy $U(x)$, is the resultant of (i) damping, proportional to velocity, (ii) a driving force due to the external field and (iii) a force taking into account the random impacts on the particle of the molecules of the medium which, due to the randomness of these collisions, is modeled as Gaussian white noise with an intensity dependent on the temperature of the medium. For the one-dimensional motion we thus have from Newton's second law the following equation of motion, called *Langevin equation*:

$$m\frac{d^2X}{dt^2} = -\beta\frac{dX}{dt} - U'(X) + \xi(t), \tag{5.80}$$

where $X = X(t)$ is the position of the particle, β is the damping constant, and $\xi(t)$ is zero-mean Gaussian white noise

$$\langle\xi(t)\rangle = 0, \qquad \langle\xi(t)\xi(t')\rangle = \beta k_B T\delta(t - t'),$$

where k_B is Boltzmann constant and T is the absolute temperature.

As we will see in Chapter 7, the Langevin equation (5.80) describes a bidimensional diffusion process (X, V) ($V = \dot{X}$ is the velocity of the particle) whose transition density, $p(x, v, t|x_0, v_0, t_0)$, obeys the following Fokker-Planck equation

$$\frac{\partial p}{\partial t} = -\frac{\partial}{\partial x}(vp) + \frac{1}{m}\frac{\partial}{\partial v}\left\{[\beta v + U'(x)]p\right\} + \frac{\beta k_B T}{2m^2}\frac{\partial^2 p}{\partial v^2}. \tag{5.81}$$

Defining the scaled variables

$$y = x\sqrt{m/k_BT}, \quad u = v\sqrt{m/k_BT}, \quad V(y) = U(x)/k_BT, \quad \gamma = \beta/m,$$

the FPE takes the form

$$\frac{\partial p}{\partial t} = -\frac{\partial}{\partial y}(up) + \frac{\partial}{\partial u}\left\{[\gamma u + V'(y)]p\right\} + \frac{1}{2}\gamma\frac{\partial^2 p}{\partial u^2}, \qquad (5.82)$$

which is called *Kramers equation*.

Defining the bidimensional variable $\mathbf{x} = (y, u)$ we see that Kramers equation is a bidimensional FPE [cf. Eq. (5.3)],

$$\frac{\partial p}{\partial t} = -\boldsymbol{\nabla}\cdot(\mathbf{f}p) + \frac{1}{2}\boldsymbol{\nabla}^2(\mathbf{D}p),$$

where[8]

$$\boldsymbol{\nabla} = \left(\frac{\partial}{\partial y}, \frac{\partial}{\partial u}\right), \quad \mathbf{f}(\mathbf{x}) = (u, -\gamma u - U'(y)), \quad \mathbf{D} = \begin{pmatrix} 0 & 0 \\ 0 & \gamma \end{pmatrix}.$$

Since under time reversal y is even and u is odd, we have

$$\epsilon(y, u) = (y, -u).$$

We can now check whether the process satisfies detailed balance. Since the diffusion matrix is constant, condition (5.70), $D_{jk}(\mathbf{x}) = \epsilon_j\epsilon_k D_{jk}(\epsilon\mathbf{x})$, is trivially satisfied. On the other hand

$$\epsilon\mathbf{f}(\epsilon\mathbf{x}) = \left(-u, \gamma u - U'(y)\right)$$

and we have

$$\left[\mathbf{f}(\mathbf{x}) + \epsilon\mathbf{f}(\epsilon\mathbf{x})\right]p_s(\mathbf{x}) = \left(0, -2\gamma u p_s(\mathbf{x})\right). \qquad (5.83)$$

Moreover

$$\boldsymbol{\nabla}\left[\mathbf{D}(\mathbf{x})p_s(\mathbf{x})\right] = \left(0, \gamma\frac{\partial p_s(\mathbf{x})}{\partial u}\right). \qquad (5.84)$$

Plugging Eqs. (5.83) and (5.84) into condition (5.69) for detailed balance, we get the following first-order partial differential equation for $p_s = p_s(y, u)$:

$$\frac{\partial p_s(\mathbf{x})}{\partial u} = -u p_s(\mathbf{x}).$$

We can easily see by direct substitution that the general solution is

$$p_s(y, u) = \phi(y)e^{-u^2/2}, \qquad (5.85)$$

where $\phi(y)$ is an arbitrary function of the scaled position. Therefore, if the stationary density of the Brownian particle has the form given by

---

[8]Note the singular character of the diffusion matrix, i.e., det $\mathbf{D} = 0$.

Eq. (5.85) then detailed balance holds. We will find the unknown $\phi(y)$ (and consequently the stationary distribution) by imposing that p_s obeys Kramers equation. Substituting Eq. (5.85) into Eq. (5.82) yields

$$0 = -u\frac{d\phi(y)}{dy} - uV'(y)\phi(y),$$

which is a first-order ordinary differential equation whose solution reads

$$\phi(y) = Ne^{-V(y)},$$

where N is an arbitrary constant. Collecting results we see that the stationary solution of Kramers equation is

$$p_s(y,u) = Ne^{-u^2/2 - V(y)},$$

(N is obtained by normalization). In the original variables we finally get the *Maxwell-Boltzmann distribution*:

$$p_s(x,v) = N\exp\left\{-\left[mv^2/2 + U(x)\right]/K_BT\right\}. \qquad (5.86)$$

Chapter 6

Linear response theory*

The theory of linear response stems from physics as the study of how physical systems in a stationary state react to external and weak perturbations.

Let us introduce the problem by an example taken from chemical physics. Suppose an electrolyte solution in equilibrium which implies that there is no net flux of electrical charge and in consequence the average electrical current is zero. At some instant of time t_0 an external electric field $\mathbf{E}$ is switch on and ions start flowing along the direction of the field. Now the system is out of equilibrium with the appearance of a non-stationary electrical current $\mathbf{J}(t, \mathbf{E})$ which will depend on the applied field. Obviously $\mathbf{J}(t, \mathbf{E})$ is such that if $\mathbf{E} = 0$ the response coincides with the equilibrium value, that is $\mathbf{J}(t, \mathbf{E} = 0) = 0$. Thus expanding the current in powers of the electrical field we have

$$\mathbf{J}(t, \mathbf{E}) = \mathbf{\Lambda}(t) \cdot \mathbf{E} + O\left(|\mathbf{E}|^2\right),$$

where $\mathbf{\Lambda}(t)$ is a matrix whose elements are $\Lambda_{kl} = \partial J_k / \partial E_l|_{\mathbf{E}=0}$. When the external field is weak we can neglect higher order terms and get a linear relation between the output $\mathbf{J}$ and the input $\mathbf{E}$. This relation is the basis of the theory of linear response. Let us observe that if as t increases the electric field tends to a constant field, $\mathbf{E}(t) \to \mathbf{E}_s$ as $t \to \infty$, then for $t \gg t_0$ the current $\mathbf{J}(t) \to \mathbf{J}_s$ tends towards a stationary state which is not an equilibrium state because $\mathbf{J}_s \neq 0$. In statistical physics this is an example of a stationary state out of equilibrium.

Linear response theory constitutes a broad field which can be addressed from the theory of random processes but also from Hamiltonian theory (classical and quantum). We here present a short review from the point of view of diffusion processes. Even within the framework of random processes, the theory of linear response can be treated with several approaches, some of them rather abstract. We will here basically follow the development in

Risken (1987). Other valuable approaches, specially addressed to statistical physics, can be found in the literature (see, for instance, Wio *et al.*, 2012).

6.1 Linear response in diffusive systems

We consider a N-dimensional diffusion process $\mathbf{X}(t)$ with vector drift $\mathbf{f}(\mathbf{x}, t)$ and diffusion matrix $\mathbf{D}(\mathbf{x}, t)$. We know from Chapter 5 that the transition density $p(\mathbf{x}, t | \mathbf{x}_0, t_0)$ and the univariate density $p(\mathbf{x}, t)$ both obey the Fokker-Planck equation, which we write in the following compact form (cf. Eq. (5.3))

$$\frac{\partial p}{\partial t} = \mathrm{L_{FP}}(\mathbf{x}, t)p, \tag{6.1}$$

where $\mathrm{L_{FP}}(\mathbf{x}, t)$ is the Fokker-Planck operator defined by

$$\mathrm{L_{FP}}(\mathbf{x}, t) = -\boldsymbol{\nabla} \cdot \mathbf{f}(\mathbf{x}, t) + \frac{1}{2}\boldsymbol{\nabla}^2 \mathbf{D}(\mathbf{x}, t). \tag{6.2}$$

In what follows we focus on the univariate density $p(\mathbf{x}, t)$ and make two assumptions:

(i) The time dependence of $\mathrm{L_{FP}}(\mathbf{x}, t)$ is through an external field which is supposed to be a weak. That is,

$$\mathbf{f}(\mathbf{x}, t) = \mathbf{f}(\mathbf{x}) + \epsilon \mathbf{f}_{\mathrm{ext}}(\mathbf{x}, t), \qquad \mathbf{D}(\mathbf{x}, t) = \mathbf{D}(\mathbf{x}) + \epsilon \mathbf{D}_{\mathrm{ext}}(\mathbf{x}, t),$$

($|\epsilon| \ll 1$). So that,

$$\mathrm{L_{FP}}(\mathbf{x}, t) = \mathrm{L_{FP}}(\mathbf{x}) + \epsilon \mathrm{L_{ext}}(\mathbf{x}, t), \tag{6.3}$$

where $\mathrm{L_{FP}}(\mathbf{x})$ is defined as in Eq. (6.2), and

$$\mathrm{L_{ext}}(\mathbf{x}, t) = -\boldsymbol{\nabla} \cdot \mathbf{f}_{\mathrm{ext}}(\mathbf{x}, t) + \frac{1}{2}\boldsymbol{\nabla}^2 \mathbf{D}_{\mathrm{ext}}(\mathbf{x}, t). \tag{6.4}$$

The Fokker-Planck equation then reads

$$\frac{\partial p}{\partial t} = [\mathrm{L_{FP}}(\mathbf{x}) + \epsilon \mathrm{L_{ext}}(\mathbf{x}, t)]\, p. \tag{6.5}$$

(ii) We also assume that the time-independent Fokker-Planck operator, $\mathrm{L_{FP}}(\mathbf{x})$, has stationary solution $p_s(\mathbf{x})$ given by (Chapter 5, Eq. (5.43))

$$\mathrm{L_{FP}}(\mathbf{x})p_s(\mathbf{x}) = 0. \tag{6.6}$$

These two assumptions allow us to look for a solution to Eq. (6.5) in the form of linear response:

$$p(\mathbf{x}, t) = p_s(\mathbf{x}) + \epsilon q(\mathbf{x}, t) + O(\epsilon^2). \tag{6.7}$$

In effect, substituting (6.7) into Eq. (6.5) we have to first order

$$\frac{\partial}{\partial t}\big[p_s(\mathbf{x}) + \epsilon q(\mathbf{x}, t) + O(\epsilon^2)\big]$$
$$= L_{FP}(\mathbf{x})p_s(\mathbf{x}) + \epsilon\big[L_{FP}(\mathbf{x})q(\mathbf{x},t) + L_{ext}(\mathbf{x},t)p_s(\mathbf{x})\big] + O(\epsilon^2).$$

But

$$\frac{\partial}{\partial t}p_s(\mathbf{x}) = 0, \qquad L_{FP}(\mathbf{x})p_s(\mathbf{x}) = 0,$$

so that $q(\mathbf{x}, t)$ obeys the inhomogeneous Fokker-Planck equation

$$\frac{\partial}{\partial t}q(\mathbf{x}, t) = L_{FP}(\mathbf{x})q(\mathbf{x}, t) + L_{ext}(\mathbf{x}, t)p_s(\mathbf{x}) + O(\epsilon). \qquad (6.8)$$

Let us see that the formal solution to this equation can be written as

$$q(\mathbf{x}, t) = \int_{-\infty}^{t} e^{(t-t')L_{FP}(\mathbf{x})}L_{ext}(\mathbf{x}, t')p_s(\mathbf{x})dt' + O(\epsilon), \qquad (6.9)$$

where the exponential in the integral is the operator defined by the power series

$$e^{(t-t')L_{FP}(\mathbf{x})} = \sum_{n=0}^{\infty}\frac{1}{n!}(t-t')^n[L_{FP}(\mathbf{x})]^n,$$

where $[L_{FP}(\mathbf{x})]^n = L_{FP}(\mathbf{x})\overset{(n)}{\cdots}L_{FP}(\mathbf{x})$.

In effect, the time derivative of Eq. (6.9) yields

$$\frac{\partial q}{\partial t} = L_{ext}(\mathbf{x}, t)p_s(\mathbf{x})$$
$$+ L_{FP}(\mathbf{x})\int_{-\infty}^{t} e^{(t-t')L_{FP}(\mathbf{x})}L_{ext}(\mathbf{x}, t')p_s(\mathbf{x})dt' + O(\epsilon),$$

but the integral on the right equals $q(\mathbf{x}, t)$ (cf. Eq. (6.9)) and we recover Eq. (6.8). In other words, Eq. (6.9) satisfies Fokker-Planck equation (6.8) which proves the statement. $\qquad \square$

We denote by $A(t) = A\big(\mathbf{X}(t)\big)$ an arbitrary dynamical function of the diffusive process $\mathbf{X}(t)$. Such kind of functions are also called "observables" of the process. If, for instance, $X(t)$ represents the velocity of a Brownian particle moving inside a fluid then an example of observable is the kinetic energy of the particle.

We define

$$\Delta A(t) \equiv A\big(\mathbf{X}(t)\big) - \langle A \rangle_s, \qquad (6.10)$$

where

$$\langle A \rangle_s = \int p_s(\mathbf{x}) A(\mathbf{x}) d\mathbf{x} \qquad (6.11)$$

is the average value of the observable A in the stationary state. We call ΔA the "fluctuation" of A around the stationary state. The average value of this fluctuation is

$$\langle \Delta A(t) \rangle = \int \left[A(\mathbf{x}) - \langle A \rangle_s \right] p(\mathbf{x}, t) d\mathbf{x}, \qquad (6.12)$$

where $p(\mathbf{x}, t)$ is the univariate distribution of $\mathbf{X}(t)$. Substituting for Eq. (6.7) we have

$$\langle \Delta A(t) \rangle = \int p_s(\mathbf{x}) A(\mathbf{x}) d\mathbf{x} - \langle A \rangle_s \int p_s(\mathbf{x}) d\mathbf{x}$$
$$+ \epsilon \int A(\mathbf{x}) q(\mathbf{x}, t) d\mathbf{x} - \epsilon \langle A \rangle_s \int q(\mathbf{x}, t) d\mathbf{x} + O(\epsilon^2). \quad (6.13)$$

Taking into account the definition (6.11) and the normalization of $p_s(\mathbf{x})$ we get

$$\int p_s(\mathbf{x}) A(\mathbf{x}) d\mathbf{x} - \langle A \rangle_s \int p_s(\mathbf{x}) d\mathbf{x} = \langle A \rangle_s - \langle A \rangle_s = 0. \qquad (6.14)$$

On the other hand, from Eq. (6.7) we see that

$$\epsilon \int q(\mathbf{x}, t) d\mathbf{x} = \int p(\mathbf{x}, t) d\mathbf{x} - \int p_s(\mathbf{x}) d\mathbf{x} + O(\epsilon^2),$$

but, due to normalization

$$\int p(\mathbf{x}, t) d\mathbf{x} = \int p_s(\mathbf{x}) d\mathbf{x} = 1,$$

hence

$$\int q(\mathbf{x}, t) d\mathbf{x} = O(\epsilon). \qquad (6.15)$$

Substituting Eqs. (6.14) and (6.15) into Eq. (6.13) we obtain

$$\langle \Delta A(t) \rangle = \epsilon \int A(\mathbf{x}) q(\mathbf{x}, t) d\mathbf{x} + O(\epsilon^2). \qquad (6.16)$$

As customary in linear response theory one sets $\epsilon = 1$. In other words, ϵ is incorporated into the (small) perturbation $q(\mathbf{x}, t)$. In what follows we will thus write Eq. (6.16) as an exact expression:

$$\langle \Delta A(t) \rangle = \int A(\mathbf{x}) q(\mathbf{x}, t) d\mathbf{x}. \qquad (6.17)$$

6.2   Linear response function

In many applications the external field leads to drift and diffusion coefficients depending on time in a multiplicative way. That is,

$$\mathbf{f}_{\text{ext}}(\mathbf{x},t) = \left(f_j^{(\text{ext})}(\mathbf{x})\phi_j(t)\right), \qquad \mathbf{D}_{\text{ext}}(\mathbf{x},t) = \left(D_{jk}^{(\text{ext})}(\mathbf{x})\phi_{jk}(t)\right),$$

where $\phi_j(t)$ and $\phi_{ij}(t)$ are given functions of time $(j,k = 1,\dots,N)$. The operator $\mathrm{L}_{\text{ext}}(\mathbf{x},t)$ given in Eq. (6.4) then reads

$$\mathrm{L}_{\text{ext}}(\mathbf{x},t) = -\sum_{j=1}^{N} \phi_j(t)\frac{\partial}{\partial x_j} f_j^{(\text{ext})}(\mathbf{x}) + \frac{1}{2}\sum_{j,k=1}^{N} \phi_{jk}(t)\frac{\partial^2}{\partial x_j \partial x_k} D_{jk}^{(\text{ext})}(\mathbf{x}).$$
(6.18)

In what follows we will write symbolically

$$\mathrm{L}_{\text{ext}}(\mathbf{x},t) \equiv F(t)\overline{\mathrm{L}}_{\text{ext}}(\mathbf{x}), \tag{6.19}$$

which provides a compact and simple way of indicating the multiplicative character of time in the external perturbation on the system. This is a convenient and formal expression but let us bear in mind that the operative form of L_{ext} is given by Eq. (6.18).

Using Eq. (6.19), the formal solution (6.9) for the external perturbation reads

$$q(\mathbf{x},t) = \int_{-\infty}^{\infty} \Theta(t-t')e^{(t-t')\mathrm{L}_{\text{FP}}(\mathbf{x})} F(t')\overline{\mathrm{L}}_{\text{ext}}(\mathbf{x})p_s(\mathbf{x})dt', \tag{6.20}$$

where we have extended the limits of the integral to the entire real line by using the Heaviside Theta function. Substituting Eq. (6.20) into Eq. (6.17) yields

$$\langle \Delta A(t)\rangle = \int_{-\infty}^{\infty} \Theta(t-t')F(t')dt' \int A(\mathbf{x})e^{(t-t')\mathrm{L}_{\text{FP}}(\mathbf{x})}\overline{\mathrm{L}}_{\text{ext}}(\mathbf{x})p_s(\mathbf{x})d\mathbf{x}.$$

Note that the fluctuation of observable A from the stationary state can then be written as

$$\langle \Delta A(t)\rangle = \int_{-\infty}^{\infty} R_A(t-t')F(t')dt' \tag{6.21}$$

where

$$R_A(t) \equiv \Theta(t)\int A(\mathbf{x})e^{t\mathrm{L}_{\text{FP}}(\mathbf{x})}\overline{\mathrm{L}}_{\text{ext}}(\mathbf{x})p_s(\mathbf{x})d\mathbf{x} \tag{6.22}$$

is the *linear response function* describing the response of the dynamical function A to the external disturbance represented by $\mathrm{L}_{\text{ext}}(\mathbf{x},t)$. Let us first observe that with the linear expression (6.21) all nonlinearities, due to inner interactions of the system, are within the expression for R_A and, therefore, integrated. Second, the Heaviside function in front of the definition of R_A indicates the principle of causality (the effect comes after the cause), that is,

$$R_A(t) = 0, \quad \text{if} \quad t < 0. \tag{6.23}$$

We present some special cases of response functions encountered in practice (Risken, 1987). These example provide physical insight on the significance of the response function.

(a) *Pulse-respnse function*

The pulse-response function is the response of A to a delta function force $F(t) = \delta(t)$. In such a case the average deviation of A from the stationary state is

$$\langle \Delta A(t) \rangle = \int_{-\infty}^{\infty} R_A(t - t')\delta(t')dt' = R_A(t),$$

which means that the response function R_A is the response of the system to an impulsive force.

(b) *Excitation function*

The excitation function is the response of the system to a step force, $F(t) = \Theta(t)$, so that the constant external field is switched on at $t = 0$. From Eq. (6.21) we have

$$\langle \Delta A(t) \rangle = \int_{-\infty}^{\infty} R_A(t - t')\Theta(t')dt' = \int_{0}^{\infty} R_A(t - t')dt'.$$

That is

$$\langle \Delta A(t) \rangle = \int_{0}^{t} R_A(t')dt',$$

and the fluctuation is given by the cumulative effect of the response function from 0 up to time t. Note that the time derivative is

$$R_A(t) = \frac{d}{dt} \langle \Delta A(t) \rangle.$$

(c) *Relaxation function*

Th relaxation function (also called after-effect function) is the response of the system to the step force $F(t) = \Theta(-t)$. In other words, we observe the system after an external and constant force has been switched off. In this case we have

$$\langle \Delta A(t) \rangle = \int_{-\infty}^{\infty} R_A(t - t')\Theta(-t')dt' = \int_{-\infty}^{0} R_A(t - t')dt'.$$

That is

$$\langle \Delta A(t) \rangle = \int_{t}^{\infty} R_A(t)dt',$$

and the fluctuation is given by the cumulative effect of the response function from t up to ∞. Now

$$R_A(t) = -\frac{d}{dt}\langle \Delta A(t) \rangle.$$

6.3 Correlation functions

We consider two observables $A(\mathbf{X}(t))$ and $B(\mathbf{X}(t))$ of the same diffusion process $\mathbf{X}(t)$ (think, for instance, of the kinetic energy and the momentum of a Brownian particle). Let us denote by $p_2(\mathbf{x}_1, t_1; \mathbf{x}_2, t_2)$ the bivariate probability density function of $\mathbf{X}(t)$. The correlation function of these two observables at different instants of time is given by

$$K_{AB}(t_1, t_2) = \langle A(\mathbf{X}(t_1))B(\mathbf{X}(t_2)) \rangle$$
$$= \int \int A(\mathbf{x}_1)p_2(\mathbf{x}_1, t_1; \mathbf{x}_2, t_2)B(\mathbf{x}_2)d\mathbf{x}_1 d\mathbf{x}_2. \quad (6.24)$$

As in the previous section we assume that $\mathbf{X}(t)$ is stationary, so that, the univariate density is independent of time (the stationary distribution) and the bivariate density p_2 is only function of time differences.

In order to evaluate an expression for the correlation function $K_{AB}(t_1, t_2)$ we have to distinguish the cases $t_1 \geq t_2$ and $t_1 \leq t_2$.

(i) We first assume that $t_1 \geq t_2$ and write

$$p_2(\mathbf{x}_1, t_1; \mathbf{x}_2, t_2) = p(\mathbf{x}_1, t_1 | \mathbf{x}_2, t_2)p_1(\mathbf{x}_2, t_2).$$

Because of the stationary character of $X(t)$ we have

$$p_2(\mathbf{x}_1, t_1; \mathbf{x}_2, t_2) = p(\mathbf{x}_1, t_1 - t_2 | \mathbf{x}_2, 0)p_s(\mathbf{x}_2), \quad (6.25)$$

where $p_s(\mathbf{x}_1)$ is the stationary density. Substituting this expression into Eq. (6.24) yields

$$K_{AB}(t_1 - t_2) = \langle A(\mathbf{X}(t_1 - t_2))B(\mathbf{X}(0)) \rangle \quad (6.26)$$
$$= \int A(\mathbf{x}_1)d\mathbf{x}_1 \int p(\mathbf{x}_1, t_1 - t_2 | x_2, 0)p_s(x_2)B(\mathbf{x}_2)d\mathbf{x}_2.$$

On the other hand, since $\mathbf{X}(t)$ is a diffusion process, the transition density $p(\mathbf{x}_1, t_1 - t_2 | \mathbf{x}_2, 0)$ is the solution to the Fokker-Planck equation

$$\frac{\partial p}{\partial t_1} = \mathrm{L}_{\mathrm{FP}}(\mathbf{x}_1)p, \qquad (6.27)$$

with the initial condition (when $t_1 = t_2$)

$$p(\mathbf{x}_1, 0 | \mathbf{x}_2, 0) = \delta(\mathbf{x}_1 - \mathbf{x}_2). \qquad (6.28)$$

Therefore, the transition density is formally given by[1]

$$p(\mathbf{x}_1, t_1 - t_2 | x_0) = e^{(t_1 - t_2)\mathrm{L}_{\mathrm{FP}}(\mathbf{x}_1)}\delta(\mathbf{x}_1 - \mathbf{x}_2). \qquad (6.29)$$

Substituting (6.29) this into (6.26) we have

$$K_{AB}(t_1 - t_2) = \int A(\mathbf{x}_1)d\mathbf{x}_1 \int e^{(t_1 - t_2)\mathrm{L}_{\mathrm{FP}}(\mathbf{x}_1)}\delta(\mathbf{x}_1 - \mathbf{x}_2)p_s(\mathbf{x}_2)B(\mathbf{x}_2)d\mathbf{x}_2$$

which, after performing the integration over $\mathbf{x}_2$, yields the final result for $t_1 \geq t_2$

$$K_{AB}(t_1 - t_2) = \int A(\mathbf{x}_1)e^{(t_1 - t_2)\mathrm{L}_{\mathrm{FP}}(\mathbf{x}_1)}p_s(\mathbf{x}_1)B(\mathbf{x}_1)d\mathbf{x}_1. \qquad (6.30)$$

(ii) When $t_1 \leq t_2$ we write $p_2(\mathbf{x}_1, t_1; \mathbf{x}_2, t_2) = p_2(\mathbf{x}_1, 0; \mathbf{x}_2, t_2 - t_1)$ and instead of Eqs. (6.25) and (6.26) we have

$$p_2(\mathbf{x}_1, t_1; \mathbf{x}_2, t_2) = p(\mathbf{x}_2, t_2 - t_1 | \mathbf{x}_1, 0)p_s(\mathbf{x}_1),$$

and

$$\begin{aligned} K_{AB}(t_2 - t_1) &= \langle A(\mathbf{X}(0))\rangle B(\mathbf{X}(t_2 - t_1))\rangle \\ &= \int B(\mathbf{x}_2)d\mathbf{x}_2 \int p(\mathbf{x}_2, t_2 - t_1 | x_1, 0)p_s(x_1)A(\mathbf{x}_1)d\mathbf{x}_1. \end{aligned}$$

Since $p(\mathbf{x}_2, t_2 - t_1 | x_1, 0)$ is the solution to

$$\frac{\partial p}{\partial t_2} = \mathrm{L}_{\mathrm{FP}}(\mathbf{x}_2)p, \qquad p(\mathbf{x}_2, 0 | \mathbf{x}_1, 0) = \delta(\mathbf{x}_2 - \mathbf{x}_1),$$

we have

$$p(\mathbf{x}_2, t_2 - t_1 | x_1, 0) = e^{(t_2 - t_1)\mathrm{L}_{\mathrm{FP}}(\mathbf{x}_2)}\delta(\mathbf{x}_2 - \mathbf{x}_1)$$

and finally $(t_1 \leq t_2)$

$$K_{AB}(t_2 - t_1) = \int B(\mathbf{x}_2)e^{(t_2 - t_1)\mathrm{L}_{\mathrm{FP}}(\mathbf{x}_2)}p_s(\mathbf{x}_2)A(\mathbf{x}_2)d\mathbf{x}_2. \qquad (6.31)$$

---

[1]Equation (6.29) is proven similarly to Eq. (6.9), i.e., taking the time derivative (with respect to t_1) of Eq. (6.29) and realizing that the formal solution satisfies Eqs. (6.1)-(6.28).

A particular case of these expressions is the *autocorrelation function*, for which $A = B$. That is

$$K_A(t_1, t_2) = \langle A(X(t_1))A(X(t_2)) \rangle.$$

From Eqs. (6.30) and (6.31) we write

$$K_A(\tau) = \int A(\mathbf{x}) e^{|\tau| L_{FP}(\mathbf{x})} p_s(\mathbf{x}) A(\mathbf{x}) d\mathbf{x}, \qquad (6.32)$$

where $\tau = t_1 - t_2$ and $-\infty < \tau < \infty$. We see from Eq. (6.32) that $K_A(\tau)$ is an even function of the time difference:

$$K_A(-\tau) = K_A(\tau).$$

6.3.1 The fluctuation-dissipation theorem

In the preceding results on correlation functions we have supposed that the diffusive system is in the stationary state, so that the evolution of the system is governed by a time independent Fokker-Planck operator $L_{FP}(\mathbf{x})$ and the stationary density is the solution to $L_{FP}(\mathbf{x})p_s(\mathbf{x}) = 0$ (cf. Sect. 6.1).

Let us now assume that at a given instant of time a weak external field has been switched on. We also assume that the time dependence of the external field is multiplicative which leads to an interaction operator $L_{ext}(\mathbf{x}, t)$ that can be written symbolically as (cf. Eq. (6.19))

$$L_{ext}(\mathbf{x}, t) = F(t)\overline{L}_{ext}(\mathbf{x}), \qquad (6.33)$$

where the operator $\overline{L}_{ext}$ is defined through Eq. (6.18). We will see that there exists a direct connection between the response function, $R_A(t)$, of any observable A due to the external perturbation and the correlation function $K_{AB}(t)$, with B conveniently defined (see below). Such a connection is known as the fluctuation-dissipation theorem.

We define the function

$$B(\mathbf{x}) \equiv [p_s(\mathbf{x})]^{-1} \overline{L}_{ext}(\mathbf{x}) p_s(\mathbf{x}) \qquad (6.34)$$

where p_s is the stationary density.[2] Suppose we have an observable A of the diffusion process $X(t)$. The correlation between $A(X(t_1))$ and the

---

[2] In many cases the stationary density is of the form $p_s(\mathbf{x}) = Ne^{-\phi(\mathbf{x})}$, where $N > 0$ and ϕ is an arbitrary function such that $\int e^{-\phi(\mathbf{x})} d\mathbf{x} = 1/N$. In these cases the function B can be written as

$$B(\mathbf{x}) = e^{\phi(\mathbf{x})} \overline{L}_{ext}(\mathbf{x}) e^{-\phi(\mathbf{x})}.$$

dynamical function $B(X(t_2))$ when $t_1 \geq t_2$ is given by Eq. (6.30) which, defining $\tau = t_1 - t_2$, we write as

$$K_{AB}(\tau) = \int A(\mathbf{x})e^{\tau L_{\mathrm{FP}}(\mathbf{x})}p_s(\mathbf{x})B(\mathbf{x})dx,$$

$(\tau \geq 0)$. Substituting for (6.34) we have

$$K_{AB}(\tau) = \int A(\mathbf{x})e^{\tau L_{\mathrm{FP}}(\mathbf{x})}\overline{L}_{\mathrm{ext}}(\mathbf{x})p_s(\mathbf{x})dx. \qquad (6.35)$$

However, for $\tau > 0$ the integral on the right hand side is precisely the linear response function of A to the time-independent part of the external perturbation $\overline{L}_{\mathrm{ext}}(\mathbf{x})$ (cf. Eq. (6.22)). The correlation function and the response function are, therefore, related by the *fluctuation-dissipation theorem*:

$$R_A(\tau) = \Theta(\tau)K_{AB}(\tau), \qquad (6.36)$$

where B is defined by Eq. (6.34).

Recall that the fluctuation around the equilibrium of observable A, due to the external field (6.33), is given by Eq. (6.21)

$$\langle \Delta A(t)\rangle = \int_{-\infty}^{\infty} R_A(t - t')F(t')dt'.$$

Substituting for Eq. (6.36) we have

$$\langle \Delta A(t)\rangle = \int_{-\infty}^{\infty} \Theta(t - t')K_{AB}(t - t')F(t')dt' = \int_{-\infty}^{t} K_{AB}(t - t')F(t')dt'$$

and performing the change of variables $t - t' \to t'$ we obtain

$$\langle \Delta A(t)\rangle = \int_{0}^{\infty} K_{AB}(t')F(t - t')dt'. \qquad (6.37)$$

Since $K_{AB}(t') = \langle A(\mathbf{X}(t'))B(\mathbf{X}(0))\rangle$, we write

$$\langle \Delta A(t)\rangle = \int_{0}^{\infty} \langle A(\mathbf{X}(t'))B(\mathbf{X}(0))\rangle F(t - t')dt'. \qquad (6.38)$$

Because of stationarity –which implies the invariance under time translations– the correlation function is also defined as $K_{AB}(t - t') = \langle A(\mathbf{X}(t))B(\mathbf{X}(t'))\rangle$. Therefore, an alternative form of writing Eq. (6.38) is

$$\langle \Delta A(t)\rangle = \int_{-\infty}^{t} \langle A(\mathbf{X}(t))B(\mathbf{X}(t'))\rangle F(t')dt'. \qquad (6.39)$$

6.4 Examples

We present some examples which, besides illustrating the above formalism, have physical relevance.

6.4.1   One dimensional diffusions

Let us consider one-dimensional diffusions with drift $f(x)$ and constant diffusion coefficient D. The Fokker-Planck operator is

$$L_{\mathrm{FP}}(x) = -\frac{\partial}{\partial x} f(x) + \frac{1}{2} D \frac{\partial^2}{\partial x^2},$$

and the stationary density, $L_{\mathrm{FP}}(x) p_s(x) = 0$, is given by the potential solution (see Chapter 4, Eq. (4.44))

$$p_s(x) = \frac{N}{D} \exp \left\{ \frac{2}{D} \int_{-\infty}^{\infty} f(x) dx \right\},$$

where N is the normalization constant.

At some instant of time an external field is switched on so that the new drift and diffusion coefficients are $f(x,t)$ and $D(x,t)$. We suppose the simple case, but rather frequent in practical applications, that

$$f(x,t) = f(x) + F(t), \qquad D(x,t) = D,$$

where $F(t)$ is an external (and weak) force.[3] Note that in this case we have

$$L_{\mathrm{ext}}(x) = -F(t) \frac{\partial}{\partial x}, \qquad \overline{L}_{\mathrm{ext}}(x) = -\frac{\partial}{\partial x}.$$

Since

$$\overline{L}_{\mathrm{ext}}(x) p_s(x) = -\frac{\partial}{\partial x} p_s(x) = -\frac{2}{D} f(x) p_s(x),$$

the correlation function $K_{AB}(\tau)$ $(\tau \geq 0)$ given in Eq. (6.35) is

$$K_{AB}(\tau) = -\frac{2}{D} \int_{-\infty}^{\infty} A(x) e^{\tau L_{\mathrm{FP}}(x)} f(x) p_s(x) dx,$$

and the fluctuation of A around the steady state, Eq, (6.37), reads

$$\langle \Delta A(t) \rangle = -\frac{2}{D} \int_{0}^{\infty} F(t - t') dt' \int_{-\infty}^{\infty} A(x) e^{t' L_{\mathrm{FP}}(x)} f(x) p_s(x) dx.$$

---

[3] In Chapter 7 we will see that (i) the unperturbed case correspond to a diffusion process $X(t)$ obeying a Langevin equation of the form $\dot{X} = f(X) + \sqrt{D} \xi(t)$, where $\xi(t)$ is Gaussian white noise, and (ii) the perturbed case corresponds to $\dot{X} = f(X) + \sqrt{D} \xi(t) + F(t)$ which means the addition of an external and deterministic time-dependent force to the random noise.

6.4.2   The Brownian harmonic oscillator

We consider a Brownian particle of mass m moving under the action of a linear harmonic force (i.e., a quadratic potential). Let us denote by $X(t)$ and $V(t) = \dot{X}(t)$ the position and the velocity of the Brownian particle at time t. Newton law leads to the following Langevin equation for the position [4]

$$\ddot{X} + \beta\dot{X} + \omega_0^2 X = \xi(t), \tag{6.40}$$

where ω_0 is the natural frequency of the oscillator, $\beta > 0$ is the damping constant (we assume the oscillator to be in the overdamped regime, so that $\beta > 2\omega_0$) and $\xi(t)$ is zero-mean Gaussian white noise,[5]

$$\langle\xi(t)\rangle = 0, \qquad \langle\xi(t_1)\xi(t_2)\rangle = D\delta(t_1 - t_2).$$

We suppose that at some initial time t_0 the position, $X(t_0) = x_0$, and the velocity, $\dot{X}(t_0) = v_0$, of the oscillator are known.

We have seen in Chapter 3, Sect. 3.2 that, even in the deterministic limit (i.e., $D = 0$), the position $X(t)$ is not Markovian. However, the bidimensional process (X, V) is Markovian. From Eq. (6.40) we see that the oscillator can be described by the following pair of (stochastic) differential equations

$$\begin{cases} \dot{X}(t) = V(t), \\ \dot{V}(t) = -\beta V(t) - \omega_0^2 X(t) + \xi(t). \end{cases} \tag{6.41}$$

In Chapter 8 we will see that this bidimensional process is a diffusion process with vector drift and diffusion matrix given by

$$\mathbf{f}(x, v) = \left(v, -\beta v - \omega_0^2 x\right), \qquad \mathbf{D} = \begin{pmatrix} 0 & 0 \\ 0 & D \end{pmatrix}. \tag{6.42}$$

As the reader can prove by direct substitution, the solution to Eq. (6.40) with the initial conditions $X(t_0) = x_0$ and $\dot{X}(t_0) = v_0$ can be written in the form

$$X(t) = e^{-\beta(t-t_0)/2}\left[x_0\cosh\gamma(t-t_0) + \frac{1}{\gamma}(v_0 + \beta x_0/2)\sinh\gamma(t-t_0)\right]$$

$$+ \frac{1}{\gamma}\int_{t_0}^{t} e^{-\beta(t-t')/2}\sinh\gamma(t-t')\xi(t')dt', \tag{6.43}$$

---

[4]The system described by Eq. (6.40) corresponds to an oscillator forced by an external random force. Other situations are possible, as the Kubo oscillator where the frequency has a random part, or even oscillators with random damping (Gitterman, 2012).

[5]In terms of physical parameters the diffusion constant is given by $D = \beta k_B T/m$, where k_B is Boltzmann constant and T is the absolute temperature of the medium.

where

$$\gamma = \frac{1}{2}(\beta^2 - 4\omega_0^2)^{1/2}.$$

Since $\langle \xi(t) \rangle = 0$, the mean value of the position is

$$\langle X(t) \rangle = e^{-\beta(t-t_0)/2}\left[x_0 \cosh \gamma(t-t_0) + \frac{1}{\gamma}(v_0 + \beta x_0/2)\sinh \gamma(t-t_0)\right]. \quad (6.44)$$

Bearing in mind that $\beta/2 > \gamma$, we see that as $t_0 \to -\infty$ (this corresponds to the stationary state, cf. Chapter 3, Sect. 3.6)

$$e^{-\beta(t-t_0)/2}\cosh \gamma(t - t_0) \to 0, \qquad e^{-\beta(t-t_0)/2}\sinh \gamma(t - t_0) \to 0,$$

and in the stationary state the average value vanishes,

$$\langle X(t) \rangle_{stat} = \lim_{t_0 \to -\infty} \langle X(t) \rangle = 0.$$

Let us now assume that the oscillator is in the stationary state (i.e., $t_0 = -\infty$) and that at $t = 0$ an external force is switched on. Now we have

$$\ddot{X} + \beta \dot{X} + \omega_0^2 X = F(t) + \xi(t), \quad (6.45)$$

where $F(t)$ is the external force per unit mass.

Since $\langle X \rangle_{stat} = 0$ the fluctuation of the position due to the external field is

$$\Delta X(t) = X(t) - \langle X \rangle_{stat} = \langle X(t) \rangle.$$

Therefore, $\Delta X(t)$ obeys Eq. (6.45) and the solution is given by (cf. Eq. (6.43))

$$\Delta X(t) = e^{-\beta(t-t_0)/2}\left[x_0 \cosh \gamma(t - t_0) + \frac{1}{\gamma}(v_0 + \beta x_0/2)\sinh \gamma(t - t_0)\right]$$

$$+ \frac{1}{\gamma}\int_{t_0}^{t} e^{-\beta(t-t')/2}\sinh \gamma(t - t')[F(t) + \xi(t')]dt'.$$

In the stationary state ($t_0 \to -\infty$) and recalling that $\langle \xi(t) \rangle = 0$, we see that the stationary average fluctuation is

$$\langle \Delta X(t) \rangle = \frac{1}{\gamma}\int_{-\infty}^{t} e^{-\beta(t-t')/2}\sinh \gamma(t - t')F(t)dt'. \quad (6.46)$$

A comparison of Eqs. (6.21) and (6.46) allows us to write

$$\langle \Delta X(t) \rangle = \int_{-\infty}^{\infty} R_X(t - t')F(t')dt',$$

where $R_X(t)$ is the response function of the position to the external force:

$$R_X(t) = \frac{1}{\gamma}\Theta(t)e^{-\beta t/2}\sinh \gamma t. \quad (6.47)$$

6.4.3   *Brownian particle in a field of force*

We now generalize the previous example and look at the problem from another angle. We consider the one-dimensional motion of a Brownian particle of unit mass inside a fluid and bounded by a conservative field with potential energy $U(x)$ and force $-U'(x)$. This is *Kramers problem* which has been discussed in Chapter 5, Sect. 5.6. Instead of the linear equation (6.40), the position of the particle obeys now the non-linear Langevin equation of second order [6]

$$\ddot{X} + \beta \dot{X} + U'(X) = \xi(t), \qquad (6.48)$$

where $\xi(t)$ is Gaussian white noise with

$$\langle \xi(t) \rangle = 0, \qquad \langle \xi(t)\xi(t') \rangle = D\delta(t - t'),$$

where $D = \beta k_B$, k_B is Boltzmann constant and T is the absolute temperature of the fluid.[7]

Denote by $V(t) = \dot{X}(t)$ the random velocity of the particle. From Eq. (6.48) we see that the dynamics of the Brownian particle can also be described by the following pair of first-order differential equations

$$\begin{cases} \dot{X}(t) = V(t), \\ \dot{V}(t) = -\beta V(t) - U'(X) + \xi(t). \end{cases} \qquad (6.49)$$

As we have seen in Sect. 5.6 of Chapter 5 the bidimensional process (X, V) is a diffusion process with vector drift and diffusion matrix given by

$$\mathbf{f}(x, v) = (v, -\beta v - U'(x)), \qquad \mathbf{D} = \begin{pmatrix} 0 & 0 \\ 0 & D \end{pmatrix}, \qquad (6.50)$$

where $D = \beta k_B T$. The operator $\mathrm{L_{FP}}(x, v)$ defined in Eq. (6.2) now reads

$$\mathrm{L_{FP}}(x, v) = -\frac{\partial}{\partial x}v + \frac{\partial}{\partial v}\left[\beta v + U'(x)\right] + \frac{1}{2}D\frac{\partial^2}{\partial v^2}, \qquad (6.51)$$

and the stationary density $p_s(x, v)$ is the solution to $\mathrm{L_{FP}}p_s = 0$. Explicitly

$$-\frac{\partial}{\partial x}(vp_s) + \frac{\partial}{\partial v}\left[((\beta v + U'(x))p\right] + \frac{1}{2}D\frac{\partial^2 p_s}{\partial v^2} = 0.$$

---

[6]The linear oscillator discussed above is a particular case of Eq. (6.48) when the potential energy is a quadratic function of the position and the damping constant β is replaced by 2β.

[7]Recall we have supposed a Brownian particle of unit mass. For an arbitrary mass the diffusion coefficient is $D = \beta k_B T/m$.

Direct substitution into this equation shows that the stationary density is given by the Maxwell-Boltzmann distribution (see also Eq. (5.86), Chapter 5)

$$p_s(x,v) = N \exp\left\{ -\frac{\beta}{D}\left[\frac{1}{2}v^2 + U(x)\right]\right\}, \qquad (6.52)$$

where N is the normalization constant.

Let us remark that in the stationary state the average value of the velocity is zero. Indeed, from Eq. (6.52), and due to the even character of p_s with respect to v, we have [8]

$$\langle V \rangle_{st} = N \int_{-\infty}^{\infty} e^{-\beta U(x)/D}dx \int_{-\infty}^{\infty} v e^{-\beta v^2/2D}dv = 0.$$

As in the previous example we suppose that the system is in the stationary state and that at $t = 0$ a time-dependent external force (per unit mass) $F(t)$ is switched on. The equation of motion becomes

$$\ddot{X} + \beta\dot{X} + U'(X) = F(t) + \xi(t). \qquad (6.53)$$

Now the diffusion matrix remains constant but the drift becomes

$$\mathbf{f}(x,v,t) = \big(v, -\beta v - U'(x) - F(t)\big).$$

With the notation of Sect. 6.1 we write $\mathbf{f}(x,v,t) = \mathbf{f}(x,v) + \mathbf{f}_{ext}(x,v,t)$, where $\mathbf{f}(x,v,t)$ is given by Eq. (6.50) and $\mathbf{f}_{ext}(x,v,t) = \big(0,-F(t)\big)$. This leads to a Fokker-Planck operator, $L_{FP}(x,v,t)$, in the form given by Eq. (6.3) (with $\epsilon = 1$):

$$L_{FP}(\mathbf{x},t) = L_{FP}(x,v) + L_{ext}(x,v,t),$$

where $L_{FP}(x,v)$ is given in Eq. (6.51) and

$$L_{ext}(x,v,,t) = -F(t)\frac{\partial}{\partial v}. \qquad (6.54)$$

Therefore (cf. Eq. (6.19))

$$\overline{L}_{ext}(x,v) = -\frac{\partial}{\partial v}.$$

[8]The stationary value of the average position, $\langle X \rangle_{st}$, will depend on the form of the potential energy $U(x)$. For a symmetric potential, $U(-x) = U(x)$, the stationary average of the position is also zero:

$$\langle X \rangle_{st} = \int_{-\infty}^{\infty} dv \int_{-\infty}^{\infty} x p_s(x,v)dx = \left(\frac{\pi D}{\beta}\right)^{1/2}\int_{-\infty}^{\infty} x e^{-\beta U(x)/D}dx = 0.$$

Note that this is the case of the harmonic potential of the previous example, which, in such a case results in a Gaussian stationary density.

Let us focus on the velocity $V(t)$ of the particle and find out how $V(t)$ responds to the external perturbation $F(t)$. The average fluctuation of the velocity around the stationary value is (cf. Eq. (6.21))

$$\langle \Delta V(t) \rangle = \int_{-\infty}^{\infty} R_V(t - t')F(t')dt', \qquad (6.55)$$

where $\Delta V(t) = V(t) - \langle V \rangle_{st} = V(t)$ and $R_V(t)$ is the linear response function defined in Eq. (6.22). From the fluctuation-dissipation theorem, Eq. (6.36), we know that $R_V(t)$ is related to the correlation function by

$$R_V(t) = \Theta(t)K_{V,B}(t), \qquad (6.56)$$

where $K_{V,B}(t) = \langle V(t)B\big(X(0), V(0)\big) \rangle$ and the function $B(x, v)$ is defined by Eq. (6.34):

$$B(x, v) = [p_s(x, v)]^{-1}\overline{L}_{\text{ext}}(x, v)p_s(x, v).$$

From Eqs. (6.52) and (6.54) we find

$$B(x, v) = \frac{\beta v}{D}.$$

Hence, $K_{AB}(\tau)$ is proportional to the autocorrelation of the velocity

$$K_V(t) = \frac{\beta}{D}\langle V(t)V(0) \rangle,$$

and the response function is also proportional to the autocorrelation of the velocity

$$R_V(t) = \frac{\beta}{D}\Theta(t)\langle V(t)V(0) \rangle.$$

Substituting Eq. (6.56) into Eq. (6.55) yields the average fluctuation of the velocity due to the external field

$$\langle \Delta V(t) \rangle = \frac{\beta}{D}\int_{-\infty}^{t} \langle V(t')V(0) \rangle F(t - t')dt'.$$

This expression can be written in a somewhat more intuitive way. Indeed, performing the change of variables $t - t' \to t'$ in the integral and realizing that $\langle V(t - t')V(0) \rangle = \langle V(t)V(t') \rangle$ due to the stationary character of $V(t)$, we get

$$\langle \Delta V(t) \rangle = \frac{\beta}{D}\int_{t}^{\infty} \langle V(t)V(t') \rangle F(t')dt'. \qquad (6.57)$$

A remarkable expression relating velocity fluctuations with the autocorrelation of the velocity.

6.5   Susceptibility

The notion of susceptibility is related with the Fourier analysis of the linear response function. We introduce the concept by the example of a simple external periodic force which we write in complex form

$$F(t) = F_0 e^{-i\omega_0 t}, \tag{6.58}$$

where F_0 is the amplitude and ω_0 the frequency of the external force. The fluctuation of an observable A from the stationary state is now given by (cf. Eq. (6.21))

$$\langle \Delta A(t) \rangle = F_0 \int_{-\infty}^{\infty} R_A(t - t') e^{-i\omega_0(t - t')} dt' = F_0 e^{-i\omega_0 t} \int_{-\infty}^{\infty} R_A(\tau) e^{i\omega_0 \tau} d\tau,$$

and the average fluctuation can be written as

$$\langle \Delta A(t) \rangle = F_0 e^{-i\omega_0 t} \chi_A(\omega_0), \tag{6.59}$$

where

$$\chi_A(\omega_0) = \int_{-\infty}^{\infty} R_A(\tau) e^{i\omega_0 \tau} d\tau, \tag{6.60}$$

is called the susceptibility of the system.

We observe that in this simple case of a periodic external force with a single frequency ω, the response of the system given in Eq. (6.59) appears with the same frequency than that of the perturbation. However, beyond the linear regime (recall that $F(t)$ must be small) we should take into account terms with F^2, F^3,..., and many important nonlinear phenomena –such as turbulence, non-linear optics, etc.– appear.

The simple formula (6.59) for the response to a periodic stimulus can be generalized to any kind of external force $F(t)$, whether periodic or not. We look at Eq. (6.60) –which we recall has been obtained assuming a single periodic force– and suppose that it defines the susceptibility to any kind of perturbation. That is to say, *susceptibility is defined as the Fourier transform of the response function*:

$$\chi_A(\omega) \equiv \int_{-\infty}^{\infty} e^{i\omega \tau} R_A(\tau) d\tau, \tag{6.61}$$

and, hence, the response function is the inverse Fourier transform of the susceptibility:

$$R_A(\tau) = \frac{1}{2\pi} \int_{-\infty}^{\infty} e^{-i\omega \tau} \chi_A(\omega) d\omega. \tag{6.62}$$

We incidentally note that from the causality principle[9] we can also write

$$\chi_A(\omega) = \int_0^\infty e^{i\omega\tau} R_A(\tau) d\tau. \tag{6.63}$$

Plugging (6.61) into (6.21) we have

$$\langle \Delta A(t) \rangle = \int_{-\infty}^\infty R_A(t - t') F(t') dt'$$

$$= \frac{1}{2\pi} \int_{-\infty}^\infty e^{-i\omega t} \chi_A(\omega) d\omega \int_{-\infty}^\infty F(t') e^{i\omega t'} dt',$$

but the last integral is the Fourier transform of the external force

$$\tilde{F}(\omega) = \int_{-\infty}^\infty e^{i\omega t} F(t) dt. \tag{6.64}$$

Therefore,

$$\langle \Delta A(t) \rangle = \frac{1}{2\pi} \int_{-\infty}^\infty e^{-i\omega t} \chi_A(\omega) \tilde{F}(\omega) d\omega, \tag{6.65}$$

which constitutes the generalization of Eq. (6.59) to arbitrary perturbations.[10]

Looking at Eq. (6.65) we realize that the response $\langle \Delta A \rangle$ is the inverse Fourier transform of the product $\chi_A(\omega)\tilde{F}(\omega)$:

$$\langle \Delta A(t) \rangle = \mathcal{F}^{-1}\left\{ \chi_A(\omega)\tilde{F}(\omega) \right\},$$

and the general and simple relation follows

$$\langle \Delta \tilde{A}(\omega) \rangle = \chi_A(\omega)\tilde{F}(\omega), \tag{6.66}$$

where

$$\langle \Delta \tilde{A}(\omega) \rangle \equiv \int_{-\infty}^\infty e^{i\omega t} \langle \Delta A(t) \rangle dt,$$

is the Fourier transform of the average fluctuation.

---

[9]That is, $R_A(\tau) = 0$ for $\tau < 0$, cf. Eq. (6.22).

[10]The simple example (6.59) is included into (6.65). Indeed, the Fourier transform of the periodic force (6.58) is

$$\tilde{F}(\omega) = F_0 \int_{-\infty}^\infty e^{i(\omega-\omega_0)t} dt = 2\pi F_0 \delta(\omega - \omega_0),$$

where we have used the following representation of the delta function

$$\delta(\omega) = \frac{1}{2\pi} \int_{-\infty}^\infty e^{\pm i\omega t} dt.$$

Substituting this expression of $\tilde{F}(\omega)$ into Eq. (6.65) immediately leads to Eq. (6.59).

We will now relate the susceptibility $\chi_A(\omega)$ with the correlation function $K_{AB}(t)$. The starting point is the fluctuation-dissipation theorem (cf. Eq. (6.36))

$$R_A(\tau) = \Theta(\tau)K_{AB}(\tau).$$

Since susceptibility is the Fourier transform of the response function, we have $\chi_A(\omega) = \mathcal{F}\{\Theta(\tau)K_{AB}(\tau)\}$; that is,

$$\chi_A(\omega) = \int_0^\infty e^{i\omega\tau} K_{AB}(\tau)d\tau. \tag{6.67}$$

However, the integral extends only on the half real line instead of on the entire line $(-\infty, \infty)$ and consequently susceptibility *is not* the Fourier transform of the correlation function. The latter, usually called "spectral density" (cf. Chapter 2, Sect. 2.5) is

$$\tilde{K}_{AB}(\omega) = \int_{-\infty}^\infty e^{i\omega\tau} K_{AB}(\tau)d\tau. \tag{6.68}$$

6.5.1 *Kramers-Kronig relations*

In order to relate susceptibilities and spectral densities -in other words, to establish the fluctuation-dissipation relation in Fourier space– we need to obtain an interesting property of the susceptibility, known as the *Kramers-Kronig relations*.[11]

Before obtaining Kramers-Kronig relations we will prove the following important formula [12]

$$\int_0^\infty e^{\pm i\omega\tau}d\tau = \pi\delta(\omega) \pm i\mathcal{P}\left(1/\omega\right), \tag{6.69}$$

where $\delta(\omega)$ is Dirac delta function and $\mathcal{P}(\cdot)$ is Cauchy principal value, a generalized function with defining properties

$$\mathcal{P}\left(\frac{1}{x-a}\right) = \frac{1}{x-a}, \qquad x \neq a, \tag{6.70}$$

and

$$\int_{-\infty}^\infty \varphi(x)\mathcal{P}\left(\frac{1}{x-a}\right)dx = \int_{-\infty}^\infty \frac{\varphi(x) - \varphi(a)}{x-a}dx, \tag{6.71}$$

where a is an arbitrary real number and $\varphi(x)$ any well-behaved function such that $\varphi(x) \to 0$ as $x \to \pm\infty$.

---

[11] From a mathematical point of view Kramers-Kronig relations are two formulas connecting the real and imaginary parts of any analytical complex function (at least in the upper-half complex plane).

[12] Equation (6.69) is sometimes called *Sochozki formula* in the Russian literature (see, for instance, Vladimirov, 1984).

Proof of Eq. (6.69). We call

$$g(\omega) = \int_0^\infty e^{\pm i\omega\tau} d\tau,$$

and write

$$g(\omega) = \lim_{\epsilon \to 0} \int_0^\infty e^{-(\epsilon \mp i\omega)\tau} d\tau = \lim_{\epsilon \to 0} \frac{1}{\epsilon \mp i\omega}.$$

Suppose that $\varphi(\omega)$ is a regular function such that $\varphi(\omega) \to 0$ as $\omega \to \pm\infty$. Then

$$\int_{-\infty}^\infty g(\omega)\varphi(\omega)d\omega = \lim_{\epsilon \to 0} \int_{-\infty}^\infty \frac{\varphi(\omega)}{\epsilon \mp i\omega} d\omega = \lim_{\epsilon \to 0} \int_{-\infty}^\infty \frac{\epsilon \pm i\omega}{\epsilon^2 + \omega^2} \varphi(\omega)d\omega$$

$$= \lim_{\epsilon \to 0} \int_{-\infty}^\infty \frac{\epsilon \pm i\omega}{\epsilon^2 + \omega^2} [\varphi(\omega) + \varphi(0) - \varphi(0)]d\omega,$$

that is

$$\int_{-\infty}^\infty g(\omega)\varphi(\omega)d\omega = \varphi(0) \lim_{\epsilon \to 0} \int_{-\infty}^\infty \frac{\epsilon \pm i\omega}{\epsilon^2 + \omega^2} d\omega$$

$$+ \lim_{\epsilon \to 0} \int_{-\infty}^\infty \frac{\epsilon \pm i\omega}{\epsilon^2 + \omega^2} [\varphi(\omega) - \varphi(0)]d\omega. \qquad (6.72)$$

However,

$$\int_{-\infty}^\infty \frac{\epsilon \pm i\omega}{\epsilon^2 + \omega^2} d\omega = \epsilon \int_{-\infty}^\infty \frac{d\omega}{\epsilon^2 + \omega^2} \pm i \int_{-\infty}^\infty \frac{\omega d\omega}{\epsilon^2 + \omega^2}.$$

The last integral vanishes because the integrand is an odd function of ω. Moreover

$$\epsilon \int_{-\infty}^\infty \frac{d\omega}{\epsilon^2 + \omega^2} = \epsilon \lim_{R \to \infty} \int_{-R}^R \frac{d\omega}{\epsilon^2 + \omega^2} = 2 \lim_{R \to \infty} \arctan(R/\epsilon) = \pi.$$

On the other hand,

$$\lim_{\epsilon \to 0} \int_{-\infty}^\infty \frac{\epsilon \pm i\omega}{\epsilon^2 + \omega^2} [\varphi(\omega) - \varphi(0)]d\omega = \pm i \int_{-\infty}^\infty \frac{\varphi(\omega) - \varphi(0)}{\omega} d\omega.$$

Collecting results into Eq. (6.72) we get

$$\int_{-\infty}^\infty g(\omega)\varphi(\omega)d\omega = \pi\varphi(0) \pm i \int_{-\infty}^\infty \frac{\varphi(\omega) - \varphi(0)}{\omega} d\omega.$$

Hence

$$g(\omega) = \pi\delta(\omega) \pm i\mathcal{P}(1/\omega),$$

which is Eq. (6.69).

$\square$

Let us now address to the derivation of Kramers-Kronig relations. Since $e^{i\omega\tau} = \cos\omega\tau + i\sin\omega\tau$, we write Eq. (6.61) as

$$\chi_A(\omega) = \chi'_A(\omega) + i\chi''_A(\omega), \tag{6.73}$$

where

$$\chi'_A(\omega) = \int_{-\infty}^{\infty} R_A(\tau)\cos\omega\tau d\tau, \qquad \chi''_A(\omega) = \int_{-\infty}^{\infty} R_A(\tau)\sin\omega\tau d\tau, \tag{6.74}$$

are the real and imaginary parts of the susceptibility. Recalling that $R_A(\tau)$ is a real function (cf. Eq. (6.22)) we see from (6.74) that

$$\chi'_A(-\omega) = \chi'_A(\omega), \qquad \chi''_A(-\omega) = -\chi''_A(\omega). \tag{6.75}$$

We will now show that $\chi'(\omega)$ and $\chi''(\omega)$ are not independent but connected by Kramers-Kronig relations. Starting from Eq. (6.63) and using Eq. (6.61) we have

$$\chi_A(\omega) = \int_0^{\infty} e^{i\omega\tau} R_A(\tau) d\tau = \frac{1}{2\pi} \int_0^{\infty} e^{i\omega\tau} d\tau \int_{-\infty}^{\infty} e^{-i\eta\tau} \chi_A(\eta) d\eta$$

and interchanging the order of the integrals, we write

$$\chi(\omega) = \frac{1}{2\pi} \int_{-\infty}^{\infty} \chi_A(\eta) d\eta \int_0^{\infty} e^{i(\omega-\eta)\tau} d\tau. \tag{6.76}$$

Using Eq. (6.69) we have

$$\chi_A(\omega) = \frac{1}{2} \int_{-\infty}^{\infty} \chi_A(\eta)\delta(\omega-\eta) d\eta + \frac{i}{2\pi} \int_{-\infty}^{\infty} \chi_A(\eta)\mathcal{P}\left(\frac{1}{\omega-\eta}\right) d\eta$$

$$= \frac{1}{2}\chi_A(\omega) - \frac{i}{2\pi} \int_{-\infty}^{\infty} \chi_A(\eta)\mathcal{P}\left(\frac{1}{\eta-\omega}\right) d\eta,$$

that is,

$$\chi_A(\omega) = -\frac{i}{\pi} \int_{-\infty}^{\infty} \chi_A(\eta)\mathcal{P}\left(\frac{1}{\eta-\omega}\right) d\eta. \tag{6.77}$$

Substituting (6.73) into (6.77) and equating real and imaginary parts we see that real and imaginary parts of the susceptibility are not independent but related to each other by the *Kramers-Kronig relations*:

$$\chi'_A(\omega) = \frac{1}{\pi} \int_{-\infty}^{\infty} \chi''_A(\eta)\mathcal{P}\left(\frac{1}{\eta-\omega}\right) d\eta, \tag{6.78}$$

$$\chi''_A(\omega) = -\frac{1}{\pi} \int_{-\infty}^{\infty} \chi'_A(\eta)\mathcal{P}\left(\frac{1}{\eta-\omega}\right) d\eta. \tag{6.79}$$

Using Eq. (6.71) we write the Kramers-Kronig relations in the following more operative form:

$$\chi'_A(\omega) = \frac{1}{\pi} \int_{-\infty}^{\infty} \frac{\chi''_A(\eta) - \chi''_A(\omega)}{\eta-\omega} d\eta, \tag{6.80}$$

$$\chi''_A(\omega) = -\frac{1}{\pi} \int_{-\infty}^{\infty} \frac{\chi'_A(\eta) - \chi'_A(\omega)}{\eta-\omega} d\eta, \tag{6.81}$$

which clearly show that the knowledge of the imaginary part of susceptibility allow us to get the real part and vice versa.

6.5.2   *Fluctuation-dissipation theorem in Fourier space*

We finally present the fluctuation-dissipation theorem in Fourier space, that is to say, the relationship between the spectral density $\tilde{K}_{AB}(\omega)$ and susceptibility $\chi_A(\omega)$. To this end we must distinguish the cases in which the correlation function $K_{AB}(\tau)$ is symmetric or antisymmetric.

(a) Symmetric correlation

Suppose that the correlation function is symmetric:

$$K_{AB}(-\tau) = K_{AB}(\tau).$$

Note that this implies

$$\int_0^\infty K_{AB}(\tau) \cos \omega\tau \, d\tau = \frac{1}{2} \int_{-\infty}^\infty K_{AB}(\tau) e^{i\omega\tau} d\tau.$$

Using this expression and the fluctuation-dissipation theorem, $R_A(\tau) = \Theta(\tau)K_{AB}(\tau)$, we obtain the following expression for the real part of susceptibility (cf. Eq. (6.74))

$$\chi'_A(\omega) = \int_{-\infty}^\infty R_A(\tau) \cos \omega\tau \, d\tau = \int_0^\infty K_{AB}(\tau) \cos \omega\tau \, d\tau$$

$$= \frac{1}{2} \int_{-\infty}^\infty e^{i\omega\tau} K_{AB}(\tau) d\tau,$$

that is,

$$\chi'_A(\omega) = \frac{1}{2}\tilde{K}_{AB}(\omega), \tag{6.82}$$

where $\tilde{K}_{AB}(\omega)$ is the power spectrum. Eq. (6.82) is, in the Fourier space, the fluctuation-dissipation theorem for symmetric correlations.

Note that the Kramers-Kronig relation (6.81) allows us to obtain the imaginary part of susceptibility as

$$\chi''_A(\omega) = -\frac{1}{2\pi} \int_{-\infty}^\infty \frac{\tilde{K}_{AB}(\eta) - \tilde{K}_{AB}(\omega)}{\eta - \omega} d\eta. \tag{6.83}$$

The complete susceptibility is therefore given by

$$\chi_A(\omega) = \frac{1}{2}\tilde{K}_{AB}(\omega) - \frac{i}{2\pi} \int_{-\infty}^\infty \frac{\tilde{K}_{AB}(\eta) - \tilde{K}_{AB}(\omega)}{\eta - \omega} d\eta. \tag{6.84}$$

(b) Antisymmetric correlation

In this case $K_{AB}(-\tau) = -K_{AB}(\tau)$ and for the imaginary part of the susceptibility we have

$$\chi_A''(\omega) = \int_{-\infty}^{\infty} R_A(\tau) \sin \omega\tau d\tau = \int_0^{\infty} K_{AB}(\tau) \sin \omega\tau d\tau$$

$$= \frac{1}{2i} \int_{-\infty}^{\infty} K_{AB}(\tau) e^{i\omega\tau} d\tau.$$

The fluctuation-dissipation theorem in Fourier space now reads

$$\chi_A''(\omega) = \frac{1}{2i}\tilde{K}_{AB}(\omega). \tag{6.85}$$

Equation (6.80) yields

$$\chi_A'(\omega) = \frac{1}{2i\pi} \int_{-\infty}^{\infty} \frac{\tilde{K}_{AB}(\eta) - \tilde{K}_{AB}(\omega)}{\eta - \omega} d\eta, \tag{6.86}$$

and the complete susceptibility $\chi_A = \chi_A' + i\chi_A''$ is

$$\chi_A(\omega) = \frac{1}{2}\tilde{K}_{AB}(\omega) - \frac{i}{2\pi} \int_{-\infty}^{\infty} \frac{\tilde{K}_{AB}(\eta) - \tilde{K}_{AB}(\omega)}{\eta - \omega} d\eta, \tag{6.87}$$

which agrees with the symmetric case (6.84).

PART 2

Stochastic calculus with some applications

Chapter 7

Introduction to stochastic calculus

In the second part of the book we address, in an intuitive and informal way, the question of integrating diffusion processes and functions of them, the so-called stochastic calculus. This is a complex and important problem and constitutes the cornerstone of stochastic differential equations which have a wide range of applications in numerous branches of science. In this chapter and the next we present the basics of stochastic calculus, including stochastic differential equations, while in the last chapter of this part we apply the methods to some economic problems.

7.1   Introductory remarks

Suppose we have an one-dimensional diffusion process $X(t)$ with drift $f(x,t)$ and diffusion coefficient $D(x,t)$. As we have seen in Chapter 4 the increment, $\Delta X(t) = X(t + \Delta t) - X(t)$, is given by [cf. Eq. (4.55)]

$$\Delta X(t) = f(x,t)\Delta t + \sqrt{D(x,t)}\Delta W(t) + O(\Delta t^2), \qquad (7.1)$$

where $x = X(t)$ and $\Delta W(t)$ is the increment of the Wiener process which, for small time increments, can be written as [cf. Eq. (4.56)]

$$\Delta W(t) = Z\sqrt{\Delta t} + O(\Delta t^2), \qquad (7.2)$$

where Z is a Gaussian random variable with zero mean and unit variance. The ratio $\Delta W(t)/\Delta t$ is thus given by

$$\frac{\Delta W(t)}{\Delta t} = Z\frac{1}{\sqrt{\Delta t}} + O(\Delta t). \qquad (7.3)$$

As we have seen in Chapter 4, Eqs. (7.2) and (7.3) show in a simple and intuitive way that the Wiener process has continuous trajectories ($\Delta W(t) \to 0$ as $\Delta t \to 0$) but it is not differentiable ($|\Delta W(t)/\Delta t| \to \infty$ as $\Delta t \to 0$).

Before proceeding further let us see that using Eq. (7.2) we can obtain a key result in the development of the stochastic calculus which is related to the square of Wiener process increments, $\Delta W(t)^2$. To this end we present the following definition: Two random processes $X(t)$ and $Y(t)$ are *equal in mean square sense* if

$$\left\langle [X(t) - Y(t)]^2 \right\rangle = 0, \qquad (7.4)$$

for all t. And the same definition applies if $Y(t)$ is not a random process but a deterministic quantity. From the equality (7.4) it follows that (Papoulis, 1984)

$$\text{Prob}\{X(t) = Y(t)\} = 1,$$

which means that the processes $X(t)$ and $Y(t)$ are also equal in probability.

Let us now prove the key result mentioned above and show that, for sufficiently small values of Δt,

$$\Delta W(t)^2 = \Delta t \qquad (7.5)$$

in mean square sense. Indeed, from Eq. (7.2) we have

$$\Delta W(t)^2 = \left[Z\sqrt{\Delta t} + O(\Delta t^2) \right]^2 = Z^2 \Delta t + O\!\left(\Delta t^{5/2}\right)$$

and

$$\left\langle \Delta W(t)^2 \right\rangle = \Delta t + O\!\left(\Delta t^{5/2}\right), \qquad (7.6)$$

because $\langle Z^2 \rangle = 1$. Therefore,

$$\left\langle \left(\Delta W^2 - \langle \Delta W^2 \rangle \right)^2 \right\rangle = \langle \Delta W^4 \rangle - \langle \Delta W^2 \rangle^2$$
$$= \left(\langle Z^4 \rangle - 1 \right)\Delta t^2 + O\!\left(\Delta t^4\right),$$

but, due to the Gaussian character of Z with zero mean and unit variance, we have $\langle Z^4 \rangle = 1$, hence

$$\left\langle \left(\Delta W^2 - \langle \Delta W^2 \rangle \right)^2 \right\rangle = O\!\left(\Delta t^4\right).$$

This implies that for sufficiently small values of Δt,

$$\Delta W^2 = \langle \Delta W^2 \rangle$$

in mean square sense. Using (7.6) we see that $\langle \Delta W^2 \rangle = \Delta t$ which proves Eq. (7.5).

$\square$

Following an analogous reasoning one can easily show that

$$\Delta W(t)^{2+n} = 0, \qquad (n = 1, 2, 3, \dots), \qquad (7.7)$$

in mean square sense and for sufficiently small values of Δt.

Despite the no differentiability of the Wiener process, these results can be symbolically written as

$$dW(t)^2 = dt, \qquad \text{and} \qquad dW(t)^{2+n} = 0, \tag{7.8}$$

$(n = 1, 2, 3, \dots)$ in mean square sense. Let us also note that, within the same symbolism, we can write Eq. (7.1) in the form

$$dX(t) = f(X, t)dt + g(X, t)dW(t), \tag{7.9}$$

where $X = X(t)$ and $g(X, t) = \sqrt{D(X, t)}$. Eq. (7.9) is an example of the so-called stochastic differential equations that we will study in the next chapter.

Squaring Eq. (7.9) and using (7.8) we have

$$dX^2 = g(x, t)^2 dt + O(dt^2) \tag{7.10}$$

and symbolically write

$$dX(t) = O(dt^{1/2}), \tag{7.11}$$

a distinctive characteristic of any diffusion process.

Returning to the stochastic differential equation (7.9), and within this formal framework, we write (7.9) as

$$\frac{dX(t)}{dt} = f(X, t) + g(X, t)\xi(t), \tag{7.12}$$

where $\xi(t)$ (symbolically the derivative of the Wiener process) is Gaussian white noise, that is, a Gaussian and stationary random process with zero mean and delta correlated [cf. Eq. (4.58)]. Differential equations in the form given by (7.12) appear frequently in the physics literature where they are called *Langevin equations*.

We remark that these expressions are purely formal because $dW(t)$ and $\xi(t)$ do not exist in the sense of ordinary differentials and functions. Even an appropriate definition of these magnitudes, would render the multiplicative term $g(x, t)\xi(t)$ undefined. Indeed, as we will see next, there are several interpretations of the multiplicative term which arise because, due to the extreme randomness of white noise, it is not clear what value of X should be used during an infinitesimal timestep dt. According to the *Itô interpretation* the value of X that appears in $g(X, t)$ is the one *before the beginning of the timestep*, i.e., $X = X(t)$, whereas the *Stratonovich interpretation* uses the value of X at the *middle of the timestep*, that is, $X = X(t + dt/2) = X(t) + dX(t)/2$. We will discuss these aspects in more detail in the forthcoming sections.

7.2   Stochastic integrals

As remarked above, the analysis of stochastic dynamical systems often leads
to stochastic differential equations of the form given by Eq. (7.9):

$$dX(t) = f(X,t)dt + g(X,t)dW(t),$$

where $X = X(t)$. This differential equation is equivalent to the following
integral equation

$$X(t) = x_0 + \int_{t_0}^{t} f(X(t'),t')dt' + \int_{t_0}^{t} g(X(t'),t')dW(t'), \qquad (7.13)$$

where t_0 is an arbitrary initial time and the initial value $x_0 = X(t_0)$ can be
either a known constant or a random variable. The first integral on the right
hand side of Eq. (7.13) can be understood as an ordinary Riemann-Stieltjes
integral. The second integral is more than a problem because $dW(t)$ does
not exists in the usual sense. Let us, however, note that if $g(X,t) = g(t)$
does not depend on $X = X(t)$ and it is, in addition, a deterministic function
of time then, integrating by parts, we have

$$\int_{t_0}^{t} g(t')dW(t') = g(t)W(t) - g(t_0)W(t_0) - \int_{t_0}^{t} \dot{g}(t')W(t')dt', \qquad (7.14)$$

and, since $W(t')$ is continuous, the last integral is an ordinary Riemann-
Stieltjes integral evaluated on the individual sample paths of $W(t')$.

Suppose next that $g(t) = G(t)$ is a random function of time (that is to
say, a random process). The problem now is obtaining a precise definition
of the integral

$$I(t) = \int_{t_0}^{t} G(t')dW(t') \qquad (7.15)$$

for as board a class of random processes $G(t)$ as possible.

As a first approximation we will try to define $I(t)$ as a kind of Riemann-
Stieltjes integral. Thus, dividing the interval $[t_0, t]$ into n subintervals with
partitioning times

$$t_0 \leq t_1 \leq t_2 \cdots \leq t_{n-1} \leq t$$

and setting the intermediate times

$$t_{k-1} \leq \tau_k \leq t_k,$$

we define the integral as the limit (in mean square sense) when $n \to \infty$ [or,
equivalently, when $\delta_n = \max(t_k - t_{k-1}) \to 0$] of the partial sums

$$S_n = \sum_{k=1}^{n} G(\tau_k)[W(t_k) - W(t_{k-1})]. \qquad (7.16)$$

We therefore define

$$\int_{t_0}^{t} G(t')dW(t') = \text{ms}- \lim_{n\to\infty} \sum_{k=1}^{n} G(\tau_k)\big[W(t_k) - W(t_{k-1})\big], \qquad (7.17)$$

where, by ms-lim we mean the *mean square limit* defined as

$$\text{ms}- \lim_{n\to\infty} X_n = X \qquad \text{if} \qquad \lim_{n\to\infty} \big\langle (X_n - X)^2 \big\rangle = 0. \qquad (7.18)$$

For the rest of this chapter *all limits are to be interpreted in mean square sense* and, when there is no confusion, we will simply use the conventional notation:

$$\lim_{n\to\infty} X_n.$$

However, for the definition of the Riemann-Stieljes integral to make sense, it is indispensable that the result of the limit $n \to \infty$ in Eq. (7.17) be independent of the choice of the intermediate points τ_k. Unfortunately, this is not in general the case for stochastic integrals of the form (7.15). Let us first show this in the particular case when $G(t)$ is the Wiener process $W(t)$.

Example

Consider the integral

$$\int_{t_0}^{t} W(t')dW(t'). \qquad (7.19)$$

If we assume the existence of the integral as an ordinary Riemann-Stieljes integral, formal application of the classical rules of integration by parts yields

$$\int_{t_0}^{t} W(t')dW(t') = \frac{1}{2}\left[W(t)^2 - W(t_0)^2\right]. \qquad (7.20)$$

As mentioned above, this would imply the mean-square convergence of the sums

$$S_n = \sum_{k=1}^{n} W(\tau_k)[W(t_k) - W(t_{k-1})],$$

with ever finer partitioning (i.e., $n \to \infty$) and for arbitrary choices of the intermediate times τ_k. Let us see, however, that the mean square limit of S_n depends on the intermediate times.

In what follows we use the shorthand notation

$$W_{t_k} = W(t_k) \qquad \text{and} \qquad W_{\tau_k} = W(\tau_k).$$

Taking into account (recall $t_n = t$)

$$\sum_{k=1}^{n}(W_{t_k}^2 - W_{t_{k-1}}^2) = W_t^2 - W_{t_0}^2,$$

we can easily, but somewhat tediously, show that

$$S_n = \frac{1}{2}W_t^2 - \frac{1}{2}W_{t_0} - \frac{1}{2}\sum_{k=1}^{n}(W_{t_k} - W_{t_{k-1}})^2 \qquad (7.21)$$

$$+ \sum_{k=1}^{n}(W_{\tau_k} - W_{t_{k-1}})^2 + \sum_{k=1}^{n}(W_{t_k} - W_{\tau_k})(W_{\tau_k} - W_{t_{k-1}}).$$

Bearing in mind that for sufficiently small Δt and in mean square sense $\Delta W^2 = \Delta t$, we have

$$\sum_{k=1}^{n}(W_{t_k} - W_{t_{k-1}})^2 = \sum_{k=1}^{n}(t_k - t_{k-1}) = t - t_0, \qquad (7.22)$$

and

$$\sum_{k=1}^{n}(W_{\tau_k} - W_{t_{k-1}})^2 = \sum_{k=1}^{n}(\tau_k - t_{k-1}). \qquad (7.23)$$

On the other hand, call

$$\Delta_n = \sum_{k=1}^{n}(W_{t_k} - W_{\tau_k})(W_{\tau_k} - W_{t_{k-1}}),$$

then

$$\langle \Delta_n \rangle = \sum_{k=1}^{n}\langle W_{t_k} - W_{\tau_k}\rangle\langle W_{\tau_k} - W_{t_{k-1}}\rangle = 0,$$

since Wiener increments are independent and have zero mean (cf. Sect. 4.5 of Chapter 4). Therefore,

$$\left\langle (\Delta_n - \langle \Delta_n \rangle)^2 \right\rangle = \langle \Delta_n^2 \rangle$$

$$= \sum_{k=1}^{n}\sum_{l=1}^{n}\langle W_{t_k} - W_{\tau_k}\rangle\langle W_{\tau_k} - W_{t_{k-1}}\rangle\langle W_{t_l} - W_{\tau_l}\rangle\langle W_{\tau_l} - W_{t_{l-1}}\rangle = 0,$$

showing that

$$\Delta_n = \sum_{k=1}^{n}(W_{t_k} - W_{\tau_k})(W_{\tau_k} - W_{t_{k-1}}) = 0 \qquad (7.24)$$

in mean square sense.

Plugging Eqs. (7.22)-(7.24) into Eq. (7.21) we get

$$S_n = \frac{1}{2}W_t^2 - \frac{1}{2}W_{t_0}^2 - \frac{1}{2}(t - t_0) + \sum_{k=1}^n (\tau_k - t_{k-1}), \qquad (7.25)$$

which clearly shows that the partial sums depend on the intermediate points τ_k. Thus, for example, if we choose

$$\tau_k = (1 - a)t_{k-1} + at_k, \qquad (0 \leq a \leq 1), \qquad (7.26)$$

then $\tau_k - t_{k-1} = a(t_k - t_{k-1})$ and

$$\sum_{k=1}^n (\tau_k - t_{k-1}) = a \sum_{k=1}^n (t_k - t_{k-1}) = a(t - t_0).$$

Substituting into (7.25) yields

$$S_n = \frac{1}{2}W_t^2 - \frac{1}{2}W_{t_0}^2 + \left(a - \frac{1}{2} \right)(t - t_0).$$

Therefore,

$$\int_{t_0}^t W(t')dW(t') = \frac{1}{2}\left[W(t)^2 - W(t_0)^2 \right] + \left(a - \frac{1}{2} \right)(t - t_0) \qquad (7.27)$$

(in mean square sense), which depends, through a, on the intermediate points and in general differs from the classical value (7.20).

Let us note before proceeding further that, since $\langle W(t)^2 \rangle = t$, the average value of the integral reads

$$\left\langle \int_{t_0}^t W(t')dW(t') \right\rangle = a(t - t_0). \qquad (7.28)$$

We therefore see that in order to uniquely define the integral (7.19) it is necessary to fix the intermediate points τ_k. In particular for $a = 0$, that is, if $\tau_k = t_{k-1}$, we obtain the *Itô integral*

$$\text{(Itô)} \quad \int_{t_0}^t W(t')dW(t') = \lim_{n \to \infty} \sum_{k=1}^n W(t_{k-1})\left[W(t_k) - W(t_{k-1}) \right]$$

$$= \frac{1}{2}\left[W(t)^2 - W(t_0)^2 - (t - t_0) \right]. \qquad (7.29)$$

In this case [cf. Eq. (7.28)]

$$\text{(Itô)} \quad \left\langle \int_{t_0}^t W(t')dW(t') \right\rangle = 0. \qquad (7.30)$$

An alternative definition of the stochastic integral was introduced by Stratonovich in which the anomalous term $t - t_0$ does not appear and the integration process follows the rules of ordinary calculus (Stratonovich, 1966). The *Stratonovich integral* consists in choosing $a = 1/2$ in Eq. (7.27) which amounts to evaluate the integral at the middle point $\tau_k = (t_{k-1} + t_k)/2$ and, since

$$W(\tau_k) = [W(t_{k-1}) + W(t_k)]/2 \qquad \text{as} \qquad t_k - t_{k-1} \to 0, \qquad (7.31)$$

we define

$$\text{(Strat)} \quad \int_{t_0}^{t} W(t')dW(t')$$

$$= \lim_{n \to \infty} \sum_{k=1}^{n} \frac{W(t_{k-1}) + W(t_k)}{2} \left[W(t_k) - W(t_{k-1})\right]$$

$$= \frac{1}{2}\left[W(t)^2 - W(t_0)^2\right]. \qquad (7.32)$$

Now

$$\text{(Strat)} \quad \left\langle \int_{t_0}^{t} W(t')dW(t') \right\rangle = \frac{1}{2}(t - t_0). \qquad (7.33)$$

7.3   Nonanticipating functions and the Itô integral

We will now extend the definition of the Itô integral, given in Eq. (7.29) for the particular example explained above, to more general stochastic integrals of the form:

$$\int_{t_0}^{t} G(t')dW(t'), \qquad (7.34)$$

where $G(t)$ is a random function of time, i.e., a stochastic process.[1] In order to do so we must restrict ourselves to a broad but special kind of functions, the so-called nonanticipating functions. This is a rather simple concept which, as Gardiner states: "can be easily made quite obscure by complex notation" (Gardiner, 1986).

A random function $G(t)$ is called a *nonanticipating function* of t if, for any $s > t$, $G(t)$ is statistically independent of $W(s) - W(t)$. In other words, *$G(t)$ is independent of the behavior of the Wiener process in the future of time t.* Note that this is a very reasonable requirement for a physical function which could be a solution of stochastic differential equations of the form (7.9) [cf. Eq. (7.13)] in which the solution $X(t)$ involves the Wiener process $W(t')$ only for $t' \leq t$.

---

[1]When $G(t) = g(t)$ is nonrandom the integral can be defined as an ordinary Riemann-Stieljes integral [see Eq. (7.14) and the comment below].

Examples of nonanticipating functions are the Wiener process itself and integrals of functions of the Wiener process (Gardiner, 1986):

$$W(t), \qquad \int_{t_0}^{t} F(W(t'))dt', \qquad \int_{t_0}^{t} F(W(t'))dW(t'),$$

and also integrals of nonanticipating functions

$$\int_{t_0}^{t} G(t')dt', \qquad \int_{t_0}^{t} G(t')dW(t'),$$

where $G(t)$ is a nonanticipating random function.

Let us stress that nonanticipating functions naturally appear in the study of stochastic differential equations where causality is expected, that is to say, the unknown future cannot influence the present.

For nonanticipating functions we can give a precise definition of the Itô stochastic integral, based in Eq. (7.17), as follows:

The Itô stochastic integral of a nonanticipating random function $G(t)$ is defined as the limit of the partial sums

$$\int_{t_0}^{t} G(t')dW(t') = \lim_{n \to \infty} \sum_{k=1}^{n} G(t_{k-1})[W(t_k) - W(t_{k-1})], \qquad (7.35)$$

where $t_0 \leq t_1 \leq \cdots \leq t_n = t$ is an arbitrary partition of the interval $[t_0, t]$ and the limit $n \to \infty$ is equivalent to the limit $\delta_n = \max |t_k - t_{k-1}| \to 0$.

It can be proven that the stochastic integral thus defined exists as long as $G(t)$ is nonanticipating and continuous (Arnold, 1974).

Another class of stochastic integral is provided by integrals of the form

$$\int_{t_0}^{t} G(t')dW(t')^{2+n} \qquad (n = 1, 2, 3, \dots),$$

where $G(t)$ is nonanticipating. In the sense of Itô, this type of integrals is defined in accordance to Eq. (7.35) as

$$\int_{t_0}^{t} G(t')dW(t')^{2+n} = \lim_{n \to \infty} \sum_{k=1}^{n} G(t_{k-1})[W(t_k) - W(t_{k-1})]^{2+n}. \qquad (7.36)$$

We have shown in Sect. 7.1 that [cf. Eqs. (7.5) and (7.7)]

$$[W(t_k) - W(t_{k-1})]^2 = t_k - t_{k-1} \quad \text{and} \quad [W(t_k) - W(t_{k-1})]^{2+n} = 0,$$

as $(t_k - t_{k-1}) \to 0$ and in mean square sense. Hence,

$$\int_{t_0}^{t} G(t')dW(t')^{2+n} = \begin{cases} \int_{t_0}^{t} G(t')dt', & n = 0, \\ 0, & n = 1, 2, 3, \dots. \end{cases} \qquad (7.37)$$

This result, that can be extended to include noninteger powers, that is, $dW(t)^{2+\alpha}$ ($\alpha \geq 0$). It can also be shown that,

$$\int_{t_0}^{t} G(t')dW(t')^n dt' = 0, \qquad (7.38)$$

($n = 1, 2, 3, \dots$). From an intuitive point of view this can be seen because $dW(t)$ is an infinitesimal of order $dt^{1/2}$ [cf. Eq. (7.8)]. Hence, $dW(t)dt$ is of order $dt^{3/2}$ and in calculating differentials, infinitesimal of order higher than dt are neglected.

7.4   Some properties of the Itô integral

We now present few relevant properties of the Itô integral such as the integration of powers of the Wiener process, general differentiation rules and correlation of integrals.

(1) Integration of powers of $W(t)$

We will prove the formula

$$\int_{t_0}^{t} W(t')^n dW(t') = \frac{1}{1+n} \left[W(t)^{n+1} - W(t_0)^{n+1} \right]$$

$$- \frac{n}{2} \int_{t_0}^{t} W(t')^{n-1} dt', \qquad (7.39)$$

($n = 1, 2, 3, \dots$), which generalizes the expression (7.29) and replaces the classical rule of integration of powers.

Indeed, using the binomial expansion and bearing in mind that $dW(t)^2 = dt$ and $dW(t)^{2+n} = 0$, we write

$$d[W(t)^m] \equiv W(t + dt)^m - W(t)^m = \left[W(t) + dW(t) \right]^m - W(t)^m$$

$$= \sum_{k=1}^{m} \binom{m}{k} W(t)^{m-k} dW(t)^k$$

$$= mW(t)^{m-1} dW(t) + \frac{1}{2}m(m-1)W(t)^{m-2}dt.$$

That is,

$$W(t)^{m-1}dW(t) = \frac{1}{m}d[W(t)^m] - \frac{1}{2}(m-1)W(t)^{m-2}dt,$$

choosing $m = 1 + n$ we have

$$W(t)^n dW(t) = \frac{1}{1+n}d[W(t)^{1+n}] - \frac{1}{2}nW(t)^{n-1}dt$$

and integrating we obtain (7.39).

$\square$

(2) Differentiation rule

As is well known from ordinary calculus the differential of a composite function $\phi(y, t)$ is

$$d\phi(y, t) = \frac{\partial \phi}{\partial t} dt + \frac{\partial \phi}{\partial y} dy.$$

Let us now generalize this expression, within Itô calculus, when y is a random function, specifically, when it is the Wiener process. Since

$$d\phi(W(t), t) = \phi(W(t) + dW(t), t + dt) - \phi(W(t), t),$$

a Taylor expansion up to second-order terms yields

$$d\phi(W(t), t) = \frac{\partial \phi}{\partial t} dt + \frac{1}{2} \frac{\partial^2 \phi}{\partial t^2} dt^2 + \frac{\partial \phi}{\partial W} dW(t)$$
$$+ \frac{1}{2} \frac{\partial^2 \phi}{\partial W^2} dW(t)^2 + \frac{\partial^2 \phi}{\partial W \partial t} dW(t) dt,$$

but $dW(t)^2 = dt$ (in mean square sense) and

$$dt^2 \to 0, \qquad dW(t) dt = O(dt^{3/2}) \to 0,$$

and all higher powers vanish. We, therefore, obtain the generalized differentiation formula

$$d\phi(W(t), t) = \left(\frac{\partial \phi}{\partial t} + \frac{1}{2} \frac{\partial^2 \phi}{\partial W^2} \right) dt + \frac{\partial \phi}{\partial W} dW(t). \qquad (7.40)$$

Comparing with the classical expression we see that it appears an extra term involving the second derivative. Let us stress once more that this is the direct consequence of the fact that $dW(t) = O(dt^{1/2})$.

(3) Average value

We have seen that when $G(t) = W(t)$ the mean value of the Itô integral, Eq. (7.34), is zero [cf. Eq. (7.30)]. Let us now see that the same result applies to any nonanticipating function. Indeed,

$$\left\langle \int_{t_0}^{t} G(t') dW(t') \right\rangle = \lim_{n \to \infty} \left\langle \sum_{k=1}^{n} G(t_{k-1}) [W(t_k) - W(t_{k-1})] \right\rangle$$

$$= \lim_{n \to \infty} \sum_{k=1}^{n} \langle G(t_{k-1}) \rangle \langle W(t_k) - W(t_{k-1}) \rangle,$$

because, by definition of nonanticipating function, $G(t_{k-1})$ is independent of the Wiener increment $W(t_k) - W(t_{k-1})$. Moreover, $\langle W(t_k) - W(t_{k-1}) \rangle = 0$ and we conclude

$$\left\langle \int_{t_0}^{t} G(t') dW(t') \right\rangle = 0. \qquad (7.41)$$

(4) Correlation function

Suppose $F(t)$ and $G(t)$ are arbitrary nonanticipating function, then the following correlation formula holds

$$\left\langle \int_{t_0}^t F(t_1)dW(t_1) \int_{t_0}^t G(t_2)dW(t_2) \right\rangle = \int_{t_0}^t \langle F(t_1)G(t_1)\rangle dt_1. \qquad (7.42)$$

Although this formula can be proved making use of the definition (7.35), we will utilize a more direct approach. Recall from Eq. (4.59) of Chapter 4 that

$$dW(t) = \xi(t)dt, \qquad (7.43)$$

where $\xi(t)$ is Gaussian white noise, that is, a Gaussian process with zero mean and delta correlated:

$$\langle \xi(t) \rangle = 0, \qquad \langle \xi(t_1)\xi(t_2)\rangle = \delta(t_1 - t_2). \qquad (7.44)$$

Substituting (7.43) into (7.42) and noting that if $F(t)$ and $G(t)$ are nonanticipating, $\xi(t)$ is independent of them, we have

$$\left\langle \int_{t_0}^t F(t_1)dW(t_1) \int_{t_0}^t G(t_2)dW(t_2) \right\rangle = \int_{t_0}^t dt_1 \int_{t_0}^t \langle F(t_1)G(t_2)\rangle \langle \xi(t_1)\xi(t_2)\rangle dt_2$$

$$= \int_{t_0}^t dt_1 \int_{t_0}^t \langle F(t_1)G(t_2)\rangle \delta(t_1 - t_2)dt_2,$$

and implementing the delta function we get

$$\left\langle \int_{t_0}^t F(t_1)dW(t_1) \int_{t_0}^t G(t_2)dW(t_2) \right\rangle = \int_{t_0}^t \langle F(t_1)G(t_1)\rangle dt_1,$$

proving Eq. (7.42).

$\square$

Chapter 8

Stochastic differential equations

One of the most important aspects of stochastic calculus is provided by stochastic differential equations. Centering on one-dimensional problems, a very general form of stochastic differential equation (SDE) is

$$\frac{dX(t)}{dt} = F\big(t, X(t); Z(t)\big), \tag{8.1}$$

where $F(t, x; z)$ is an arbitrary function and $Z(t)$ is a random process with known statistical properties that is often called "input noise" or just "noise". Obviously, the solution of Eq. (8.1) is a random process $X(t)$ and the main objective is to know its statistical properties.

Two simplifications are usually found in practice. The first one consists in assuming that $Z(t) = \xi(t)$, where $\xi(t)$ is Gaussian white noise with zero mean and delta correlation. A second simplification supposes that F is a linear function of the noise. These two assumptions lead to the Langevin equation

$$\frac{dX(t)}{dt} = f(X, t) + g(X, t)\xi(t), \tag{8.2}$$

which, bearing in mind that $\xi(t)$ is the derivative of the Wiener process,[1] can be written in the mathematical standard form of a SDE

$$dX(t) = f(X, t)dt + g(X, t)dW(t). \tag{8.3}$$

The noise term appearing in (8.2) and (8.3) is called *additive* when $g(X, t)$ is independent of X and *multiplicative* otherwise. As we have seen in the previous chapter these equations have no meaning unless we provide an interpretation of the noise term $g(X, t)dW(t)$ in the multiplicative case.

---

[1] In the sense of generalized functions, as discussed in Chapter 4.

8.1   Itô stochastic differential equations

Assuming that $X(t_0) = X_0$ the SDE (8.3) is equivalent to [2]

$$X(t) = X_0 + \int_{t_0}^{t} f\big(X(t'),t'\big)dt' + \int_{t_0}^{t} g\big(X(t'),t'\big)dW(t'). \tag{8.4}$$

Let us first observe that for an additive noise term, where $g = g(t)$ is nonrandom, the last integral in Eq. (8.4) is defined as an ordinary Riemann-Stieljes integral (cf. Eq. (7.14)). However, for the multiplicative case the integral needs to be defined. Thus, If the SDE (8.3) is interpreted in the sense of Itô and g is a nonanticipating function, the integral of the noise terms is defined as the Itô integral (cf. Eq. (7.35)) [3]

$$\int_{t_0}^{t} g\big(X(t'),t'\big)dW(t') = \lim_{n\to\infty} \sum_{k=1}^{n} g\big(X(t_{k-1}),t_{k-1}\big)\big[W(t_k) - W(t_{k-1})\big],$$

$$\tag{8.5}$$

and Eq. (8.3) is called the Itô stochastic differential equation.

8.1.1   *Existence and uniqueness*

We now present some relevant properties of Itô SDE's. We start with the conditions on the existence and uniqueness of the solution of Eq. (8.3), which are essentially the same than those corresponding to nonrandom differential equations. The main result is provided by the following theorem:

Theorem.

Suppose we have the Itô SDE

$$dX(t) = f\big(X,t\big)dt + g\big(X,t\big)dW(t), \qquad X(t_0) = X_0, \tag{8.6}$$

where f and g are nonanticipating functions, X_0 is a random variable independent of the Wiener process $W(t)$ and $\langle X_0^2 \rangle$ finite. If the following assumptions are satisfied

(i) (Lipschitz condition) For all $t \in [t_0, T]$, $x, y \in \mathbb{R}$, there exists a constant $K > 0$ such that

$$|f(x,t) - f(y,t)| + |g(x,t) - g(y,t)| \leq K|x - y|. \tag{8.7}$$

(ii) (Restriction on growth) For all $t \in [t_0, T]$ and $x \in \mathbb{R}$

$$|f(x,t)|^2 + |g(x,t)|^2 \leq K^2\big(1 + |x|^2\big). \tag{8.8}$$

---

[2]The initial value X_0 may or may not be random, if it is, X_0 must be considered nonanticipating, i.e., independent of the Wiener noise.

[3]In Eq. (7.35) we have defined the Itô integral of a random function $G(t)$ with respect to the Wiener measure $dW(t)$. Here $G(t) = g\big(X(t),t\big)$ and randomness is through $X(t)$.

Then Eq. (8.6) has on the interval $[t_0, T]$ a continuous and unique solution $X(t)$ (with probability 1) that satisfies the initial condition $X(t_0) = X_0$.

The proof is rather technical and basically consists in applying the Picard iteration method adapted to random variables. We refer the interested reader to Arnold's book (1974) for details.[4]

Remark 1

The Lipschitz condition ensures that $f(x, t)$ and $g(x, t)$ do not change faster in x than does x itself. Hence, discontinuous functions of x are excluded as coefficients but also continuous functions of the type $f(x, t) = |x|^\alpha$ $(0 < \alpha < 1)$ for which condition (8.7) does not hold. Thus, for example, it can be proved that the SDE

$$dX(t) = |X(t)|^\alpha dW(t),$$

has only one solution for $\alpha \geq 1/2$ but infinitely many for $0 < \alpha < 1/2$ (Arnold, 1974).

It can also be shown that for the Lipschitz condition to be satisfied it suffices that $f(x, t)$ and $g(x, t)$ have continuous and bounded partial derivatives of first order with respect to x for every $t \in [t_0, T]$ (Arnold, 1974).

Remark 2

The restriction on growth allows at most linear increases of these functions. If this condition does not hold, an "explosion" takes place, that is, $X(t)$ becomes infinite in finite time. As a simple example consider the nonrandom equation

$$dx(t) = \frac{1}{2}x^3 dt, \qquad x(0) = x_0.$$

The solution for $x_0 \neq 0$ is

$$x(t) = \left(x_0^{-2} - t\right)^{-1/2}$$

and "explodes" at $t = x_0^{-2}$. Note that, for a given time interval $[0, T]$, there always are initial values (those such that $x_0 \geq T^{-1/2}$) for which the solution does not exist on the entire interval $[0, T]$.

---

[4]See also Oksendal (2010) for more formal aspects of SDE's and their applications.

Remark 3

If $f(x,t) \equiv f(x)$ and $g(x,t) \equiv g(x)$ do not depend on time, the corresponding SDE

$$dX(t) = f(X)dt + g(X)dW(t),$$

is called *autonomous*. We can easily see that this equation is invariant under the time translation $t \to t - t_0$,[5] so that we can take $t_0 = 0$ without loss of generality.

For the existence and uniqueness of autonomous SDE's we only need the Lipschitz condition,

$$|f(x) - f(y)| + |g(x) - g(y)| \le K|x - y|,$$

because it is shown that the restriction on growth follows from this condition (Arnold, 1974). Moreover, the solution is *global* in the sense that it exists on the entire half line $[0, \infty)$.

8.1.2 *Numerical integration*

Note that from the SDE (8.6) we can write

$$\Delta X(t) = f\big(X(t),t\big)\Delta t + g\big(X(t),t\big)\Delta W(t) + O\big(\Delta t^2\big), \qquad (8.9)$$

where $\Delta X(t) = X(t + \Delta t) - X(t)$ and $\Delta W(t) = W(t + \Delta t) - W(t)$. From this equation we see that the term $f(x,t)$ represents the mean velocity of the random motion described by $X(t)$, whereas the noise intensity $g(x,t)$ is a measure of the magnitude of the fluctuations of $\Delta X(t)$ about the mean value.

Recall that the increment of the Wiener process can be written as (cf. Eq. (4.56) of Chapter 4)

$$\Delta W(t) = Z\sqrt{\Delta t} + O(\Delta t^2),$$

where Z is a Gaussian random variable with zero mean and unit variance. We can thus write Eq. (8.9) as

$$X(t + \Delta t) = X(t) + f\big(X(t),t\big)\Delta t + Zg\big(X(t),t\big)\sqrt{\Delta t} + O(\Delta t^2), \quad (8.10)$$

which allows to calculate $X(t + \Delta t)$, for sufficiently small Δt, once we know $X(t)$.

More explicitly, assuming $X(t_0) = X_0$ and discretizing the time interval $[t_0, t]$ by a lattice of intermediate times:

$$t_0 < t_1 < \cdots < t_{n-1} < t_n = t,$$

---

[5]Recall that the Wiener process is time homogeneous.

Eq. (8.10) can be written as

$$X_{k+1} = X_k + f(t_k, X_k)(t_{k+1} - t_k)$$
$$+ Zg(t_k, X_k)\sqrt{t_{k+1} - t_k} + O(|t_{k+1} - t_k|^2), \qquad (8.11)$$

where $X_k = X(t_k)$ $(k = 0, 1, 2, \cdots, n - 1)$.

Therefore, the numerical procedure for solving the Itô SDE consists in calculating X_{k+1} from the knowledge of X_k by adding a deterministic term $f(X_k, t_k)\Delta t_k$ and a fluctuating term $Zg(X_k, t_k)\sqrt{\Delta t_k}$, where $\Delta t_k = t_{k+1} - t_k$ is the lattice size. Note that by letting the lattice size going to zero we obtain the formal solution of the SDE such that $X(t_0) = X_0$.

8.2 Diffusion processes and the Fokker-Planck equation

Under rather general circumstances the solution of the Itô SDE is a diffusion process. This is the content of the following theorem:

Theorem.

If the conditions of existence snd uniqueness for the SDE (8.6) are satisfied and if $f(x,t)$ and $g(x,t)$ are continuous functions of time, then the solution $X(t)$ of the SDE is a diffusion processes on $[t_0, T]$ with drift $f(x,t)$ and diffusion coefficient $g^2(x,t)$.

Proof. We present an intuitive and heuristic proof and refer the interested reader to Arnold (1974) for a more formal proof.

We know that diffusion processes are Markovian random processes with continuous sample paths. Moreover, the fact that the trajectories of $X(t)$, solution to the SDE (8.6), are continuous is readily seen from the discretization of the SDE given in Eq. (8.9):

$$\Delta X(t) = f(X, t)\Delta t + g(X, t)\Delta W(t) + O(\Delta t^2), \qquad (8.12)$$

which implies

$$\Delta X(t) \to 0 \qquad \text{as} \qquad \Delta t \to 0,$$

meaning that $X(t)$ has continuous sample paths.

The Markovian character of $X(t)$ can also be seen from Eq. (8.12) since $X(t + \Delta t)$ $(\Delta t > 0)$ only depends on the previous value, $X(t)$, but not on earlier values. Hence, $X(t)$ is Markovian and, therefore, a diffusion process.

Let us now see that the drift is given by $f(x,t)$ and the diffusion coefficient by $g^2(x,t)$. Indeed, we know that (cf. Eq. (3.23) of Chapter 3) the drift is given by the first infinitesimal conditional moment defined as

$$\lim_{\Delta t \to 0} \frac{1}{\Delta t} \Big\langle \Delta X(t) | X(t) = x \Big\rangle.$$

Substituting for Eq. (8.12) and taking into account that $\langle \Delta W \rangle = 0$, we get

$$\frac{1}{\Delta t}\Big\langle \Delta X(t)\big|X(t) = x \Big\rangle = f(x,t) + O(\Delta t),$$

which proves that $f(x,t)$ is the drift of the diffusion process.

On the other hand, the second infinitesimal moment, i.e. the diffusion coefficient $D(x,t)$, is

$$D(x,t) = \lim_{\Delta t \to 0} \frac{1}{\Delta t}\Big\langle \Delta X(t)^2\big|X(t) = x \Big\rangle.$$

Squaring Eq. (8.9) and bearing in mind that $\Delta W(t) = O(\Delta t^{1/2})$ and $\Delta W(t)^2 = \Delta t$ (in mean square sense) we have

$$\Big\langle \Delta X(t)^2\big|X(t) = x \Big\rangle = g^2(x,t)\Delta t + O\big(\Delta t^{3/2}\big).$$

Hence,

$$D(x,t) = g^2(x,t). \tag{8.13}$$

$\square$

We finally establish the connection between Itô SDE's and Fokker-Planck equations. Using the above result this is rather straightforward, for the solution $X(t)$ of the SDE

$$dX(t) = f(X,t)dt + g(X,t)dW(t), \qquad X(t_0) = x_0,$$

is a diffusion process with drift $f(x,t)$ and diffusion coefficient $g^2(x,t)$. Therefore, the transition probability density function of that solution,

$$p(x,t|x_0,t_0)dx = \text{Prob}\big\{x < X(t) < x + dx|X(t_0) = x_0\big\},$$

will satisfy the Fokker-Planck equation (FPE)

$$\frac{\partial p}{\partial t} = -\frac{\partial}{\partial x}\big[f(x,t)p\big] + \frac{1}{2}\frac{\partial^2}{\partial x^2}\big[g^2(x,t)p\big], \tag{8.14}$$

with initial condition

$$p(x,t_0|x_0,t_0) = \delta(x - x_0). \tag{8.15}$$

8.2.1 Changing variables: Itô formula

Suppose we have the SDE

$$dX(t) = f(X,t)dt + g(X,t)dW(t) \qquad (8.16)$$

and want to make a change of variable $Y = \phi(X)$. What is the SDE followed by $Y(t)$? For nonrandom equations, the change of variable would follow the laws of ordinary calculus. Unfortunately, this is not the case for the Itô SDE. Indeed, by Taylor expanding and using Eq. (8.16), we have

$$
\begin{aligned}
d\phi(X) &= \phi(X+dX) - \phi(X) \\
&= \phi'(X)dX + \frac{1}{2}\phi''(X)dX^2 + O(dX^3) \\
&= \phi'(X)\Big[f(X,t)dt + g(X,t)dW(t)\Big] \\
&\quad + \frac{1}{2}\phi''(X)\Big[f(X,t)dt + g(X,t)dW(t)\Big]^2 + O(dX^3),
\end{aligned}
$$

but

$$
\begin{aligned}
\Big[f(X,t)dt + g(X,t)dW(t)\Big]^2 &= g^2(x,t)dW(t)^2 \\
+ 2f(X,t)g(X,t)dtdW(t) + O(dt^2) &= g^2(x,t)dt + O(dt^{3/2}),
\end{aligned}
$$

where we have taken into account that $dW(t) = O(dt^{1/2})$ and that $dW(t)^2 = dt$ (in mean square sense). Substituting this into the expression for $d\phi$ and discarding infinitesimals of order higher than 1, we get the *Itô formula*:

$$
\begin{aligned}
d\phi(X) = \Big[f(X,t)\phi'(X) + \frac{1}{2}g^2(X,t)\phi''(X) \Big]dt \\
+ g(X,t)\phi'(X)dW(t), \qquad (8.17)
\end{aligned}
$$

which, because of the term $\phi''(X)$, clearly shows that changing variables does not follow the rules of ordinary calculus unless $\phi(X)$ is a linear function of X for which $\phi''(X) = 0$.

8.2.2 Multidimensional equations

Let $\mathbf{X}(t) = \big(X_1(t), \dots, X_N(t)\big)$ be a N dimensional random process whose time evolution obeys the following set of N SDE's

$$dX_j(t) = f_j(\mathbf{X},t)dt + \sum_{k=1}^{N} g_{jk}(\mathbf{X},t)dW_k(t), \qquad (8.18)$$

$(j = 1, 2, \ldots, N)$, where

$$\mathbf{f}(\mathbf{X}, t) = \Big(f_j(x, t)\Big), \qquad \mathbf{g}(\mathbf{X}, t) = \Big(g_{jk}(x, t)\Big)$$

are the vector drift and the noise intensity matrix respectively. Let us denote by $\mathbf{W}(t) = \big(W_1(t), \ldots, W_N(t)\big)$ the N dimensional Wiener process, with correlation matrix

$$\langle W_j(t_1) W_k(t_2) \rangle = \delta_{jk} \min(t_1, t_2),$$

$(j, k = 1, 2, \ldots, N)$. Proceeding similarly as in Chapter 7 (Sect. 7.1) we can easily show that (cf. Eq. (7.8))

$$dW_j(t) dW_k(t) = \delta_{jk} dt, \qquad dW_j(t)^{2+n} = 0 \ (n > 0), \tag{8.19}$$

(in mean square sense). Note that this means that, as in one dimension, $dW_j(t)$ and, hence, $d\mathbf{W}(t)$ are infinitesimals of order $1/2$.

As an alternative way, the N-dimensional SDE (8.18) can be written as a set on N Langevin equations

$$\frac{dX_j(t)}{dt} = f_j(\mathbf{X}, t) + \sum_{k=1}^{N} g_{jk}(\mathbf{X}, t) \xi_k(t), \tag{8.20}$$

where $\xi_j(t)$ $(j = 1, 2, \ldots, N)$ are Gaussian white noises with correlation matrix

$$\langle \xi_j(t_1) \xi_k(t_2) \rangle = \delta_{jk} \delta(t_1 - t_2).$$

In vector form Eqs. (8.18) and (8.20) are written as

$$d\mathbf{X}(t) = \mathbf{f}(\mathbf{X}, t) dt + \mathbf{g}(\mathbf{X}, t) d\mathbf{W}(t), \tag{8.21}$$

and

$$\frac{d\mathbf{X}(t)}{dt} = \mathbf{f}(\mathbf{X}, t) + \mathbf{g}(\mathbf{X}, t) \boldsymbol{\xi}(t), \tag{8.22}$$

where $\boldsymbol{\xi}(t) = (\xi_1(t), \xi_2(t), \ldots, \xi_N(t))$.

Similarly to the one-dimensional form, the discretized form the SDE (8.18) reads

$$\Delta X_j(t) = f_j(\mathbf{X}, t) \Delta t + \sum_{k=1}^{N} g_{jk}(\mathbf{X}, t) \Delta W_k(t) + O(\Delta t^2), \tag{8.23}$$

where $\Delta X_j(t) = X_j(t + \Delta t) - X(t_j)$ $(j = 1, 2, \ldots, N)$. As in one dimension, the multidimensional process $\mathbf{X}(t)$ is Markovian since, from Eq. (8.23) one clearly sees that $\mathbf{X}(t + \Delta t)$ $(\Delta t > 0)$ solely depends on $\mathbf{X}(t)$ but not on earlier values. Moreover, trajectories are continuous functions of time for

$\Delta X_j(t) \to 0$ as $\Delta t \to 0$ for all $j = 1, 2, \ldots, N$. Consequently, $\mathbf{X}(t)$, solution to the vector SDE (8.21), is a N dimensional diffusion process. Let us see which are the drift and diffusion matrix.

For a N dimensional diffusion process, the components of the vector drift are defined as (cf. Eq. (3.47) of Chapter 3)

$$A_j(\mathbf{x}, t) = \lim_{\Delta t \to 0} \frac{1}{\Delta t} \Big\langle \Delta X_j(t) \big| \mathbf{X}(t) = \mathbf{x} \Big\rangle$$

and substituting for (8.23) we see that the drift coincides with $\mathbf{f}(\mathbf{x}, t)$ as otherwise expected.

As to the diffusion matrix, let us recall that its elements are defined as (cf. Eq. (3.48) of Chapter 3)

$$D_{jk}(\mathbf{x}, t) = \lim_{\Delta t \to 0} \frac{1}{\Delta t} \Big\langle \Delta X_j(t) \Delta X_k(t) \big| \mathbf{X}(t) = \mathbf{x} \Big\rangle.$$

Using (8.23) we have

$$\Delta X_j(t) \Delta X_k(t) = \sum_{j=1}^{N} \sum_{k=1}^{N} g_{jl}(\mathbf{x}, t) g_{km}(\mathbf{x}, t) \Delta W_l(t) \Delta W_m(t) + O(\Delta t^2),$$

but (see Eq. (8.19))

$$\Delta W_l(t) \Delta W_m(t) = \delta_{lm} \Delta t$$

(in mean square sense). Hence

$$\Delta X_j(t) \Delta X_k(t) = \sum_{l=1}^{N} g_{jl}(\mathbf{x}, t) g_{kl}(\mathbf{x}, t) \Delta t + O(\Delta t^2),$$

and

$$D_{jk}(\mathbf{x}, t) = \sum_{l=1}^{N} g_{jl}(\mathbf{x}, t) g_{kl}(\mathbf{x}, t), \tag{8.24}$$

or in matrix form

$$\mathbf{D}(\mathbf{x}, t) = \mathbf{g}(\mathbf{x}, t) \mathbf{g}^{\mathrm{T}}(\mathbf{x}, t), \tag{8.25}$$

where $g^{\mathrm{T}}(\mathbf{x}, t)$ is the transposed matrix.

The Fokker-Planck equation for the transition PDF $p(\mathbf{x}, t | \mathbf{x}_0, t_0)$ is (cf. Eq. (5.1), Chapter 5)

$$\frac{\partial p}{\partial t} = -\sum_{j=1}^{N} \frac{\partial}{\partial x_j} \big[f_j(\mathbf{x}, t) p(\mathbf{x}, t | \mathbf{x}_0, t_0) \big]$$

$$+ \frac{1}{2} \sum_{j,k=1}^{N} \frac{\partial^2}{\partial x_j \partial x_k} \bigg[\sum_{l=1}^{N} g_{jl}(\mathbf{x}, t) g_{kl}(\mathbf{x}, t) p(\mathbf{x}, t | \mathbf{x}_0, t_0) \bigg], \tag{8.26}$$

with the initial condition

$$p(\mathbf{x}, t_0 | \mathbf{x}_0, t_0) = \delta(\mathbf{x} - \mathbf{x}_0). \tag{8.27}$$

8.2.3   Itô formula in several dimensions

Let us finally generalize Itô formula to several dimensions. We want to evaluate the differential, $d\phi(\mathbf{X})$, of an arbitrary function of the solution, $\mathbf{X} = \mathbf{X}(t)$, to the N dimensional SDE (8.21). We proceed as in one dimension and expanding in Taylor series we have

$$d\phi(\mathbf{X}) = \phi(\mathbf{X} + d\mathbf{X}) - \phi(\mathbf{X}) \tag{8.28}$$

$$= \sum_j \frac{\partial\phi}{\partial x_j}dX_j + \frac{1}{2}\sum_{j,k}\frac{\partial^2\phi}{\partial x_j\partial x_j}dX_j dX_k + O(|d\mathbf{X}|^3),$$

but from Eq. (8.18) we get

$$dX_j dX_k = \sum_{l,m}g_{jl}(\mathbf{X},t)g_{km}(\mathbf{X},t)dW_l(t)dW_m(t) + O(dt^{3/2})$$

which, after using $dW_l(t)dW_m(t) = \delta_{lm}dt$ (cf. Eq. (8.19)), yields

$$dX_j dX_k = \sum_l g_{jl}(\mathbf{X},t)g_{kl}(\mathbf{X},t)dt + O(dt^{3/2}). \tag{8.29}$$

Substituting Eqs. (8.18) and (8.29) into Eq. (8.28) we obtain the N *dimensional Itô formula*

$$d\phi(\mathbf{X}(t)) = \left[\sum_j f_j(\mathbf{X},t)\frac{\partial\phi}{\partial x_j} + \sum_{j,k,l}g_{jl}(\mathbf{X},t)g_{kl}(\mathbf{X},t)\frac{\partial^2\phi}{\partial x_j\partial x_k}\right]dt$$

$$+ \sum_{j,l}g_{jl}(\mathbf{X},t)\frac{\partial\phi}{\partial x_j}dW_l(t). \tag{8.30}$$

8.3   Stratonovich stochastic integrals and differential equations

Itô integrals and equations have the peculiarity that do not follow the traditional rules of calculus and this makes their handling more intricate than ordinary equations. Stratonovich (1966) avoided such an inconvenience by introducing an alternative integration procedure which obeys the usual rules of calculus.

Itô interpretation has some advantages from a formal point of view – such as holding the martingale property (Arnold, 1974)– as well as being particularly suited for the study of financial systems, as we will show in the next chapter. On the other hand, the Stratonovich SDE is more appropriate for the representation of many processes of physics and other natural sciences. This is due to the fact that white noise is only an idealization which does not exist in nature where real noises are approximated to white noise under certain circumstances. We will see in Sect. 8.5 that, if we consider such a limiting procedure, the appropriate integral is that of Stratonovich.

8.3.1 The Stratonovich integral

Recall that in Sect. 7.2 we have studied a particular example of stochastic integral,

$$\int_{t_0}^t W(t')dW(t'),$$

which is defined as a stochastic limit of Riemannian sums:

$$\int_{t_0}^t W(t')dW(t') = \lim_{n\to\infty} \sum_{k=1}^n W(\tau_k)\big[W(t_k) - W(t_{k-1})\big],$$

$(t_0 \le t_1 \le \cdots \le t_n = t, \tau_k \in [t_{k-1}, t_k])$. We have shown that this limit depended on the choice of the intermediate points τ_k.

Itô's choice,

$$\tau_k = t_{k-1},$$

has, as remarked above, some advantages because in this case the increments of the Wiener process $\Delta W(t_k) = W(t_k) - W(t_{k-1})$ "point to the future". However, the resulting limiting procedure leads to a value which does not conform with the rules of ordinary calculus:

$$\int_{t_0}^t W(t')dW(t') = \frac{1}{2}[W(t)^2 - W(t_0)^2] - \frac{1}{2}(t - t_0).$$

This inconvenience (but, unfortunately, also some advantages of Itô's choice) is removed by Stratonovich's choice in which the intermediate points of the partitioning are taken to be at the center of each infinitesimal interval, that is,

$$\tau_k = \frac{1}{2}(t_k + t_{k+1}).$$

With this choice the result of the integral agrees with ordinary calculus:

$$\text{(S)} \int_{t_0}^t W(t')dW(t') = \lim_{n\to\infty} \sum_{k=1}^n \frac{W(t_{k-1}) + W(t_k)}{2}[W(t_k) - W(t_{k-1})]$$

$$= \frac{1}{2}\lim_{n\to\infty}\sum_{k=1}^n [W(t_k)^2 - W(t_{k-1})^2] = \frac{1}{2}[W(t)^2 - W(t_0)^2].$$

Let us now turn to more general forms of stochastic integrals and define

$$\text{(S)} \int_{t_0}^t G\big(W(t'), t'\big)dW(t') \tag{8.31}$$

$$= \lim_{n\to\infty} \sum_{k=1}^n G\left(\frac{W(t_{k-1}) + W(t_k)}{2}, \frac{t_k + t_{k-1}}{2}\right)[W(t_k) - W(t_{k-1})],$$

where $G(x,t)$ is a continuous function of t with finite partial derivative with respect to x and such that the second moment $\langle G[t,W(t)]^2 \rangle$ is finite.

The Itô representation of this kind of integral is given by (see also Eq. (7.35))

$$\int_{t_0}^{t} G(W(t'),t') dW(t') = \lim_{n\to\infty} \sum_{k=1}^{n} G(W(t_{k-1}),t_{k-1}) [W(t_k) - W(t_{k-1})].$$
(8.32)

The relationship with Stratonovich's integral (8.31) is given by the following result:[6]

$$(S) \int_{t_0}^{t} G(W(t'),t') dW(t') = \int_{t_0}^{t} G(W(t'),t') dW(t')$$

$$+ \frac{1}{2} \int_{t_0}^{t} \frac{\partial G(W(t'),t')}{\partial W(t')} dt'. \quad (8.33)$$

Proof. In effect, calling

$$\Delta t_k = t_k - t_{k-1}, \qquad \Delta W_k = W(t_k) - W(t_{k-1}) \qquad (8.34)$$

we can write

$$\frac{t_k + t_{k-1}}{2} = t_{k-1} + \frac{1}{2} \Delta t_k, \qquad \frac{W(t_{k-1}) + W(t_k)}{2} = W_{k-1} + \frac{1}{2} \Delta W_k, \quad (8.35)$$

where $W_{k-1} \equiv W(t_{k-1})$.

With this shorthand notation we write Eq. (8.31) as

$$(S) \int_{t_0}^{t} G(W(t'),t') dW(t') = \lim_{n\to\infty} \sum_{k=1}^{n} G\left(W_{k-1} + \frac{\Delta W_k}{2}, t_{k-1} + \frac{\Delta t_k}{2}\right) \Delta W_k$$

and, since $\Delta W_k = O(\Delta t_k^{1/2})$, a Taylor series expansion of G yields

$$G\left(W_{k-1} + \frac{\Delta W_k}{2}, t_{k-1} + \frac{\Delta t_k}{2}\right)$$

$$= G(W_k, t_{k-1}) + \frac{1}{2} \frac{\partial G(W_{k-1}, t_{k-1})}{\partial W_{k-1}} \Delta W_k + O(\Delta t_k).$$

Thus,

$$(S) \int_{t_0}^{t} G(W(t'),t') dW(t') = \lim_{n\to\infty} \sum_{k=1}^{n} \Big[G(W_{k-1}, t_{k-1}) \Delta W_k$$

$$+ \frac{1}{2} \frac{\partial G(W_{k-1}), t_{k-1}}{\partial W_{k-1}} (\Delta W_k)^2 + O(\Delta t_k^{3/2}) \Big],$$

---

[6]For the rest of this chapter, Stratonovich integrals are indicated by the symbol (S) in front of them. Itô integrals bear no symbol.

but $(\Delta W_k)^2 = \Delta t_k + O(\Delta t_k^2)$ and $\Delta t_k \to 0$ as $n \to \infty$; hence

$$(\text{S}) \int_{t_0}^{t} G(W(t'), t') dW(t') = \lim_{n \to \infty} \left[\sum_{k=1}^{n} G(W_{k-1}, t_{k-1}) \Delta W_k \right.$$

$$\left. + \frac{1}{2} \sum_{k=1}^{n} \frac{\partial G(W_{k-1}), t_{k-1}}{\partial W_{k-1}} \Delta t_k \right].$$

In the first sum on the right we recognize Itô integral (8.32) while the second sum corresponds to an ordinary Riemann integral. Therefore,

$$(\text{S}) \int_{t_0}^{t} G(W(t'), t') dW(t') = \int_{t_0}^{t} G(W(t'), t') dW(t').$$

$$+ \frac{1}{2} \int_{t_0}^{t} \frac{\partial G(W(t'), t')}{\partial W(t')} dt'$$

which is Eq. (8.33). $\qquad\qquad\qquad\qquad\qquad\qquad\qquad\qquad\qquad\qquad\Box$

Let us now generalize the definition of the Stratonovich integral of $G(W(t), t)$ given in Eq. (8.31) to include functions of the form $G(X(t), t)$, where $X(t)$ is the solution of a stochastic differential equation.

We have seen that the generic one-dimensional SDE

$$dX(t) = f(X, t) + g(X, t) dW(t), \tag{8.36}$$

with $X(t_0) = X_0$, can be written as an integral equation

$$X(t) = X_0 + \int_{t_0}^{t} f(X(t'), t') dt' + \int_{t_0}^{t} g(X(t'), t') dW(t'), \tag{8.37}$$

in which the second integral has been interpreted in the sense of Itô. We have also seen that Itô integrals have an equivalent representation in the sense of Stratonovich. Then the Itô SDE (8.36) will have an equivalent Stratonovich SDE. However, the Itô integral appearing in Eq. (8.37) does not coincide with the stochastic integral defined in Eq. (8.32), because the integrand in Eq. (8.37) is a *functional* of the Wiener process (through $X(t)$ solution to (8.36)) while the integrand in Eq. (8.31) is an ordinary function (not a functional) of $W(t)$. Hence, in order to build a theory of SDE's on the basis of Stratonovich's idea we must generalize the above definitions.

Definition. Let $g(X, t)$ be a differentiable function where $X = X(t)$ is a solution of the SDE (8.36), then the Stratonovich integral of g with respect to the Wiener process is

$$\text{(S)} \int_{t_0}^{t} g\big(X(t'),t'\big)dW(t') \tag{8.38}$$

$$= \lim_{n\to\infty} \sum_{k=1}^{n} g\left(\frac{X(t_{k-1})+X(t_k)}{2},t_{k-1}\right)\big[W(t_k)-W(t_{k-1})\big].$$

Let us observe that, as in Eq. (8.31), on the right hand side of (8.38) we could also have written

$$g\left(\frac{X(t_{k-1})+X(t_k)}{2},\frac{t_{k-1}+t_k}{2}\right),$$

however, this essentially alters nothing.[7]

Using the short-hand notation defined in Eqs. (8.34)-(8.35) we can write Eq. (8.38) as

$$\text{(S)} \int_{t_0}^{t} g\big(X(t'),t'\big)dW(t') = \lim_{n\to\infty} \sum_{k=1}^{n} g\left(X_{k-1}+\frac{\Delta X_k}{2},t_{k-1}\right)\Delta W_k. \tag{8.39}$$

Note that Itô integral can be likewise defined as

$$\int_{t_0}^{t} g\big(X(t'),t'\big)dW(t') = \lim_{n\to\infty} \sum_{k=1}^{n} g\big(X_{k-1},t_{k-1}\big)\Delta W_k. \tag{8.40}$$

8.3.2   *Relation with Itô integral*

The connection between these two definitions of the stochastic integral of $g(t,X(t))$ is similar to that of the integral of $G(t,W(t))$ given in Eq. (8.33). In the present case we have

$$\text{(S)} \int_{t_0}^{t} g\big(X(t'),t'\big)dW(t') = \int_{t_0}^{t} g\big(X(t'),t'\big)dW(t') \tag{8.41}$$

$$+ \frac{1}{2}\int_{t_0}^{t} g\big(X(t'),t'\big)\frac{\partial g\big(X(t'),t'\big)}{\partial X(t')}dt'.$$

---

[7]One can easily see that the difference between both expressions for g is of order Δt_k and since $\Delta W(t_k) = O(\Delta t_k^{1/2})$, this difference contributes to the integral with negligible terms of order $\Delta t_k^{3/2}$.

Proof. We proceed as before and expand g on the right hand side of Eq. (8.39) up to first order in ΔX_k.[8] We have

$$g\left(X_{k-1} + \frac{\Delta X_k}{2}, t_{k-1}\right) = g(X_{k-1}, t_{k-1})$$

$$+ \frac{1}{2}\frac{\partial g(X_{k-1}, t_{k-1})}{\partial X_{k-1}}\Delta X_k + O(\Delta X_k^2).$$

On the other hand, from Eq. (8.37) we see

$$X(t_k) = X(t_{k-1}) + \int_{t_{k-1}}^{t_k} f(X(t'), t')dt' + \int_{t_{k-1}}^{t_k} g(X(t'), t')dW(t')$$

and as $t_{k-1} \to t_k$ (i.e., as $\Delta t_k \to 0$) we get (recall $\Delta W_k = O(\Delta t_k^{1/2})$)

$$\Delta X_k = f(X_{k-1}, t_{k-1})\Delta t_k + g(X_{k-1}, t_{k-1})\Delta W_k + O(\Delta t_k^{3/2}). \qquad (8.42)$$

Hence

$$g\left(X_{k-1} + \frac{\Delta X_k}{2}, t_{k-1}\right) = g(X_{k-1}, t_{k-1})$$

$$+ \frac{1}{2}\frac{\partial g(X_{k-1}, t_{k-1})}{\partial X_{k-1}}\Big[f(X_{k-1}, t_{k-1})\Delta t_k$$

$$+ g(X_{k-1}, t_{k-1})\Delta W_k\Big] + O(\Delta t_k^{3/2}),$$

thus

$$g\left(X_{k-1} + \frac{\Delta X_k}{2}, t_{k-1}\right)\Delta W_k = g(X_{k-1}, t_{k-1})\Delta W_k$$

$$+ \frac{1}{2}\frac{\partial g(X_{k-1}, t_{k-1})}{\partial X_{k-1}}g(X_{k-1}, t_{k-1})(\Delta W_k)^2 + O(\Delta t_k^{3/2}).$$

Recalling that $(\Delta W_k)^2 = \Delta t_k$ (in mean square sense) and substituting the resulting expression into Eq. (8.39) we obtain

$$(\text{S})\int_{t_0}^t g(X(t'), t')dW(t') = \lim_{n\to\infty}\sum_{k=1}^n g(X_{k-1}, t_{k-1})\Delta W_k \qquad (8.43)$$

$$+ \frac{1}{2}\left[\lim_{n\to\infty}\sum_{k=1}^n \frac{\partial g(X_{k-1}, t_{k-1})}{\partial X_{k-1}}g(X_{k-1}, t_{k-1})\right].\Delta t_k$$

However,

$$\lim_{n\to\infty}\sum_{k=1}^n g(X_{k-1}, t_{k-1})\Delta W_k = \int_{t_0}^t g(X(t'), t')dW(t')$$

---

[8] Bear in mind that in Eq. (8.39) Δt_k and ΔX_k both vanish as $n \to \infty$.

is the Itô integral of g with respect to $W(t)$ (cf. Eq. (8.40)) and

$$\lim_{n\to\infty} \sum_{k=1}^{n} \frac{\partial g(X_{k-1}, t_{k-1})}{\partial X_{k-1}} g(X_{k-1}, t_{k-1})\Delta t_k = \int_{t_0}^{t} \frac{\partial g(X(t'), t')}{\partial X(t')} g(X(t'), t')dt'$$

is an ordinary Riemann-Stilejes integral. Substituting this two equations into (8.43) proves Eq. (8.41).

□

8.3.3 Stratonovich stochastic differential equations

We know that (Itô) SDE (8.36) is equivalent to the integral form given in Eq. (8.37). This equivalence leads us to define Stratonovich SDE as

$$X(t) = X_0 + \int_{t_0}^{t} f(X(t'), t')dt' + (S)\int_{t_0}^{t} g(X(t'), t')dW(t'), \qquad (8.44)$$

which we write symbolically by

$$(S) \quad dX(t) = f(X, t) + g(X, t)dW(t), \qquad (8.45)$$

where the (S) means that the equation has to be integrated in the sense of Stratonovich.

The connection between Itô and Stratonovich SDE's is obtained through Eq. (8.41):

$$(S)\int_{t_0}^{t} g(X(t'), t')dW(t') = \int_{t_0}^{t} g(X(t'), t')dW(t')$$
$$+ \frac{1}{2}\int_{t_0}^{t} g(X(t'), t')\frac{\partial g(X(t'), t')}{\partial X(t')}dt',$$

by the replacement

$$(S)\, g(X, t)dW(t) \longrightarrow g(X, t)dW(t) + \frac{1}{2}g(X, t)\frac{\partial g(X, t)}{\partial X(t)}dt. \qquad (8.46)$$

Therefore, the Itô equation

$$dX(t) = \left[f(X, t) + \frac{1}{2}g(X, t)\frac{\partial g(X, t)}{\partial X(t)} \right]dt + g(X, t)dW(t) \qquad (8.47)$$

corresponds to the Stratonovich SDE (8.45),

$$(S) \quad dX(t) = f(X, t) + g(X, t)dW(t),$$

in the sense of coincidence of solutions. Conversely, the Stratonovich equation

$$(S) \quad dX(t) = \left[f(X, t) - \frac{1}{2}g(X, t)\frac{\partial g(X, t)}{\partial X(t)} \right]dt + g(X, t)dW(t) \qquad (8.48)$$

corresponds to the Itô equation

$$dX(t) = f(X,t) + g(X,t)dW(t).$$

Let us remark that for additive equations in which $g(X,t) = g(t)$ is a nonrandom function of time (or a constant), Itô and Stratonovich SDE's and integrals coincide. However, in the general multiplicative case, one obtains two different Markov processes as solutions that *differ in the drift but not in the fluctuating term* which, in particular, implies that both Itô and Stratonovich solutions are non differentiable.

8.3.4 *Changing variables*

The Stratonovich convention has the advantage of following the rules of ordinary calculus and this is very useful, for instance, if we have to change variables in the SDE in order to obtain a simpler and more manageable form of the differential equation and one does not have to resort to the awkward Itô formula. We have the following result.

Theorem.

The rule for the change of variables in Stratonovich differential equations is the same as in ordinary calculus. That is, if

$$(S) \quad dX = f(X,t)dt + g(X,t)dW(t), \tag{8.49}$$

then

$$(S) \quad d\phi(X) = \left[f(X,t)dt + g(X,t)dW(t)\right]\phi'(X). \tag{8.50}$$

Proof. Let us start from the Stratonovich SDE (8.49) and convert it to Itô (cf. Eq. (8.47))

$$dX = \left[f(X,t) + \frac{1}{2}g(X,t)\frac{\partial g(X,t)}{\partial X}\right]dt + g(X,t)dW(t). \tag{8.51}$$

We now make the change of variable, $X \to Z$, given by $\phi(X)$ with inverse, $Z \to X$, given by $\psi(Z)$; that is

$$Z = \phi(X) \quad \text{and} \quad X = \psi(Z). \tag{8.52}$$

It follows

$$\phi'(X) = [\psi'(Z)]^{-1} \quad \text{and} \quad \phi''(X) = -\psi''(Z)[\psi'(Z)]^{-3}. \tag{8.53}$$

Now Itô formula (8.17) for Itô SDE (8.51) reads

$$d\phi(X) = \left[f(X,t)\phi'(X) + \frac{1}{2}g(X,t)\frac{\partial g(X,t)}{\partial X} \right.$$
$$\left. + \frac{1}{2}g^2(X,t)\phi''(X) \right] dt + g(X,t)\phi'(X)dW(t),$$

that is to say,

$$dZ = \left[f(Z,t)[\psi'(Z)]^{-1} + \frac{1}{2}g(Z,t)\frac{g(Z,t)}{\partial Z}[\psi'(Z)]^{-2} \right. \tag{8.54}$$
$$\left. - \frac{1}{2}g^2(Z,t)\frac{g(Z,t)}{\partial Z}[\psi'(Z)]^{-3}\psi''(Z) \right] dt + g(Z,t)[\psi'(Z)]^{-1}dW,$$

where $f(Z,t) \equiv f(X(Z),t)$ and similarly for $g(Z,t)$. We have also taken into account Eq. (8.53) and

$$\frac{\partial g(X,t)}{\partial X} = \frac{\partial g(t,Z)}{\partial Z}\frac{dZ}{dX} = \frac{\partial g(t,Z)}{\partial Z}[\psi'(Z)]^{-1}.$$

Equation (8.54) is the Itô SDE for $Z(t)$, with drift

$$\tilde{f}(Z,t) \equiv f(Z,t)[\psi'(Z)]^{-1} + \frac{1}{2}g(Z,t)\frac{g(Z,t)}{\partial Z}[\psi'(Z)]^{-2}$$
$$- \frac{1}{2}g^2(Z,t)\frac{g(Z,t)}{\partial Z}[\psi'(Z)]^{-3}\psi''(Z)$$

and noise intensity

$$\tilde{g}(Z,t) \equiv g(Z,t)[\psi'(Z)]^{-1}.$$

The corresponding Stratonovich equation, Eq. (8.48), will be given by

$$(S) \quad Z = \left[\tilde{f}(t,Z) - \frac{1}{2}\tilde{g}(t,Z)\frac{\partial \tilde{g}(t,Z)}{\partial Z} \right] dt + \tilde{g}(t,Z)dW(t).$$

However,

$$\tilde{f}(t,Z) - \frac{1}{2}\tilde{g}(t,Z)\frac{\partial \tilde{g}(t,Z)}{\partial Z}$$
$$= f(Z,t)[\psi'(Z)]^{-1} + \frac{1}{2}g(Z,t)\frac{g(Z,t)}{\partial Z}[\psi'(Z)]^{-2}$$
$$- \frac{1}{2}g^2(Z,t)\frac{g(Z,t)}{\partial Z}[\psi'(Z)]^{-3}\psi''(Z)$$
$$- \frac{1}{2}g(Z,t)[\psi'(Z)]^{-1}\frac{\partial}{\partial Z}\left[g(Z,t)[\psi'(Z)]^{-1} \right]$$
$$= f(Z,t)[\psi'(Z)]^{-1}$$

and the Stratonovich SDE for $Z(t)$ reads

$$\text{(S)} \quad dZ = \left[f(Z,t)dt + g(Z,t)dW(t) \right] [\psi'(Z)]^{-1}, \qquad (8.55)$$

or in terms of the original process $X(t)$

$$\text{(S)} \quad d\phi(X) = \left[f(X,t)dt + g(X,t)dW(t) \right] \phi'(X)$$

which is Eq. (8.50).

$\square$

8.3.5 The Fokker-Planck equation

We know that to Itô SDE

$$dX = f(X,t)dt + g(X,t)dW(t),$$

corresponds the FPE (cf. Eq. (8.14))

$$\frac{\partial p}{\partial t} = -\frac{\partial}{\partial x}[f(x,t)p] + \frac{1}{2}\frac{\partial^2}{\partial x^2}[g^2(x,t)p].$$

On the other hand the Stratonovich SDE

$$\text{(S)} \quad dX = f(X,t)dt + g(X,t)dW(t), \qquad X(t_0) = x_0$$

is equivalent to Itô SDE

$$dX = \left[f(X,t) + \frac{1}{2}g(X,t)\frac{\partial g(X,t)}{\partial X} \right] dt + g(X,t)dW(t).$$

Therefore, the FPE according to Stratonovich rule will be given by

$$\frac{\partial p}{\partial t} = -\frac{\partial}{\partial x}\left\{ \left[f(x,t) + \frac{1}{2}g(x,t)\frac{\partial g(x,t)}{\partial x} \right] p \right\}$$

$$+ \frac{1}{2}\frac{\partial^2}{\partial x^2}[g^2(x,t)p]. \qquad (8.56)$$

Realizing that

$$\frac{\partial}{\partial x}\left\{ g(x,t)\frac{\partial}{\partial x}[g(x,t)p] \right\} = \frac{\partial^2}{\partial x^2}[g^2(x,t)p] - \frac{\partial}{\partial x}\left[g(x,t)\frac{\partial g(x,t)}{\partial x}p \right],$$

that is

$$\frac{\partial^2}{\partial x^2}[g^2(x,t)p] = \frac{\partial}{\partial x}\left[g(x,t)\frac{\partial g(x,t)}{\partial x}p \right] + \frac{\partial}{\partial x}\left\{ g(x,t)\frac{\partial}{\partial x}[g(x,t)p] \right\},$$

we see that Eq. (8.56) can be written in the following more compact form

$$\frac{\partial p}{\partial t} = -\frac{\partial}{\partial x}[f(x,t)p] + \frac{1}{2}\frac{\partial}{\partial x}\left[g(x,t)\frac{\partial}{\partial x}[g(x,t)p] \right]. \qquad (8.57)$$

which is called the *Stratonovich form* of the Fokker-Planck equation and is standard in the physics literature (van Kampen, 1992).

The extra term in Eq. (8.56), with respect to the Itô form of the FPE,

$$\frac{1}{2}g(x,t)\frac{\partial g(x,t)}{\partial x}$$

is sometimes called *spurious drift*. Let us finally note that the initial condition is obviously the same for both forms; that is

$$p(x,t_0|x_0,t_0) = \delta(x - x_0).$$

8.3.6 Multidimensional equations

The above results can be easily generalized to N dimensional processes $\mathbf{X}(t) = (X_1(t), \ldots, X_N(t))$. Then the system SDE's

$$\text{(S)} \quad dX_j = f_j(\mathbf{X},t)dt + \sum_{k=1}^{N} g_{jk}(\mathbf{X},t)dW_k(t), \tag{8.58}$$

$(j = 1, 2, \ldots, N)$, or equivalently

$$X_j(t) = X_j(t_0) + \int_{t_0}^{t} f_j(t, \mathbf{X}(t'))dt' + \text{(S)} \sum_{k=1}^{N} \int_{t_0}^{t} g_{jk}(\mathbf{X}(t'),t')dW_k(t')dt',$$

has the Stratonovich form if the sum on the right hand side is defined as the sum of Stratonovich integrals (cf. Eq. (8.38))

$$\text{(S)} \int_{t_0}^{t} g_{jk}(\mathbf{X}(t'),t')dW_k(t')dt' \tag{8.59}$$

$$= \lim_{n\to\infty} \sum_{l=1}^{n} g_{jk}\left(\frac{\mathbf{X}(t_{l-1}) + \mathbf{X}(t_l)}{2}, t_{l-1}\right) [W_k(t_l) - W_k(t_{l-1})],$$

where $t_0 \le t_1 \le \cdots \le t_n = t$ is a partition of the interval $[t_0, t]$.

Likewise for the Itô form

$$dX_j = f_j(\mathbf{X},t)dt + \sum_{k=1}^{N} g_{jk}(\mathbf{X},t)dW_k(t),$$

$(j = 1, 2, \ldots, N)$, we have

$$\int_{t_0}^{t} g_{jk}(\mathbf{X}(t'),t')dW_k(t')dt' = \lim_{n\to\infty} \sum_{l=1}^{n} g_{jk}(\mathbf{X}(t_{l-1}),t_{l-1}) [W_k(t_l) - W_k(t_{l-1})].$$

Following an identical reasoning as in one dimension we can see that the connection between Stratonovich and Itô forms is given by (cf. Eq. (8.41))

$$(S) \int_{t_0}^{t} g_{jk}\big(\mathbf{X}(t'),t'\big)dW_k(t') = \int_{t_0}^{t} g_{jk}\big(\mathbf{X}(t'),t'\big)dW_k(t') \qquad (8.60)$$

$$+ \frac{1}{2}\sum_{i=1}^{N} \int_{t_0}^{t} g_{ik}\big(\mathbf{X}(t'),t'\big)\frac{\partial g_{jk}\big(\mathbf{X}(t'),t'\big)}{\partial X_i}dt'.$$

That is,

$$(S) \sum_{k=1}^{N} g_{jk}\big(\mathbf{X},t\big)dW(t) \longrightarrow \sum_{k=1}^{N} g_{jk}\big(\mathbf{X},t\big)dW(t)$$

$$+ \frac{1}{2}\sum_{i,k=1}^{N} g_{ik}\big(\mathbf{X},t\big)\frac{\partial g_{jk}\big(\mathbf{X},t\big)}{\partial X_i}dt.$$

Therefore, the Stratonovich SDE

$$(S) \quad dX_j = f_j\big(\mathbf{X},t\big)dt + \sum_{k=1}^{N} g_{jk}\big(\mathbf{X},t\big)dW_k(t),$$

is equivalent to Itô SDE

$$dX_j = \left[f_j\big(\mathbf{X},t\big) + \frac{1}{2}\sum_{i,k=1}^{N} g_{ik}\big(\mathbf{X},t\big)\frac{\partial g_{jk}\big(\mathbf{X},t\big)}{\partial X_i} \right]dt$$

$$+ \sum_{k=1}^{N} g_{jk}\big(\mathbf{X},t\big)dW_k(t).$$

Conversely, the Itô SDE

$$dX_j = f_j\big(\mathbf{X},t\big)dt + \sum_{k=1}^{N} g_{jk}\big(\mathbf{X},t\big)dW_k(t),$$

is equivalent to the Stratonovich SDE

$$(S) \quad dX_j = \left[f_j\big(\mathbf{X},t\big) - \frac{1}{2}\sum_{i,k=1}^{N} g_{ik}\big(\mathbf{X},t\big)\frac{\partial g_{jk}\big(\mathbf{X},t\big)}{\partial X_i} \right]dt$$

$$+ \sum_{k=1}^{N} g_{jk}\big(\mathbf{X},t\big)dW_k(t).$$

Finally, the Fokker-Planck equation corresponding to the multidimensional Stratonovich SDE is

$$\frac{\partial p}{\partial t} = - \sum_{i=1}^{N} \frac{\partial}{\partial x_i} \left[f_i(\mathbf{X}, t) p \right]$$

$$+ \frac{1}{2} \sum_{i,j,k=1}^{N} \frac{\partial}{\partial x_i} \left[g_{ik}(\mathbf{X}, t) \frac{\partial}{\partial x_j} \left[g_{jk}(\mathbf{X}, t) p \right] \right], \qquad (8.61)$$

as follows from an identical reasoning than that of the one dimensional case.

8.4 Examples

We present some examples of stochastic differential equations that are found in many practical applications. We begin with the historical example of the Brownian motion that results in a second-order stochastic differential equation for the position of the Brownian particle which is equivalent to a bidimensional equation for position and velocity. In the case of no external field driving the motion, the velocity of the particle follows a first-order linear equation whose generalization is addressed in the subsequent examples. The second example is an additive SDE with linear drift that has, as a particular case, the Ornstein-Uhlenbeck process, surely the most widely used SDE since it represents countless numbers of phenomena. A final and simple example is a multiplicative linear noise process that exemplifies in an easy manner the differences between additive and multiplicative processes.

8.4.1 *Brownian motion*

Albert Einstein in 1905 and, independently, Marian von Smoluchowski in 1906 gave the explanation for the Brownian motion, the random motion of little macroscopic particles moving in fluids. Both assumed that the motion of the Brownian particle was a diffusion process and derived a Fokker-Planck equation for the transition probability. Shortly after, Paul Langevin in 1908 gave a complementary explanation based on Newton laws of dynamics which resulted in the first stochastic differential equation also known as the Langevin equation.

In the original Langevin's approach the Brownian particle is assumed to be unboundedly moving inside a fluid without any external field and only subjected to two forces: (i) a viscous drag proportional to the velocity, and (ii) a fluctuating force due to the continual collisions of the molecules of the fluid on the particle and that Langevin described as "extremely irregular" (usually this fluctuating force is supposed to be Gaussian white noise at first approximation). Assuming, for simplicity, one-dimensional motion and denoting by $X = X(t)$ the position of the particle at time t, Newton second law directly leads to the Langevin equation

$$m\ddot{X} + \gamma\dot{X} = \kappa\xi(t), \tag{8.62}$$

where m is the mass of the particle, $\gamma > 0$ represents damping, $\kappa > 0$ measures the strength of the random force which is assumed to be zero-mean Gaussian white noise, and dots represent time derivatives.[9]

Equation (8.62) is a second-order SDE which reduces to a first-order SDE for the velocity $V = \dot{X}$:

$$\dot{V} + \beta V = k\xi(t), \tag{8.63}$$

where $\beta = \gamma/m$ is the damping constant and $k = \kappa/m$ is the noise intensity. The Langevin equation (8.63) is a first-order linear equation[10] that can be readily solved . As we see by direct substitution, the solution with initial condition $V(t_0) = V_0$ is

$$V(t) = V_0 e^{-\beta(t-t_0)} + k \int_{t_0}^{t} e^{-\beta(t-t')}\xi(t')dt'.$$

In the stationary state $t_0 \to -\infty$ and this solution reads

$$V(t) = k \int_{-\infty}^{t} e^{-\beta(t-t')}\xi(t')dt', \tag{8.64}$$

and the (stationary) autocorrelation of the volatility is

$$\langle V(t_1)V(t_2) \rangle = k^2 \int_{-\infty}^{t_1} e^{-\beta(t_1-t_1')}dt_1' \int_{-\infty}^{t_2} e^{-\beta(t_2-t_2')} \langle \xi(t_1')\xi(t_2') \rangle dt_2'.$$

Since $\langle \xi(t_1')\xi(t_2') \rangle = \delta(t_1' - t_2')$, we have

$$\int_{-\infty}^{t_2} e^{-\beta(t_2-t_2')} \langle \xi(t_1')\xi(t_2') \rangle dt_2' = \int_{-\infty}^{t_2} e^{-\beta(t_2-t_2')}\delta(t_1' - t_2')dt_2'$$

$$= \int_{-\infty}^{\infty} \theta(t_2 - t_2')e^{-\beta(t_2-t_2')}\delta(t_1' - t_2')dt_2' = \Theta(t_2 - t_1')e^{-\beta(t_2-t_1')},$$

---

[9]Equation. (8.62) represents the simplest case of an unbounded particle. The problem was generalized by Kramers in 1940, who considered a Brownian particle moving in a field of force represented by a potential energy $U(x)$. The Langevin equation in this case reads (Gardiner, 1986; Risken, 1987)

$$m\ddot{X} + \gamma\dot{X} + U'(X) = k\xi(t).$$

We have already encountered this equation in others parts of this book (cf. Chapters 5 and 6).

[10]In fact, it is an Ornstein-Uhlenbeck process, see Sect. 8.4.3 below

where $\Theta(\cdot)$ is the Heaviside step function. Hence

$$\langle V(t_1)V(t_2) \rangle = k^2 e^{-\beta(t_1+t_2)} \int_{-\infty}^{t_1} \Theta(t_2 - t_1')e^{2\beta t_1'} dt_1'.$$

The result of the integral depends on whether $t_1 < t_2$ or $t_1 > t_2$. We can easily see that

$$\int_{-\infty}^{t_1} \Theta(t_2 - t_1')e^{2\beta t_1'} dt_1' = \frac{1}{2\beta} \begin{cases} e^{2\beta t_1}, & \text{if } t_1 < t_2: \\ e^{2\beta t_2}, & \text{if } t_1 > t_2. \end{cases}$$

Therefore, the stationary autocorrelation of the velocity reads

$$\langle V(t_1)V(t_2) \rangle = \frac{k^2}{2\beta} e^{-\beta|t_1 - t_2|}, \tag{8.65}$$

and the stationary variance is

$$\langle V^2(t) \rangle = \frac{k^2}{2\beta}. \tag{8.66}$$

We know from statistical mechanics that the average kinetic energy of the Brownian particle is related to the absolute temperature T of the medium through Boltzmann constant k_B by the equipartition law (Risken, 1987)

$$\left\langle \frac{1}{2}mV^2 \right\rangle = \frac{1}{2}k_B T,$$

so that

$$\langle V^2 \rangle = \frac{k_B}{m}T,$$

and substituting for Eq. (8.66) we obtain

$$k^2 = \frac{2\beta k_B T}{m}, \tag{8.67}$$

which relates the noise intensity k with the damping constant β and the absolute temperature and it constitutes one of the simplest form of the fluctuation-dissipation theorem (cf. Chapter 6, Sect. 6.3).

Mean-squared displacement

From an experimental point of view, it is rather difficult to measure the velocity autocorrelation of the Brownian particle. The reason is the extremely irregular evolution of the velocity because, as is apparent from the Langevin equation (8.63) and due to white noise, $V(t)$ is not a differentiable function of time (at least in ordinary sense). It is more accessible the mean squared displacement, $\langle [X(t) - X_0]^2 \rangle$, where X_0 is the position of the particle at the time when the experimenter starts observing the motion. This initial time, because of the stationary character of the Brownian motion, can be taken to be 0 without loss of generality.

Let us evaluate the mean square displacement of the Brownian particle. Since

$$X(t) = X_0 + \int_0^t V(t')dt',$$

we have

$$\langle [X(t) - X_0]^2 \rangle = \int_0^t \int_0^t \langle V(t')V(t'') \rangle dt' dt''.$$

In the stationary state $V(t)$ is given by (8.64) and

$$\langle [X(t) - X_0]^2 \rangle = \frac{k^2}{2\beta} \int_0^t dt' \int_0^t e^{-\beta|t'-t''|} dt'',$$

but

$$\int_0^t dt' \int_0^t e^{-\beta|t'-t''|} dt'' = 2 \int_0^t dt' \left[\int_0^{t'} e^{-\beta(t'-t'')} dt'' + \int_{t'}^t e^{-\beta(t''-t')} dt'' \right]$$

$$= \frac{2t}{\beta} - \frac{2}{\beta^2} \left(1 - e^{-\beta t} \right).$$

Hence,

$$\langle [X(t) - X_0]^2 \rangle = \frac{k^2}{\beta^2} t - \frac{k^2}{\beta^3} \left(1 - e^{-\beta t} \right).$$

If we observe the Brownian particle for a sufficiently long time the mean-squared displacement shows the well known linear growth:

$$\langle [X(t) - X_0]^2 \rangle \sim Dt, \qquad (t \gg \beta^{-1}), \tag{8.68}$$

where $D = k^2/\beta^2$ is the diffusion coefficient. Let us finally note from Eq. (8.67) that

$$D = \frac{2k_B T}{m\beta}, \tag{8.69}$$

which relates the diffusion coefficient (fluctuation) with the damping constant β (dissipation) and the temperature. Eq. (8.69) is the original fluctuation-dissipation theorem, also known as Einstein relation.

8.4.2   *Additive linear equations*

We next consider the following additive SDE with linear drift

$$dX(t) = [a(t)X(t) + b(t)]dt + g(t)dW(t), \qquad (8.70)$$

where $a(t), b(t)$ and $g(t)$ are nonrandom functions. The SDE is, therefore, additive and there is no difference between Itô and Stratonovich conventions and, in particular, the rules of ordinary calculus hold.

One can easily check by direct substitution that the solution to Eq. (8.70) with $X(t_0) = X_0$ is

$$X(t) = \exp\left[\int_{t_0}^{t} a(t')dt'\right] \left\{ X_0 + \int_{t_0}^{t} \exp\left[-\int_{t_0}^{t'} a(t'')dt''\right] \right.$$

$$\left. \times \left[b(t')dt' + g(t')dW(t')\right] \right\}. \qquad (8.71)$$

When X_0 is nonrandom, or random but Gaussian, the solution $X(t)$ is a Gaussian process since (8.71) is a linear superposition of Gaussian variables. The probability density function of $X(t)$ will thus be determined solely by first and second moments.

Bearing in mind that $\langle dW(t)\rangle = 0$ we see from (8.71) that

$$\langle X(t)\rangle = \exp\left[\int_{t_0}^{t} a(t')dt'\right] \left\{ \langle X_0\rangle + \int_{t_0}^{t} \exp\left[-\int_{t_0}^{t'} a(t'')dt''\right] b(t')dt' \right\}.$$
$$(8.72)$$

Note that the average $m(t) \equiv \langle X(t)\rangle$ is the solution to the deterministic equation

$$\dot{m}(t) = a(t)m(t) + b(t), \qquad m(t_0) = \langle X_0\rangle,$$

as can be seen taking the average of SDE (8.70).

Let us now evaluate the correlation function

$$K(t_1, t_2) = \left\langle [X(t_1) - \langle X(t_1)\rangle][X(t_2) - \langle X(t_2)\rangle] \right\rangle. \qquad (8.73)$$

To this end we call

$$\Phi(t) \equiv \exp\left[\int_{t_0}^{t} a(t')dt'\right] \qquad (8.74)$$

and write Eqs. (8.71) as

$$X(t) = \Phi(t)\left\{ X_0 + \int_{t_0}^{t} \frac{1}{\Phi(t')}\left[b(t')dt' + g(t')dW(t')\right] \right\}. \qquad (8.75)$$

Hence

$$\langle X(t)\rangle = \Phi(t)\left[\langle X_0\rangle + \int_{t_0}^{t}\frac{b(t')}{\Phi(t')}dt'\right]. \tag{8.76}$$

Substituting (8.75) and (8.76) into (8.73) yields

$$K(t_1,t_2) = \Phi(t_1)\Phi(t_2)\left[\left\langle\left(X_0 - \langle X_0\rangle\right)^2\right\rangle\right.$$

$$\left. + \int_{t_0}^{t_1}\frac{g(t_1')}{\Phi(t_1')}\int_{t_0}^{t_2}\frac{g(t_2')}{\Phi(t_2')}\langle dW(t_1)dW(t_2)\rangle\right].$$

Recalling that

$$\left\langle dW(t_1')dW(t_2')\right\rangle = \left\langle\xi(t_1')\xi(t_2')\right\rangle dt_1' dt_2' = \delta(t_1 - t_2)dt_1' dt_2',$$

we get

$$\int_{t_0}^{t_1}\frac{g(t_1')}{\Phi(t_1')}dt_1'\int_{t_0}^{t_2}\frac{g(t_2')}{\Phi(t_2')}\delta(t_1' - t_2')dt_1' = \int_{t_0}^{\min(t_1,t_2)}\frac{g^2(t')}{\Phi^2(t')}dt'.$$

Therefore,

$$K(t_1,t_2) = \Phi(t_1)\Phi(t_2)\left[\left\langle\left(X_0 - \langle X_0\rangle\right)^2\right\rangle + \int_{t_0}^{\min(t_1,t_2)}\frac{g^2(t')}{\Phi^2(t')}dt'\right] \tag{8.77}$$

or, in the original notation (cf. Eq. (8.74)),

$$K(t_1,t_2) = \exp\left[\int_{t_0}^{t_1}a(t')dt' + \int_{t_0}^{t_2}a(t')dt'\right]\left\{\left\langle\left(X_0 - \langle X_0\rangle\right)^2\right\rangle\right.$$

$$\left. + \int_{t_0}^{\min(t_1,t_2)}\exp\left[-2\int_{t_0}^{t'}a(t'')dt''\right]g^2(t')dt'\right\}. \tag{8.78}$$

8.4.3 The Ornstein-Uhlenbeck process

One particular case of the linear equation (8.70) is the Ornstein-Uhlenbeck (OU) process which, we recall from Chapter 4 (Sect. 4.5), is a diffusion process with linear drift and constant diffusion coefficient. This immediately leads to the following SDE

$$dX(t) = -\alpha x dt + k dW(t), \tag{8.79}$$

where $\alpha > 0$ and $k > 0$ are constant parameters.[11] Now function $\Phi(t)$ defined in (8.74) takes the simple form

$$\Phi(t) = e^{-\alpha(t-t_0)}$$

---

[11]In many of its countless applications the OU process has the slightly more general form

$$dX(t) = -\alpha(X - b)dt + k dW(t),$$

where $b \in \mathbb{R}$ is constant. However, the spatial translation $X \to X - b$ reduces it to the standard form (8.79) without loss of generality.

and the OU process explicitly is (cf. cf Eq. (8.75))

$$X(t) = e^{-\alpha(t-t_0)} \left[X_0 + k \int_{t_0}^{t} e^{\alpha(t'-t_0)} dW(t') \right],$$

which we write in a more convenient form as

$$X(t) = X_0 e^{-\alpha(t-t_0)} + k \int_{t_0}^{t} e^{-\alpha(t-t')} dW(t'). \qquad (8.80)$$

We have proved in Chapter 4 that the OU process is stationary which means that the transition density of the process, $p(x, t|x_0, t_0)$, goes to a finite and non vanishing value as $t - t_0 \to \infty$.[12] Therefore, letting $t_0 \to -\infty$ (that is, assuming that the process began at the infinite past) in Eq. (8.80) we get the stationary value:

$$X_s(t) = k \int_{-\infty}^{t} e^{-\alpha(t-t')} dW(t'), \qquad (8.81)$$

which obviously does not depend on the initial value X_0.

From Eq. (8.76) we see that the mean value is

$$\langle X(t) \rangle = \langle X_0 \rangle e^{-\alpha(t-t_0)}, \qquad (8.82)$$

while from (8.78) we obtain the correlation function

$$K(t_1, t_2) = \text{Var}(X_0) e^{-\alpha(t_1+t_2-2t_0)} + k \int_{t_0}^{\min(t_1, t_2)} e^{-\alpha(t_1+t_2-2t')} dt'.$$

That is,

$$K(t_1, t_2) = \left[\text{Var}(X_0) - \frac{k}{2\alpha} \right] e^{-\alpha(t_1+t_2-2t_0)} + \frac{k}{2\alpha} e^{-\alpha|t_1-t_2|}. \qquad (8.83)$$

Finally, in the stationary state ($t_0 \to -\infty$) we have

$$\langle X_s(t) \rangle = 0, \qquad K_s(t_1 - t_2) = \frac{k}{2\alpha} e^{-\alpha|t_1-t_2|}. \qquad (8.84)$$

---

[12]Note that the drift and the diffusion coefficient are time independent which makes the process invariant under time translations and ultimately leads to the stationary character of the process.

8.4.4 *A multiplicative linear process*

We finally consider the following multiplicative process defined by the SDE:

$$dX(t) = kX(t)dW(t) \qquad (8.85)$$

which constitutes one of the simplest examples about the differences between Itô and Stratonovich rules. Indeed, Eq. (8.85) can be easily solved by the change of variable

$$Z(t) = \ln X(t) \qquad (8.86)$$

and proceeding further depends on the interpretation chosen.

(a) *Itô interpretation*

In this case the change of variable has to be implemented by using Itô formula (8.17). We thus have

$$dZ(t) = \frac{1}{X(t)}dX(t) - \frac{1}{2X^2(t)}dX(t)^2,$$

but from (8.85) we see (recall that $dW(t)^2 = dt$)

$$\frac{dX(t)}{X(t)} = kdW(t), \qquad \frac{dX(t)^2}{X^2(t)} = k^2 dt.$$

Hence

$$dZ(t) = -\frac{1}{2}k^2 dt + kdW(t). \qquad (8.87)$$

Equation that can be readily integrated with the result

$$Z(t) = Z(t_0) - \frac{1}{2}k^2(t - t_0) + k[W(t) - W(t_0)].$$

In the original variable we write

$$X(t) = X(t_0)\exp\left\{ -\frac{1}{2}k^2\Delta t + k\Delta W(t) \right\}, \qquad (8.88)$$

where $\Delta t = t - t_0$ and $\Delta W(t) = W(t) - W(t_0)$.

The average value is

$$\langle X(t) \rangle = e^{-k^2\Delta t/2}\langle X(t_0) \rangle \left\langle e^{k\Delta W(t)} \right\rangle,$$

where we have taken into account that $X(t_0)$ is independent of $\Delta W(t)$. In order to evaluate the average on the right hand side we will use the formula proved in Chapter 1 (cf. Eq. (1.49)) that for any Gaussian variable Z with zero mean the following equality holds

$$\left\langle e^Z \right\rangle = e^{\langle Z^2 \rangle/2}. \qquad (8.89)$$

Since $\Delta W(t)$ is Gaussian with zero mean, we have (see also Eq. (7.5))

$$\left\langle e^{k\Delta W(t)} \right\rangle = e^{k^2 \langle \Delta W(t)^2 \rangle / 2} = e^{k^2 \Delta t / 2}.$$

Hence,

$$\langle X(t) \rangle = \langle X(t_0) \rangle, \tag{8.90}$$

and the average value of the process is constant.

Likewise, and after using Eq. (8.89), we can calculate the correlation function

$$\langle X(t_1)X(t_2) \rangle = e^{-k^2(\Delta t_1 + \Delta t_2)/2} \left\langle X^2(t_0) \right\rangle$$
$$\times \exp\left\{ \frac{k^2}{2} \left\langle [\Delta W(t_1) + \Delta W(t_2)]^2 \right\rangle \right\}. \tag{8.91}$$

However,

$$\left\langle [\Delta W(t_1) + \Delta W(t_2)]^2 \right\rangle = \Delta t_1^2 + \Delta t_2^2 + 2\langle \Delta W(t_1)\Delta W(t_2)\rangle.$$

Since

$$\langle \Delta W(t_1)\Delta W(t_2) \rangle = \langle W(t_1)W(t_2) \rangle - \langle W(t_1)W(t_0) \rangle$$
$$- \langle W(t_2)W(t_0) \rangle + \langle W(t_0)^2 \rangle$$

and (cf. Eq. (4.54))

$$\langle W(t_1)W(t_2) \rangle = \min(t_1, t_2),$$

we write

$$\langle \Delta W(t_1)\Delta W(t_2) \rangle = \min(t_1, t_2) - t_0 = \min(t_1 - t_0, t_2 - t_0).$$

Hence,

$$\left\langle [\Delta W(t_1) + \Delta W(t_2)]^2 \right\rangle = \Delta t_1^2 + \Delta t_2^2 + 2\min(t_1 - t_0, t_2 - t_0).$$

Substituting into (8.91) finally yields

$$\langle X(t_1)X(t_2) \rangle = \left\langle X^2(t_0) \right\rangle \exp\left\{ k^2 \min(t_1 - t_0, t_2 - t_0) \right\}. \tag{8.92}$$

(b) *Stratonovich interpretation*

In this case the solution of the multiplicative equation (8.85) is obtained using the rules of ordinary calculus. Thus the change of variables (8.86) turns Eq. (8.85) into (compare with Eq. (8.87))

$$dZ(t) = k\,dW(t),$$

which is readily integrated yielding

$$X(t) = X(t_0) \exp\left\{ k\Delta W(t) \right\}. \tag{8.93}$$

The mean value and the correlation function can be obtained following the above line of reasoning, that is, by using (8.89). The final results read

$$\langle X(t) \rangle = \langle X(t_0) \rangle e^{k^2 \Delta t/2} \tag{8.94}$$

and

$$\langle X(t_1) X(t_2) \rangle = \langle X^2(t_0) \rangle \exp\left\{ \frac{k^2}{2} \Big[t_1 + t_2 - 2t_0 \right. \tag{8.95}$$

$$\left. + \min(t_1 - t_0, t_2 - t_0) \Big] \right\}$$

and we see a clear difference between Itô and Stratonovich results (Gardiner, 1986).

8.5 Representation of physical systems *

A great number of phenomena in natural sciences, engineering and economics are described by continuous dynamical systems under the influence of random disturbances. Such systems can be in many cases represented by differential equations of the form (see the introductory remarks to this chapter)[13]

$$\frac{dX(t)}{dt} = F\big(t, X(t)\, Z(t)\big). \tag{8.96}$$

Here $X(t)$ is the random process representing the dynamical system and $Z(t)$ is a random disturbance, sometimes called "input noise", with known statistical characteristics.

Suppose $X(t) = x$ is known, the approximate solution of (8.96) for sufficiently small Δt is

$$X(t + \Delta t) \approx x + F\big(t, x\, Z(t)\big)\Delta t.$$

Let us observe that $X(t)$ is a Markovian process only if $X(t + \Delta t)$ is statistically independent on what took place prior to instant t which implies that $Z(t)$ has to be a process with independent values at every point (a

---

[13]For simplicity we restrict ourselves to one-dimensional systems, the generalization to N dimensions is not difficult (Arnold, 1974).

completely uncorrelated process). This is the case when $Z(t)$ is white noise
for which $\langle Z(t) \rangle = 0$ and

$$\langle Z(t_1)Z(t_2) \rangle = \delta(t_1 - t_2).$$

If, in addition, the input noise $Z(t)$ is also Gaussian then $Z(t) = \xi(t)$ is
Gaussian white noise (i.e., the derivative of the Wiener process). Although
this is the most usual assumption, $Z(t)$ does not have to be necessarily
Gaussian, for instance, $Z(t)$ may be Poisson white noise also called "shot
noise". In any case, for completely uncorrelated processes, the sample
trajectories of $Z(t)$ are extremely irregular: they are discontinuous and
unbounded and their sample paths are "point clouds" randomly scattered
on the half plane spanned by $t \geq t_0$ and $X(t) \in \mathbb{R}$. Moreover, and since
$\delta(0) = \infty$, the variance of such processes, $\langle Z^2(t) \rangle = \infty$, is infinite. From a
physical point of view, the variance is related to the average energy of the
random disturbance and, thus, *for white noise the mean energy is infinite.*
This is completely unrealistic and *white noise is an idealization of real input
noises.*

On the other hand, when $Z(t)$ is white noise the solution $X(t)$ of Eq.
(8.96) is Markovian which is very convenient from a mathematical point of
view, since all results on Markov processes are applicable. If, in addition,
$Z(t) = \xi(t)$ is Gaussian then as we have seen in Sect. 7.1 of Chapter 7,
$X(t)$ is a diffusion process and its sample paths are continuous but not
differentiable (see also the discussion in Sect. 8.2). The non smoothness
of trajectories is a consequence of the Markovian character of the process
since the Markov property is equivalent to say that "for a known present,
it is forbidden to transmit information from the past into future" (Arnold,
1974). The irregular behavior of $X(t)$ arises from this.

However, natural processes, which are truly physically realizable, are
always smooth and are at most only approximately Markov (or diffusion)
processes. In such practical cases the input noise in Eq. (8.96) would only
be white in an approximate way. In other words,

$$K(t_1, t_2) = \langle Z(t_1)Z(t_2) \rangle \approx \delta(t_1 - t_2),$$

that is, $Z(t)$ is only an approximate delta-correlated process. This is the
case when, for instance

$$K(t_1, t_2) = \begin{cases} 1/\epsilon, & -\epsilon/2 \leq |t_1 - t_2| \leq \epsilon/2, \\ 0, & \text{otherwise,} \end{cases}$$

since

$$K(t_1, t_2) \longrightarrow \delta(t_1 - t_2), \qquad \text{as} \quad \epsilon \to 0.$$

We therefore conclude that white noise is a convenient mathematical idealization and real input noises are always "colored" in some degree.

In representing real systems, one frequently runs into differential equations of the form

$$\frac{dX(t)}{dt} = f(X,t) + g(X,t)Z(t), \tag{8.97}$$

where the input noise $Z(t)$, which enters linearly in the equation, is the stationary Ornstein-Uhlenbeck process, that is, the solution to the differential equation (cf. Sect. 8.4 and Eq. (8.84))

$$\frac{dZ(t)}{dt} = -\alpha Z + \sigma \xi(t), \tag{8.98}$$

with $\alpha, \sigma > 0$ and $\xi(t)$ Gaussian white noise. In the stationary state ($t_0 \to -\infty$) the correlation function of $Z(t)$ is

$$K(|t_1 - t_2|) = \frac{\sigma^2}{2\alpha} e^{-\alpha|t_1 - t_2|}. \tag{8.99}$$

The magnitude α has units (time)$^{-1}$ and its inverse, $\tau_c = 1/\alpha$ is called *correlation time* of the OU process. Moreover, in the joint limit $\alpha \to \infty$ (i.e., $\tau_c \to 0$) and $\sigma \to \infty$ with $\sigma^2/2\alpha \to D > 0$ (finite), we have (Vladimirov, 1984)

$$\frac{\sigma^2}{2\alpha} e^{-\alpha|t_1 - t_2|} \longrightarrow D\delta(t_1 - t_2)$$

and $Z(t)$ becomes Gaussian white noise.

Due to the continuity of $Z(t)$ (recall that OU noise is a diffusion process), Eq. (8.97) can be regarded as an ordinary differential equation and, consequently, yields a process $X(t)$ which now is differentiable *although not Markovian*. However, as discussed in Sect.3.2 of Chapter 3 for deterministic processes, we can recover the Markov property by enlarging phase space and considering the vector process $(X(t), Z(t))$ whose time evolution is governed by the bidimensional SDE formed by Eqs. (8.97) and (8.98).

One question arises at once, since in the limit of small correlation time the input noise $Z(t)$ becomes white noise $\xi(t)$, does Eq. (8.97) turn into a Stratonovich SDE or to an Itô SDE as $\tau_c \to 0$? The answer is Stratonovich and we have the following result:

Theorem (Wong-Zakai)

The one-dimensional differential equation

$$dX(t) = f(x,t)dt + g(x,t)Z(t)dt,$$

where $Z(t)$ is Gaussian colored noise with correlation time τ_c, becomes, as $\tau_c \to 0$, the Stratonovich SDE:

$$(S) \quad dX(t) = f(x,t)dt + g(x,t)dW(t),$$

or the equivalent Itô equation:

$$dX(t) = \left[f(x,t) + \frac{1}{2}g(x,t)\frac{\partial g(x,t)}{\partial X} \right] dt + g(x,t)dW(t).$$

Proof. This theorem was proved by Wong and Zakai (1965). We here follow van Kampen (1992) for a simple and intuitive verification of the result for autonomous equations. We refer the interested reader to the original work for a more general and formal proof.

We restrict ourselves to autonomous systems and suppose that $X(t)$ obeys an autonomous differential equation of the type (8.97), in which $f(x)$ and $g(x)$ do not depend explicitly of time. We also assume that $Z(t)$ is OU noise and, hence, Gaussian and continuos. We write Eq. (8.97) in the form

$$\frac{1}{g(X)}\frac{dX(t)}{dt} = \frac{f(X)}{g(X)} + Z(t). \tag{8.100}$$

Since $Z(t)$ is continuous, Eq. (8.100) is like an ordinary differential equation where the familiar rules of calculus hold. We thus define a new unknown function Y by the indefinite integral

$$Y = \int^X \frac{dx'}{g(x')}. \tag{8.101}$$

This change turns Eq. (8.100) into

$$\frac{dY(t)}{dt} = h(Y) + Z(t), \tag{8.102}$$

where

$$h(Y) = \frac{f(X(Y))}{g(X(Y))}, \tag{8.103}$$

and $X(Y)$ is obtained by inverting Eq. (8.101) which gives the original unknown X as a function of Y.

Suppose that the correlation time τ_c of the OU noise $Z(t)$ is very small.[14] Then $Z(t)$ approaches to Gaussian white noise $\xi(t)$ and $Y(t)$ will approximately obey the SDE

$$dY(t) = h(Y)dt + \xi(t)dt. \tag{8.104}$$

---

[14]In practice, this means that we are observing the system over time scales much greater than τ_c.

Note this is an additive equation and there are no differences between Itô and Stratonovich interpretations.

Let $q(y, t|y_0)$ and $p(x, t|x_0)$ be the PDF's of $Y(t)$ and $X(t)$ respectively. Bearing in mind that $X \to Y$ is a one-to-one correspondence, we have

$$q(y, t|y_0)dy \equiv \text{Prob}\{y \leq Y(t) < y + dy\}$$
$$= \text{Prob}\{x \leq X(t) < x + dx\} \equiv p(x, t|x_0)dx,$$

and, since (cf. (8.101))

$$\frac{dy}{dx} = \frac{1}{g(x)}, \tag{8.105}$$

we write

$$q(y, t|y_0) = p(x, t|x_0)\frac{dx}{dy} = p(x, t|x_0)g(x). \tag{8.106}$$

The Fokker-Planck equation corresponding to Langevin equation (8.104) is

$$\frac{\partial q}{\partial t} = -\frac{\partial}{\partial y}\left[h(y)q(y, t|y_0)\right] + \frac{1}{2}\frac{\partial^2}{\partial y^2}q(y, t|y_0).$$

We return to the original variable x and since (cf. (8.105))

$$\frac{\partial}{\partial y} = g(x)\frac{\partial}{\partial x}, \qquad \frac{\partial^2}{\partial y^2} = g(x)\frac{\partial}{\partial x}g(x)\frac{\partial}{\partial x},$$

we write

$$\frac{\partial q}{\partial t} = -g(x)\frac{\partial}{\partial x}\left[\frac{f(x)}{g(x)}q\right] + \frac{1}{2}g(x)\frac{\partial}{\partial x}g(x)\frac{\partial q}{\partial x}.$$

Substituting for Eq. (8.106) and some cancelations yield

$$\frac{\partial p}{\partial t} = -\frac{\partial}{\partial x}(f(x)p) + \frac{1}{2}\frac{\partial}{\partial x}\left[g(x)\frac{\partial}{\partial x}(g(x)p)\right],$$

which is the Stratonovich form of the FPE. We have therefore shown that *shifting from Gaussian colored noise to white noise one arrives at the Stratonovich SDE.* □

Chapter 9

Some financial applications

Financial dynamics is one of the fields where stochastic calculus has been widely applied. In this chapter we present a short and elementary introduction to the subject of financial dynamics and review some applications.

9.1 Market dynamics

The fact that financial prices and indices are random magnitudes has been known since long. One of the first mathematical analysis of price changes was given more than a century ago by the French mathematician Louis Bachelier who, studying the option pricing problem, proposed a model for market dynamics which assumed that price changes behave as the ordinary Brownian motion (Bachelier, 1900; reprinted in Cootner, 1964). In the technical parlance of the previous chapter, this means that prices were governed by a SDE with constant drift and constant diffusion coefficient. The model was not satisfactory because with this representation prices can attain negative values. This contradicts one of the most basic tenets of economics, the so-called principle of limited liability, that is to say, prices cannot be negative. This flaw was avoided by Osborne almost six decades later by assuming that prices obey the geometric Brownian motion; that is, prices are the exponential of an ordinary Brownian motion and, hence, they never attain negative values (Osborne, 1959; reprinted in Cootner, 1964).

Let us denote by $S(t)$ a speculative price at time t (it can also represent a financial index). The geometric Brownian motion model assumes that

$$S(t) = S_0 e^{r(t)}, \tag{9.1}$$

where $S_0 = S(t_0)$ is the initial price and $r(t)$, called *return*, is described by an ordinary Brownian motion of the form given by the following SDE:

$$dr(t) = m\,dt + \sigma\,dW(t), \tag{9.2}$$

219

where $W(t)$ is the standard Wiener process. Note that the return $r(t)$ is thus assumed to be a diffusion process with constant drift m and constant diffusion coefficient σ^2.

Let us also observe that there is no need to specify an interpretation (Itô or Stratonovich) for Eq. (9.2) because σ is constant and we are dealing with an additive equation. However, in order to proceed further and obtain, for instance, a differential equation for the price $S(t)$ we need to specify an interpretation. For the rest of this chapter we will use the Itô interpretation because, due to the peculiarities of trading, the Itô convention is more convenient for finance. Indeed, recall that the Stratonovich interpretation requires the knowledge of the price at the middle of the interval $(t, t + dt)$, that is, $S(t + dt/2)$. Since trading takes place at time t this would imply *the knowledge of price before trading took place.* On the contrary, in the Itô interpretation we only need to know S at time t which is conceptually more satisfactory from a financial point of view.

We will next obtain the SDE satisfied by $S(t)$. To this end we can either apply Itô formula [cf. Eq. (8.17) of Chapter 8] or follow a more direct way. We find more illustrating the latter approach and, thus, bearing in mind that $S(t) = S[r(t)]$, by Taylor expanding and taking into account Eq. (9.1), we have

$$dS \equiv S(r + dr) - S(r)$$
$$= \frac{dS}{dr}dr + \frac{1}{2}\frac{d^2S}{dr^2}(dr)^2 + \cdots = S_0 e^r dr + \frac{1}{2}S_0 e^r (dr)^2 + \cdots$$
$$= S(t)dr + \frac{1}{2}S(t)(dr)^2 + \cdots = S(mdt + \sigma dW) + \frac{1}{2}\sigma^2 S(dW)^2 + \cdots$$

However, as discussed in Chapter 7, $(dW)^2 = dt$ (in mean square sense) and we finally see that $S(t)$ obeys the following multiplicative SDE (in Itô sense)

$$\frac{dS(t)}{S(t)} = \left(m + \frac{1}{2}\sigma^2\right)dt + \sigma dW(t). \tag{9.3}$$

Let us note that the solution of the SDE (9.3) is given by Eq. (9.1) with $r(t)$ given by the solution of Eq. (9.2) (which, being additive, its solution is obtained using the ordinary rules of calculus). We thus have[1]

$$S(t) = S_0 \exp\left\{m(t - t_0) + \sigma[W(t) - W(t_0)]\right\}. \tag{9.4}$$

---

[1]We can alternatively obtain Eq. (9.4) integrating (9.3) directly, although, in this case, we are dealing with a multiplicative equation and must apply Itô rules of calculus which makes that the term of the spurious drift, $(1/2)\sigma^2 t$, disappears. The final result is obviously given by Eq. (9.4).

This expression shows that, due to the Gaussian character of the Wiener process, the price $S(t)$ is a log-normal random process and the geometric Brownian motion is also termed as the *log-normal model*.

The log-normal model is widely used in countless number of financial settings and applications. It has, however, several shortcomings. In order to remedy some of these deficiencies, the geometric Brownian motion, given by Eqs. (9.1)-(9.2), can be generalized in several ways. One of these generalizations assumes that the return is a more complex diffusion process obeying a stochastic differential equation of the form (which we interpret in the sense of Itô)

$$dr(t) = f(r,t)dt + g(r,t)dW(t). \qquad (9.5)$$

In this case returns, and hence prices, are driven by an external "force" $f(r,t)$ and by multiplicative noise $g(r,t)$ which, in the most general case, depend explicitly of time. Function f drives prices and function g modulates the intensity of the fluctuations around the deterministic motion set by f. In any case, and regardless the values taken by $r(t)$, prices given by Eq. (9.1) are always nonnegative thus keeping the principle of limited liability.

Another shortcoming of the geometric Brownian motion, perhaps the most important one, is the absence of "fat tails"[2] and skewness in the distribution of prices. Empirical price distributions clearly show fat tails, which means that extreme losses and profits have higher probabilities than that of the log-normal model. Empirical distributions are also biassed in the sense that losses are usually more probable than profits. In order to address these, and other problems, an intense research has aroused both in mathematics and physics (the latter is sometimes called "econophysics") which, among others, may involve the use of the Lévy process as driving noise (instead of the Wiener process), or models in which the variance σ^2 (or the noise intensity g) is a random process as in the so-called "stochastic volatility models". We will treat some of these models here and also in Chapter 12 and refer the interested reader to the literature for more information (Bouchaud and Potters, 2011; Fouque *et al.*, 2000, 2011; Mantegna and Stanley, 2007).

9.2 Valuing the future

One important quantitative procedure in economics and finance is that of "discounting". This process tries to answer a key question: How can we

---

[2]Tails of the probability distribution that are much higher than the ones of the log-normal distribution of the geometric Brownian motion.

value the future? The discounting mechanism weights the future relative to the present and the weighting method is carried through a discount function which usually takes the form of a decreasing exponential. Indeed, under a steady rate of interest r, a dollar inverted today, at time $t = 0$, will yield e^{rt} at time $t > 0$. That is to say, a dollar in any future time t is worth e^{-rt} today. In this simple example r is fixed but in practice rates are uncertain and it is not realistic to represent discounting by a deterministic function of time such as the decreasing exponential with a fixed rate and some kind of average over all possible interest rate paths must be taken. We will next develop these ideas into a consistent framework.

9.2.1 *The process of discounting*

In economics the increment at a given time of the quantity of wealth, exemplified by some magnitude $M = M(t)$, is assumed to depend linearly on the quantity itself and the duration of the variation. For a continuous and instantaneous variation one then writes:

$$dM(t) \propto M(t)dt. \tag{9.6}$$

This is a phenomenological law based on the empirical fact that the bigger $M(t)$, the greater its variation at a given time, but also on the simplifying assumption that the increment is linear in $M(t)$ and not, for instance, quadratic. Let us incidentally note that linearity is equivalent to assuming that the interest rate, i.e., the relative time derivative

$$r = \frac{1}{M(t)} \frac{dM(t)}{dt}, \tag{9.7}$$

is independent of $M(t)$. Note that this definition can be written as

$$r = \frac{d\ln M}{dt}, \tag{9.8}$$

so that the rate is the derivative of the logarithm of wealth.

In the simplest situation the growth law (9.6) represents a completely linear law with direct proportionality in which r is constant:

$$dM(t) = rM(t)dt, \tag{9.9}$$

where r is the rate and is measured in units of $1/(\text{time})$. Now the growth law is readily integrated, giving

$$M(t) = e^{r(t-t_0)} M(t_0), \tag{9.10}$$

which yields an exponential growth connecting wealth at some initial time t_0, that is to say, the present time (which, in our case and without loss of generality, can be taken equal to zero) and wealth at any future time t.

Before proceeding further we recall that the growth law (9.6), often in the simplest version (9.9), appears in numerous branches of physical and social sciences. Thus, for example, in radioactivity if $N(t)$ is the number of active nuclei at time t, the usual hypothesis is that this number decreases as

$$dN(t) = -\lambda N(t)dt,$$

where $\lambda > 0$ is the decay constant. Similar considerations apply to other situations, as they are found in chemical reactions, population dynamics, as in many other places.

In economics, discounting refers to the process of connecting wealth at different times. Specifically the discount function, which we denote by $\delta(t)$, is defined by

$$\delta(t) \equiv \frac{M(0)}{M(t)}, \tag{9.11}$$

so that $M(0) = \delta(t)M(t)$ in accordance with the fact that discounting specifically refers to weighting the future (at some time $t > 0$) relative to the present (at $t = 0$).

In the simplest case of Eq. (9.10) the discount function is given by the decreasing exponential

$$\delta(t) = e^{-rt}, \tag{9.12}$$

where $r > 0$ is the interest rate. However, as we have mentioned above, this simple form of discount, in which the interest rate is always constant, is unrealistic. A first generalization consists in assuming rates to be deterministic functions of time $r(t)$. In such a case the growth law (9.9) is replaced by

$$dM(t) = r(t)M(t)dt \tag{9.13}$$

and discount is given by

$$\delta(t) = \exp\left[-\int_0^t r(t')dt'\right]. \tag{9.14}$$

Obviously if $r(t) = r$ is constant we recover the simple exponential decay of Eq. (9.12).

However, the assumption of rates being given by constants or by deterministic functions of time is unreasonable, at least over long periods

of time. Financial interest rates are typically described as random, as the
many models for stochastic interest rates appearing in the literature show.[3]
Population dynamics are subject to random influences, as are chemical reactions and other physical processes where rates appear.

We therefore assume that $r(t)$ is a random function of time. This naturally means that discounting $\delta(t)$ is also random, as is clearly seen in Eq.
(9.14). In these circumstances the effective discount function is defined as
the average of $\delta(t)$,

$$D(t) = \left\langle \exp\left[-\int_0^t r(t')dt'\right]\right\rangle, \qquad (9.15)$$

taken over all possible realization of $r(t)$. The function $r(t)$ can, in principle, be any random process. However, the most common assumption is that
rates are diffusion processes (Brigo and Mercurio, 2006). Indeed, a natural
simplifying assumption is that $r(t)$ is a Markovian processes with continuous paths, that is, a diffusion processes. Therefore, rates are solutions to
stochastic differential equations of the form

$$dr = f(r)dt + g(r)dW(t), \qquad (9.16)$$

where $W(t)$ is the Wiener process and the equation is interpreted in the
Itô sense. Note that we assume that drift $f(r)$ and noise intensity $g(r)$ do
not depend explicitly on time, that is to say, the time dependence is only
implicit through $r = r(t)$ which means that the interest rate process is time
homogeneous and stationary. This is certainly an idealization because real
markets do not seem to be stationary, at least over long periods of time.

Defining the auxiliary random process

$$x(t) = \int_0^t r(t')dt', \qquad (9.17)$$

the equivalent discount function can be written as

$$D(t) = \left\langle e^{-x(t)}\right\rangle.$$

Therefore,

$$D(t|r_0) = \int_{-\infty}^{\infty} dr \int_{-\infty}^{\infty} e^{-x}p(x,r,t|r_0)dx, \qquad (9.18)$$

where $p(x,r,t|r_0)$ is the probability density function (PDF) of the bidimen-
sional diffusion process $(x(t),r(t))$ and we have included the dependence on
the initial rate, $r_0 = r(0)$, in the discount function $D(t|r_0)$.

---

[3]A very complete account on the theory of interest rates is given in Brigo and Mercurio
(2006).

From Eqs. (9.16)-(9.17) we see that $(x(t), r(t))$ is defined by the following pair of stochastic differential equations

$$dx = r\,dt,$$
$$dr = f(r)dt + g(r)dW(t). \tag{9.19}$$

Therefore, the joint density obeys the (forward) Fokker-Planck equation

$$\frac{\partial p}{\partial t} = -r\frac{\partial p}{\partial x} - \frac{\partial}{\partial r}[f(r)p] + \frac{1}{2}\frac{\partial^2}{\partial r^2}[g^2(r)p], \tag{9.20}$$

with the initial condition

$$p(x, r, 0|r_0) = \delta(x)\delta(r - r_0). \tag{9.21}$$

There are two different approaches for obtaining the discount function $D(t)$. One of them, which is standard in the financial literature, is based on the backward Fokker-Planck equation and it is called the *Feynman-Kac approach* (Brigo and Mercurio, 2006). A second procedure is based on Fourier analysis (Farmer *et al.*, 2015). In the next two sections we will explain both approaches.

9.2.2 *Feynman-Kac approach* *

This method obtains a partial differential equation for the discount function $D(t|r_0)$ which is based on the backward FPE for the joint density $p(x, r, t|r_0)$. In what follows we will assume that $t_0 \neq 0$ and denote $x_0 = x(t_0)$. Note that by definition $x_0 = 0$ (cf. Eq. (9.17)). However, we temporally keep $x_0 \neq 0$ and set $x_0 = 0$ at the end of the calculation if needed.

The backward FPE for the PDF $p(x, r, t|x_0, r_0, t_0)$ that corresponds to the bidimensional process (9.19) is [cf. Chapter 8 and Eq. (5.14) of Chapter 5]

$$\frac{\partial p}{\partial t_0} = -r_0\frac{\partial p}{\partial x_0} - f(r_0)\frac{\partial p}{\partial r_0} - \frac{1}{2}g^2(r_0)\frac{\partial^2 p}{\partial r_0^2}, \tag{9.22}$$

with final condition as $t_0 \to t$,

$$p(x, r, t|x_0, r_0, t) = \delta(x - x_0)\delta(r - r_0). \tag{9.23}$$

Let us observe that the problem (9.22)-(9.23) is invariant under translations of both time and variable x_0. We thus define the new variables

$$t' = t - t_0, \qquad x' = x - x_0, \tag{9.24}$$

so that

$$\frac{\partial p}{\partial t_0} = -\frac{\partial p}{\partial t'}, \qquad \frac{\partial p}{\partial x_0} = -\frac{\partial p}{\partial x'},$$

and (9.22) reads

$$\frac{\partial p}{\partial t'} = -r_0 \frac{\partial p}{\partial x'} + f(r_0)\frac{\partial p}{\partial r_0} + \frac{1}{2}g^2(r_0)\frac{\partial^2 p}{\partial r_0^2}. \tag{9.25}$$

Note that under this change of variables, we also have

$$p = p(x, r, t|x_0, r_0, t_0) = p(x, r, t|x - x', r_0, t - t') = p(x', r, t'|r_0),$$

where the last equality comes from the invariance under time and space translations, i.e., $p(x, r, t|x_0, r_0, t_0) = p(x - x_0, r, t - t_0|r_0)$. Consequently the "final condition" (9.23) becomes the initial condition

$$p(x', r, t' = 0|r_0) = \delta(x')\delta(r - r_0). \tag{9.26}$$

We multiply Eq. (9.25) by $e^{-x'}$ and integrate over x' and r, we have

$$\frac{\partial}{\partial t'}\int_{-\infty}^{\infty} dr \int_{-\infty}^{\infty} e^{-x'}p\,dx' = -r_0 \int_{-\infty}^{\infty} dr \int_{-\infty}^{\infty} e^{-x'}\frac{\partial p}{\partial x'}dx'$$

$$+ \left[f(r_0)\frac{\partial}{\partial r_0} + \frac{1}{2}g^2(r_0)\frac{\partial^2}{\partial r_0^2} \right]\int_{-\infty}^{\infty} dr \int_{-\infty}^{\infty} e^{-x'}p\,dx'. \tag{9.27}$$

But from Eq. (9.18) we see that

$$\int_{-\infty}^{\infty} dr \int_{-\infty}^{\infty} e^{-x'}p\,dx' = D(t'|r_0). \tag{9.28}$$

On the other hand, integrating by parts the first integral on the right hand side of Eq. (9.27) and using (9.28) yield

$$\int_{-\infty}^{\infty} dr \int_{-\infty}^{\infty} e^{-x'}\frac{\partial p}{\partial x'}dx' = \int_{-\infty}^{\infty} dr \int_{-\infty}^{\infty} e^{-x'}p\,dx' = D(t'|r_0), \tag{9.29}$$

where we have taken into account the boundary condition [otherwise implicit in the definition of D given in Eq. (9.18)]

$$\lim_{x' \to \pm\infty} \left[e^{-x'}p(x', r, t'|r_0) \right] = 0.$$

Substituting Eqs. (9.28) and (9.29) into Eq. (9.27) and setting $t_0 = 0$ which implies $t' = t$ [cf. Eq. (9.24)] we finally get

$$\frac{\partial D}{\partial t} = -r_0 D + f(r_0)\frac{\partial D}{\partial r_0} + \frac{1}{2}g^2(r_0)\frac{\partial^2 D}{\partial r_0^2}, \tag{9.30}$$

with the initial condition [cf. Eqs. (9.26) and (9.28)]

$$D(0|r_0) = 1. \tag{9.31}$$

The method for obtaining the discount function $D(t|r_0)$ by solving the initial-value problem (9.30)-(9.31), it is called the *Feynman-Kac approach* and Eq. (9.30) the *Feynman-Kac equation*.

In some applications (see, for instance, Sec. 9.3) it is convenient to consider $t_0 \neq 0$ so that $t' = t - t_0 \neq t$. In these cases it is appropriate to denote $D = D(t|r_0, t_0)$ and the Feynman-Kac equation (9.30) reads

$$\frac{\partial D}{\partial t_0} = r_0 D - f(r_0)\frac{\partial D}{\partial r_0} - \frac{1}{2}g^2(r_0)\frac{\partial^2 D}{\partial r_0^2}, \qquad (9.32)$$

with the final condition $D(t|r_0, t) = 1$.

9.2.3 *Fourier transform approach. The Vasicek model* *

An alternative method of obtaining the discount function, which turns out to be quite advantageous in linear cases, is based on the joint characteristic function. The latter defined as the Fourier transform of the joint density:

$$\tilde{p}(\omega_1, \omega_2, t|r_0) = \int_{-\infty}^{\infty} e^{-i\omega_2 r}dr \int_{-\infty}^{\infty} e^{-i\omega_1 x}p(x, r, t|r_0)dx. \qquad (9.33)$$

One of the chief advantages of working with the characteristic function is that obtaining the effective discount is straightforward. Indeed, comparison of Eq. (9.18),

$$D(t|r_0) = \int_{-\infty}^{\infty} dr \int_{-\infty}^{\infty} e^{-x}p(x, r, t|r_0)dx,$$

with Eq. (9.33) shows that

$$D(t|r_0) = \tilde{p}(\omega_1 = -i, \omega_2 = 0, t|r_0). \qquad (9.34)$$

Therefore, in order to obtain the discount function we only need to know the joint characteristic function of (x, r).

The Vasicek model

As an example of this procedure, we will apply the Fourier method to the Ornstein-Uhlenbeck (OU) process. In the theory of financial interest rates the OU model was proposed by Oldrich Vasicek in 1977 and it is sometimes referred to as the Vasicek model. The model is a diffusion process with linear drift and constant noise intensity (Vasicek, 1977)

$$dr = -\alpha(r - m) + kdW(t), \qquad (9.35)$$

where $r = r(t)$ is the interest rate and $W(t)$ is the standard Wiener process with zero mean and unit variance. The parameter m, sometimes referred to as *normal level*, is an average value to which rates return, $k > 0$ is the intensity of the fluctuations, and $\alpha > 0$ is the strength of the return to the mean value.

The OU process has been treated in Chapter 8 and in the stationary regime[4] the solution to Eq. (9.35) is [cf. Eq. (8.80) of Chapter 8]

$$r(t) = m + k \int_{-\infty}^{t} e^{-\alpha(t-t')} dW(t'), \tag{9.36}$$

from which it follows that the normal level is the (stationary) mean value of the process:

$$\langle r(t) \rangle = m,$$

where the average must be taken in the stationary regime ($t_0 \to -\infty$). We have also shown in Chapter 8 [cf. Eq. (8.84)] that the (stationary) correlation function reads

$$K_s(t_1 - t_2) = \frac{\sigma}{2\alpha} e^{-\alpha|t_1-t_2|},$$

showing that α^{-1} is the correlation time and

$$\sigma^2 = \frac{k^2}{2\alpha}$$

is the stationary variance which in finance it is sometimes referred to as *volatility*.

Let us next focus on the joint PDF $p(x, r, t|r_0)$. Since now $f(r) = -\alpha(r - m)$ and $g(r) = k$ the Fokker Planck equation (9.20) reads

$$\frac{\partial p}{\partial t} = -r \frac{\partial p}{\partial x} + \alpha \frac{\partial}{\partial r}[(r - m)p] + \frac{1}{2}k^2 \frac{\partial^2 p}{\partial r^2}, \tag{9.37}$$

with the initial condition

$$p(x, r, 0|r_0) = \delta(x)\delta(r - r_0). \tag{9.38}$$

The equation obeyed by the joint characteristic function $\tilde{p}(\omega_1, \omega_2, t|r_0)$ defined in Eq. (9.33) is simpler than the FPE (9.37). Indeed, Fourier transforming Eqs. (9.37) and (9.38) results in the following first-order partial differential equation for the characteristic function

$$\frac{\partial \tilde{p}}{\partial t} = (\omega_1 - \alpha\omega_2)\frac{\partial \tilde{p}}{\partial \omega_2} - \left(i\alpha m\omega_2 + \frac{1}{2}k^2\omega^2\right)\tilde{p}, \tag{9.39}$$

---

[4]Recall that for a time homogeneous process, the stationary regime is achieved as $t - t_0 \to \infty$. That is, if t_0 is finite and $t \to \infty$ or else if t is finite and $t_0 \to -\infty$ (the process begun in the infinite past).

with initial condition

$$\tilde{p}(\omega_1, \omega_2, 0 | r_0) = e^{i\omega_2 r_0}. \tag{9.40}$$

In writing Eq. (9.39) we have taken into account the following properties of the Fourier transform $\mathcal{F}\{\cdot\}$:

$$\mathcal{F}\left\{r\frac{\partial p}{\partial x}\right\} = -\omega_1 \frac{\partial \tilde{p}}{\partial \omega_2}, \qquad \mathcal{F}\left\{\frac{\partial}{\partial r}(rp)\right\} = -\omega_2 \frac{\partial \tilde{p}}{\partial \omega_2},$$

and

$$\mathcal{F}\left\{\frac{\partial p}{\partial r}\right\} = i\omega_2 \tilde{p}, \qquad \mathcal{F}\left\{\frac{\partial^2 p}{\partial r^2}\right\} = -\omega^2 \tilde{p}.$$

To solve problem (9.39)-(9.40) we proceed as in Sec. 4.5 of Chapter 4 and seek the solution in the form of a Gaussian function:

$$\tilde{p}(\omega_1, \omega_2, t | r_0) = \exp\left[-A(\omega_1, t)\omega_2^2 - B(\omega_1, t | r_0)\omega_2 - C(\omega_1, t | r_0)\right], \tag{9.41}$$

were $A(\omega_1, t)$, $B(\omega_1, t | r_0)$ and $C(\omega_1, t | r_0)$ are unknown functions. Substituting the ansatz (9.41) into Eq. (9.39) we get

$$(\dot{A} + 2\alpha A + k^2/2)\omega_2^2 + (\dot{B} + \alpha B - 2\omega_1 A + i\alpha m)\omega_2 + \dot{C} - \omega_1 B = 0,$$

(the dot denotes the derivative with respect to time). Since this is an identity valid for all values ω_2 we obtain

$$\dot{A} + 2\alpha A + k^2/2 = 0, \quad \dot{B} + \alpha B - 2\omega_1 A + i\alpha m = 0, \quad \dot{C} - \omega_1 B = 0.$$

That is,

$$\dot{A} = -2\alpha A - k^2/2, \qquad A(\omega_1, 0) = 0,$$

$$\dot{B} = -\alpha B + 2\omega_1 A - i\alpha m, \qquad B(\omega_1, 0 | r_0) = ir_0,$$

$$\dot{C} = \omega_1 B, \qquad C(\omega_1, 0 | r_0) = 0,$$

where the initial conditions stem from the combination of (9.40) and (9.41). These equations are a set of linear ordinary differential equations which can be sequentially integrated, giving

$$A(\omega_1, t) = \frac{k^2}{4\alpha}\left(1 - e^{-2\alpha t}\right),$$

$$B(\omega_1, t | r_0) = ir_0 e^{-\alpha t} + im\left(1 - e^{-\alpha t}\right) + \frac{k^2 \omega_1}{2\alpha^2}\left(1 - 2e^{-\alpha t} + e^{-2\alpha t}\right),$$

and

$$C(\omega_1, t|r_0) = i\omega_1 \frac{r_0}{\alpha} \left(1 - e^{-\alpha t}\right) + im\omega_1 \left[t - \frac{1}{\alpha}\left(1 - e^{-\alpha t}\right)\right]$$
$$+ \frac{k^2\omega_1^2}{2\alpha^3}\left[\alpha t - 2\left(1 - e^{-\alpha t}\right) + \frac{1}{2}\left(1 - e^{-2\alpha t}\right)\right]. \quad (9.42)$$

The discount function is now obtained by comparing Eqs. (9.34) with Eq. (9.41). We thus see that

$$D(t|r_0) = \exp\left[-C(\omega_1 = -i, t|r_0)\right],$$

which, after substituting for $C(-i, t|r_0)$, yields

$$\ln D(t) = -\frac{r_0}{\alpha}\left(1 - e^{-\alpha t}\right) - m\left[t - \frac{1}{\alpha}\left(1 - e^{-\alpha t}\right)\right]$$
$$+ \frac{k^2}{2\alpha^3}\left[\alpha t - 2\left(1 - e^{-\alpha t}\right) + \frac{1}{2}\left(1 - e^{-2\alpha t}\right)\right]. \quad (9.43)$$

The exponential terms in Eq. (9.43) are only significant for times smaller than the autocorrelation time of the rate, $t < \alpha^{-1}$. At longer times the exponential terms are negligible. Moreover as time increases even constant terms are negligible. Asymptotically we thus have the usual exponential law:

$$D(t) \simeq e^{-(m-k^2/2\alpha^2)t}, \quad (9.44)$$

$(t \gg \alpha^{-1})$ which shows that the long-run discount rate given by

$$r_\infty \equiv m - k^2/2\alpha^2, \quad (9.45)$$

is always smaller than the (stationary) average interest rate m by an amount that depend on the product, $k\alpha^{-1}$, of the noise intensity k and the correlation time α^{-1} of the process, the latter indicating the persistence of the fluctuations.

9.3   Pricing bonds. The term structure of interest rates *

Another application of stochastic calculus to finance, and very closely related to discounting, is bond pricing. This is a broad subject with countless studies (many of them rather abstract) which have appeared in the literature during the last decades. We will here present a short and intendedly simple introduction and refer the interested reader to more specialized works for more information (Brigo and Mercurio, 2006).

A bond is an instrument that one purchases now and delivers a payment in the future. From a more technical point of view, we say that a (discount) bond is a default-free claim on a specified sum of money to be delivered at a given future date called maturity time. Such claims are bought and issued by investors. Let us denote by $B(t_0, t)$ the price at time t_0 of a discount bond maturing at time $t \geq t_0$, with unit maturity value[5]

$$B(t, t) = 1.$$

Bonds are classified according to the *time interval to maturity* τ defined as

$$\tau = t - t_0.$$

Thus, if $\tau = 10$ years we talk about a 10 year bond that is traded initially at t_0 (for instance, today) with price $B(t_0, t_0 + 10)$ and which after 10 years has unit value. Similarly for a 3 month bond, 3 year bond etc. The central question is to know the *backward evolution* of the bond price, from unit maturity to the initial price $B(t_0|t)$.

To this end we define the *instantaneous rate of return* $r(t_0, t)$ (also called *forward rate*) as the relative time variation of the bond price [see Eq. (9.7)]

$$r(t_0, t) \equiv \frac{1}{B(t_0, t)} \frac{dB(t_0, t)}{dt_0}, \qquad (9.46)$$

or equivalently

$$r(t_0, t) = \frac{d \ln B(t_0, t)}{dt_0}. \qquad (9.47)$$

The knowledge of the forward rate $r(t_0, t)$ allows us to relate the initial price $B(t_0, t)$ and the maturing price $B(t, t) = 1$. Indeed the integration of (9.47) directly leads to

$$B(t_0, t) = \exp \left\{ -\int_{t_0}^{t} r(t_0, t') dt' \right\}. \qquad (9.48)$$

The comparison of Eq. (9.48) with Eq. (9.14) shows that $B(t_0, t)$ is the equivalent of the discount function $\delta(t)$ as well as the forward rate $r(t_0, t)$ is the equivalent of the discount rate $r(t)$. However, in what follows, we will use the notation $r(t)$ for the so-called spot rate which we will define below.

Another quantity of interest is the *yield to maturity* $y(t_0, \tau)$ defined by

$$y(t_0, \tau) \equiv -\frac{1}{\tau} \ln B(t_0, t_0 + \tau), \qquad (9.49)$$

and from (9.48) we see that

$$y(t_0, \tau) = \frac{1}{\tau} \int_{t_0}^{t_0 + \tau} r(t_0, t') dt'.$$

---

[5]If the final maturity price is not 1 (say, $B(t|t) = \beta$) then the (initial) bond price would be $\beta B(t_0, t)$.

That is to say, the yield is the time average of the forward rate over the maturity period τ.

A final quantity is needed, the *spot rate*, which is defined as the limit of the yield when maturity tends to 0,

$$r(t_0) \equiv \lim_{\tau \to 0} y(t_0, \tau) = \lim_{\tau \to 0} \left\{ \frac{1}{\tau} \int_{t_0}^{t_0+\tau} r(t_0, t')dt' \right\}. \qquad (9.50)$$

Solving the indeterminacy by expanding the integral in powers of τ, we see that the spot rate is given in terms of the forward rate by

$$r(t_0) = r(t_0, t_0). \qquad (9.51)$$

In other words, the spot rate is the instantaneous forward rate.

Let us finally note that a loan of amount M subscribed at time t_0 with an interest rate $r(t_0)$ (the spot rate) will, at time $t_0 + dt_0$, increase in value to $M + dM$, where

$$dM = r(t_0)M dt_0. \qquad (9.52)$$

Indeed, at any time t_0, the value of the spot rate $r(t_0)$ is the instantaneous increase of the loan value, that is, $r(t_0) = d \ln M(t_0)/dt$ (compare with Eq. (9.47)).[6] However, subsequent values of the spot rate are not necessarily certain. We will see next the consequences of this fact on the time evolution of the bond price $B(t_0, t)$.

9.3.1 Dynamics of the bond price

Suppose the spot rate $r(t_0)$ is random. In such a case, and analogously to discounting, the usual assumption is that $r_0 = r(t_0)$ is a Markovian random process with continuous trajectories; that is, a diffusion process obeying a stochastic differential equation of the form

$$dr_0 = f(r_0)dt_0 + g(r_0)dW(t_0), \qquad (9.53)$$

where $W(t_0)$ is the standard Wiener process. We have assumed that the drift and the noise intensity are independent of time (as is the case in most applications) thus the time dependence of these coefficients is implicit through $r_0 = r(t_0)$. We know that this implies invariance under time translations and we can set $t_0 = 0$ when needed without loss of generality.

We will now follow Vasicek (1977) and obtain the time evolution of the bond price $B(t_0, t)$ at maturity $t \geq t_0$. To this end, let us first observe that

---

[6]Let us remark again the close similarities with discounting as described in Sec. 9.2 (see Eq. (9.13)).

the most natural hypothesis consists in assuming that the bond price B is a function of the initial spot rate $r(t_0)$, i.e.,

$$B = B\big[t_0, t | r(t_0)\big]. \tag{9.54}$$

In this way $B(t_0, t | r_0)$ represents the price of a bond issued at time t_0 and maturing at time t, given that the initial interest rate is $r_0 = r(t_0)$.

The infinitesimal variation of the bond price is then defined by

$$dB = B\big[t_0 + dt_0, t | r(t_0 + dt_0)\big] - B\big[t_0, t | r(t_0)\big].$$

We expand in Taylor series up to second order

$$B\big[t_0 + dt_0, t | r(t_0 + dt_0)\big] = B\big[t_0, t | r(t_0)\big] + \frac{\partial B}{\partial t_0} dt_0 + \frac{\partial B}{\partial r_0} dr_0$$

$$+ \frac{1}{2}\left[\frac{\partial^2 B}{\partial t_0^2} dt_0^2 + \frac{\partial^2 B}{\partial r_0^2} dr_0^2 + \frac{\partial^2 B}{\partial t_0 \partial r_0} dt_0 dr_0\right] + \cdots .$$

Substituting for Eq. (9.53) and recalling from Chapter 8 that $dW(t_0) = O(dt_0^{1/2})$ and $dr_0^2 = dt_0^2$ we obtain, up to first order in dt_0,

$$dB = \left[\frac{\partial B}{\partial t_0} + f(r_0)\frac{\partial B}{\partial r_0} + \frac{1}{2}g^2(r_0)\frac{\partial^2 B}{\partial r_0^2}\right] dt_0$$

$$+ g(r_0)\frac{\partial B}{\partial r_0} dW(t_0). \tag{9.55}$$

Defining

$$\mu(t_0, t | r_0) \equiv \frac{1}{B}\left[\frac{\partial B}{\partial t_0} + f(r_0)\frac{\partial B}{\partial r_0} + \frac{1}{2}g^2(r_0)\frac{\partial^2 B}{\partial r_0^2}\right], \tag{9.56}$$

and

$$\sigma(t_0, t | r_0) \equiv -\frac{1}{B}g(r_0)\frac{\partial B}{\partial r_0}, \tag{9.57}$$

we see from (9.55) that the bond price satisfies the stochastic differential equation

$$\frac{dB}{B} = \mu(t_0, t | r_0)dt_0 - \sigma(t_0, t | r_0)dW(t_0), \tag{9.58}$$

showing that the bond price is a diffusion process as well.

Averaging Eq. (9.58) and recalling that $\langle dW(t_0)\rangle = 0$ we see that

$$\mu(t_0, t | r_0) = \left\langle \frac{1}{B}\frac{dB}{dt_0}\right\rangle,$$

which proves that $\mu(t_0, t | r_0)$ is the average of the instantaneous rate of return [cf. Eq. (9.46)] at time t_0 on a bond with maturing date t, given that the current spot rate is r_0. In an analogous way one can easily show that $\sigma^2(t_0, t|r_0)$ is the variance.

We therefore see from the above development that the bond price is a random quantity. The question is: what is the price that an investor has to buy (or sell) a bond at time t_0 maturing at time $t = t_0 + \tau$ with the current spot rate r_0? One possible answer would be proceeding as in discounting and take the average over all possible realizations of the bond price. However, this procedure implies that the expected rate of return of a bond is invariant under risk variation –i.e., under changes of the variance $\sigma^2(t_0, t|r_0)$– a fact that investors always have in mind. In this section we explain a procedure resulting in a deterministic bond price which takes into account the risk aversion of investors.[7]

9.3.2   *The market price of risk*

Consider an investor who, at time t_0, sells an amount M_1 of a bond maturing at time t_1. and at the same time buys an amount M_2 of another bond with a different maturing date t_2. The total worth of the *portfolio* thus constructed is $M = M_2 - M_1$. Note that the prices M_i $(i = 1, 2)$ are multiples of the bond prices $B(t_0, t_i|r_0)$ and, hence, obey the SDE (9.58). That is to say,

$$\frac{dM_i}{M_i} = \mu(t_0, t_i|r_0)dt_0 + \sigma(t_0, t_i|r_0)dW(t_0).$$

In consequence the infinitesimal variation $dM = dM_2 - dM_1$ of the portfolio changes over time according to

$$dM = \left[\mu(t_0, t_2|r_0)M_2 - \mu(t_0, t_1|r_0)M_1\right]dt_0$$
$$+ \left[\sigma(t_0, t_2|r_0)M_2 - \sigma(t_0, t_1|r_0)M_1\right]dW(t_0). \qquad (9.59)$$

Suppose we choose the amounts M_1 and M_2 such that

$$M_1 = \frac{M}{\sigma_2 - \sigma_1}\sigma_2, \qquad M_1 = \frac{M}{\sigma_2 - \sigma_1}\sigma_1, \qquad (9.60)$$

where $M = M_2 - M_1$ and $\sigma_i = \sigma(t_0, t_i|r_0)$ $(i = 1, 2)$. Hence M_1 is proportional to σ_2 while M_2 is proportional to σ_1. With this choice we have

$$\sigma_2 M_2 - \sigma_1 M_1 = \sigma_2 \frac{\sigma_1 M}{\sigma_2 - \sigma_1} - \sigma_1 \frac{\sigma_2 M}{\sigma_2 - \sigma_1} = 0,$$

---

[7]In practice this is true only to some extend since the theoretical procedure assumes that the market is driven by Gaussian white noise (i.e., the Wiener process) which is an idealized noise presenting –among other shortcomings– no fat tails, a key characteristic of real markets.

and the random term in Eq. (9.59) vanishes. This renders the portfolio composed of such amounts of the two bonds instantaneously riskless:

$$dM = \frac{M}{\sigma_1 - \sigma_2}(\mu_2\sigma_1 - \mu_1\sigma_2)dt_0, \qquad (9.61)$$

where $\mu_i = \mu(t_0, t_i|r_0)$. The rate of return r_M of this portfolio is

$$r_M \equiv \frac{1}{M}\frac{dM}{dt_0} = \frac{\mu_2\sigma_1 - \mu_1\sigma_2}{\sigma_1 - \sigma_2}.$$

In order to avoid *arbitrage opportunities* –that is, making profits without taking any risk– the rate r_M must be equal to the spot rate r_0. If not, the portfolio can be purchased by taking funds borrowed at the spot rate, or otherwise sold and the profits lent out to accomplish a riskless arbitrage (see Vasicek, 1977). Therefore [compare also Eq. (9.52) with Eq. (9.61)]

$$r_0 = \frac{\mu_2\sigma_1 - \mu_1\sigma_2}{\sigma_1 - \sigma_2}.$$

Rearranging terms we get $(\mu_1 - r_0)/\sigma_1 = (\mu_2 - r_0)/\sigma_2$, so that

$$\frac{\mu(t_0, t_1|r_0) - r_0}{\sigma(t_0, t_1|r_0)} = \frac{\mu(t_0, t_2|r_0) - r_0}{\sigma(t_0, t_2|r_0)}.$$

This equation is valid for arbitrary maturities $t_1, t_2, \ldots$, it then follows that the ratio $[\mu(t_0, t|r_0) - r_0]/\sigma(t_0, t|r_0)$ *must be independent of the maturity time* t.

Let us denote by $q(t_0|r_0)$ the common value of such a ratio for a bond of any maturity date, given that the current spot rate is r_0,

$$q(t_0|r_0) \equiv \frac{\mu(t_0, t|r_0) - r_0}{\sigma(t_0, t|r_0)}, \qquad (t \geq t_0). \qquad (9.62)$$

The quantity $q(t_0|r_0)$ is called the *market price of risk*, as it gives the variation of the expected rate of return on a bond (specified by the *risk premium* $\mu - r_0$) per an additional unit risk (specified by the standard deviation σ).

Note that if $q = 0$ the spot rate $r_0 = r(t_0)$ and the average rate of return μ coincide.

$$\mu(t_0, t|r_0) = r(t_0)$$

$(t = t_0 + \tau)$ meaning that the expected instantaneous rates of return on bonds are the same for all maturities.

9.3.3   The term structure equation

The above development on the market price of risk allows us to obtain a deterministic equation for the bond price $B = B(t_0, t|r_0)$. In effect, writing Eq. (9.62) as

$$\mu(t_0, t|r_0) - r_0 = \sigma(t_0, t|r_0)q(t_0|r_0),$$

and substituting μ and σ by their definitions given in Eqs. (9.56) and (9.57), we have

$$\frac{1}{B}\left[\frac{\partial B}{\partial t_0} + f(r_0)\frac{\partial B}{\partial r_0} + \frac{1}{2}g^2(r_0)\frac{\partial^2 B}{\partial r_0^2}\right] - r_0 = -q(t_0|r_0)\frac{1}{B}g(r_0)\frac{\partial B}{\partial r_0},$$

which, after rearranging terms, yields

$$\frac{\partial B}{\partial t_0} = r_0 B - [f(r_0) + g(r_0)q(t_0|r_0)]\frac{\partial B}{\partial r_0} - \frac{1}{2}g^2(r_0)\frac{\partial B}{\partial r_0^2}. \qquad (9.63)$$

This equation, called the term structure equation, is a partial differential equation for $B(t_0, t|r_0)$, once the random character of the spot rate process $r(t)$ (through f and g) is known and the market price of risk $q(t_0|r_0)$ is specified. Bond prices are obtained after solving (9.63) with the final condition

$$B(t, t|r_0) = 1. \qquad (9.64)$$

The term structure equation (9.63) for the bond price is identical to the Feynman-Kac equation (9.32) for the discount as long as we make the following change of drift

$$f(r_0) \longrightarrow f(r_0) + g(r_0)q(t_0|r_0). \qquad (9.65)$$

On the other hand, as we have seen in Sec. 9.2, the solution of the Feynman-Kac equation (9.32) for the discount $D(t|r_0)$ is written as the average [cf. Eq. (9.18)]

$$D(t|r_0, t_0) = \int_{-\infty}^{\infty} dr \int_{-\infty}^{\infty} e^{-x} p(x, r, t|r_0, t_0) dx,$$

where $p(x, r, t|r_0, t_0)$ is the probability density function of the bidimensional diffusion process $(x(t), r(t))$ defined by Eq. (9.19),

$$dx = r dt, \qquad dr = f(r)dt + g(r)dW(t).$$

Now the analogy between the term structure equation (9.63) and the Feynman-Kac equation (9.32) suggests that we can write the bond price $B(t_0, t|r_0)$ as an average over the different realizations of the spot rate

$r(t_0)$. However, this averaging procedure is taken using a modified PDF called the *risk-free measure*. Thus it can be proved in a more rigorous way that (Vasicek, 1977; Duffie, 2001)

$$B(t_0, t|r_0) = \int_{-\infty}^{\infty} dr \int_{-\infty}^{\infty} e^{-x} p^*(x, r, t|r_0, t_0) dx, \qquad (9.66)$$

where $p^*(x, r, t|r_0, t_0)$ is the risk-free measure which is the PDF of the bidimensional process $(x(t_0), r(t_0))$ defined by the following pair of SDEs which include the market price of risk [see Eq. (9.65)]:

$$dx = r dt,$$
$$dr = [f(r) + g(r)q(t|r)]dt + g(r)dW(t). \qquad (9.67)$$

That is, p^* is the solution to the FPE

$$\frac{\partial p^*}{\partial t} = -r\frac{\partial p^*}{\partial x} - \frac{\partial}{\partial r}\Big[[f(r) + g(r)q(t|r)]p^*\Big] + \frac{1}{2}\frac{\partial^2}{\partial r^2}[g^2(r)p^*], \qquad (9.68)$$

with the initial condition

$$p^*(x, r, t_0|r_0, t_0) = \delta(x)\delta(r - r_0). \qquad (9.69)$$

Since, as we have shown in Sect. 9.2.2, the Feynman-Kac approach to discounting is equivalent to the Fourier method described in Sect. 9.2.3, we can apply the latter to obtain directly the bond price knowing only the risk neutral PDF, without having to solve the Feynman-Kac equation (9.63) with condition (9.26). Indeed, the characteristic function of the risk neutral density p^* is the joint Fourier transform

$$\tilde{p}^*(\omega_1, \omega_2, t|r_0, t_0) = \int_{-\infty}^{\infty} e^{-i\omega_2 r} dr \int_{-\infty}^{\infty} e^{-i\omega_1 x} p^*(x, r, t|r_0, t_0) dx$$

which after comparing with Eq. (9.66) yields

$$B(t_0, t|r_0) = \tilde{p}^*(\omega_1 = -i, \omega_2 = 0, t|r_0, t_0). \qquad (9.70)$$

Finally, once we know the bond price, the yield to maturity $y(t_0, \tau|r_0)$ (also called *the term structure of interest rates*) is readily evaluated from Eq. (9.49):

$$y(t_0, \tau|r_0) = -\frac{1}{\tau}\ln B(t_0, t_0 + \tau|r_0). \qquad (9.71)$$

The graphic representations of $y(t_0, \tau|r_0)$ as a function of t_0 and for different values of the maturity interval τ are called *yield curves* and are of prime importance for practitioners.

9.4　Pricing options. Black-Scholes theory *

Options are financial instruments designed to protect investors of the stock market randomness. In 1973, F. Black, M. Scholes and R. Merton proposed a very popular method for obtaining a fair price for options. The pricing method is based on stochastic calculus and we will here review some of the basic aspects of the theory.

An Option is a special type of "derivative" which is a financial asset whose price depends on the price of another asset, called the "underlying". There are options of many kinds. The simplest is the *European option*, a financial instrument giving to its owner the right but not the obligation to buy (European call) or to sell (European put) stock at a fixed future date, the maturing time, and at a certain price called *exercise* or *striking price*. Other type of options are more sophisticated. One possible extension is the *American option* which gives the right to exercise the option at any time before maturity.

The main purpose in option studies is to find a fair and presumably riskless price for these instruments. The first solution to the problem was given by Bachelier in 1900 and during several years various option prices were proposed without being completely satisfactory. However, in the early ninety seventies it was finally developed a complete option valuation based on equilibrium theoretical hypothesis for speculative prices. The works of Fisher Black, Myron Scholes and Robert Merton were the culmination of this effort, and left the doors open for extending the option pricing theory in many ways (Black and Sholes, 1973; Merton, 1973).

We will here present a short summary on the Black-Scholes theory of option pricing centering only in the European call. We refer the interested reader to the literature for more information (see, for instance, Hull, 2010).

As mentioned above, a call option on a given asset is an instrument purchased at time t giving the possibility –but not the obligation– of buying a share of that asset at a future maturing time $T \geq t$ with a fixed striking price K. Let us denote by C the price of such a contract. The call price C will obviously depend on the time of purchasing t and the share price $S(t)$ at that time but also on the maturing time T and the exercise price K. That is, $C = C\big(t, S(t)|T, K\big)$, although we will usually denote it by $C(t, S)$. Since the share price $S(t)$ is a random process, obtaining a fair price for the option is not an easy task and constitutes the central problem in option pricing.

The final price of a call option maturing at T with striking price K is

$$C(T, S) = \max[S(T) - K, 0].$$ (9.72)

Indeed, if the price of the share at time T is greater than K the investor earns the difference between $S(T)$ and K, by purchasing the share at price K and immediately selling it at a bigger price $S(T)$. The price of the call purchased at T then should be given by that difference. On the other hand, if the share price at T is smaller than K, the call is worthless which proves Eq. (9.72) as the final price of the call. We thus see that option pricing is a backward problem, in the sense that we have to find the initial price of the call $C(t, S)$ knowing that the final price is given by Eq. (9.72).

Let us suppose for a moment that the price of the underlying asset evolves deterministically at a given fixed rate r, then the striking price, discounted at $t \leq T$ (see Sec. 9.2) is $Ke^{-r(T-t)}$. Therefore, the fair price of the call would be

$$C(t, S) = \max[S(t) - Ke^{-r(T-t)}, 0].$$

However, share prices and rates are random. Hence some kind of average has to be implemented. Nevertheless, the direct average of this equation on all possible realizations of $S(t)$ leads to arbitrage opportunities (i.e., making profits without risk) and the call price thus obtained is unfair (Smith, 1976). After the work of Black, Scholes and Merton it was shown that such an average must be taken with a special probability density function, the so-called *martingale measure* (Baxter and Rennie, 1998).

9.4.1 *The Black-Scholes equation*

The Black-Scholes equation is a partial differential equation satisfied by the price C of any derivative (a call or a put) dependent on a non-dividend paying stock. There are many different approaches for deriving the Black-Sholes equation. We here present a simple and intuitive derivation which is close to the original one.

The basic assumption in Black-Scholes theory is that share prices are described by diffusion processes. That is, by Markovian processes with continuous trajectories.[8] We therefore assume that stock prices evolve according to the SDE

$$\frac{dS(t)}{S(t)} = f(t, S)dt + g(t, S)dW(t),$$ (9.73)

---

[8]The continuity of the trajectories of the share price has been relaxed in several occasions. Thus, for instance, Merton (1976) included random jumps in the price evolution.

which generalizes the log-normal model presented in Sec. 9.1 [cf. Eq. (9.3)].

We consider an European option whose underlying evolves as in Eq. (9.73). In what follows we assume that (i) the underlying stock pays no dividends and (ii) the market is "frictionless" meaning that there are no transaction costs and no commissions.

Let us first prove that the price of the option, $C(t, S)$, obeys a SDE like (9.73). In other words, that the price of the derivative is also a diffusion process. Indeed, by Taylor expanding (i.e., by the Itô lemma) we have

$$dC(t) \equiv C(t + dt, S + dS) - C(t, S)$$
$$= \frac{\partial C}{\partial t} dt + \frac{\partial C}{\partial S} dS + \frac{1}{2} \frac{\partial^2 C}{\partial S^2} + O(dt dS),$$

which, after using (9.73) and recalling that $dW(t) = O(dt^{1/2})$ and $dW(t)^2 = dt$, yields (up to order dt)

$$dC = \left(\frac{\partial C}{\partial t} + Sf \frac{\partial C}{\partial S} + \frac{1}{2} S^2 g^2 \frac{\partial^2 C}{\partial S^2} \right) dt + Sg \frac{\partial C}{\partial S} dW. \qquad (9.74)$$

Defining

$$\mu_c(t) \equiv \frac{1}{C} \left(\frac{\partial C}{\partial t} + Sf \frac{\partial C}{\partial S} + \frac{1}{2} S^2 g^2 \frac{\partial^2 C}{\partial S^2} \right), \qquad \sigma_c(t) \equiv \frac{1}{C} Sg \frac{\partial C}{\partial S}, \quad (9.75)$$

we see from (9.74) that the option price obeys the SDE

$$\frac{dC(t)}{C(t)} = \mu_c(t) dt + \sigma_c(t) dW(t). \qquad (9.76)$$

The option price is therefore a diffusion process of the same kind than that of the underlying.

Consider an investor who builds up a portfolio compounded by a certain number of derivatives with price $C(t)$ and a number of underlying assets with price $S(t)$. It is also assumed that short-selling, or borrowing, is permitted without transaction costs and commissions. Specifically, we own 1 call worth $C(t)$ dollars and owe $n_s(t)$ shares, so that we borrow $n_s(t)S(t)$ dollars. The monetary value (also called nominal value) of the portfolio is

$$M(t) = C(t) - n_s(t)S(t). \qquad (9.77)$$

Before proceeding further let us note that $n_s(t)$ is the number of shares per option and $M(t)$ is the nominal value of the portfolio per option. Now for a portfolio strategy in which $n_s(t)$ is adjusted slowly relative to the changes in C, S and t we may assume that $dn_s = 0$.[9] Hence, from Eq. (9.77) we see that the change in the nominal value of the portfolio is

$$dM = dC - n_s dS. \tag{9.78}$$

Substituting for Eqs. (9.73) and (9.74) yields

$$dM = \left[\frac{\partial C}{\partial t} + Sf\left(\frac{\partial C}{\partial S} - n_s \right) + \frac{1}{2}S^2 g^2 \frac{\partial^2 C}{\partial S^2} \right] dt$$
$$+ Sg\left(\frac{\partial C}{\partial S} - n_s \right) dW. \tag{9.79}$$

This is a stochastic equation because of the Wiener noise term on the right hand side. However, and recalling that we can continuously modify our portfolio by varying the number of shares per call, we may choose n_s as

$$n_s = \frac{\partial C}{\partial S}, \tag{9.80}$$

so that the random term in (9.79) vanishes and we get a riskless portfolio. Condition (9.80) is called "delta hedging".[10] We thus have the deterministic equation [notice that with Eq. (9.80) the term involving drift f in Eq. (9.79) also vanishes]

$$dM = \left(\frac{\partial C}{\partial t} + \frac{1}{2}S^2 g^2 \frac{\partial^2 C}{\partial S^2} \right) dt. \tag{9.81}$$

If we want to avoid arbitrage opportunities the variation dM in the monetary value of the portfolio must be the same as the variation we would have had if we had put the same amount of money in a risk-free bond with interest rate r:

$$dM = rM dt. \tag{9.82}$$

Indeed, if the portfolio earned more than the risk-free rate r, one could borrow from the bank, paying interest at rate r, invest in the riskless portfolio, and make a profit without taking any risk. If, on the other hand, the portfolio earned less than r, we could short-selling the portfolio and invest the cash obtained to buy a risk-free bond.

---

[9] Note that we have not anticipated the change in S and C, i.e., $n_s(t)$ is a nonanticipating function of S and C (see Chapter 8). That is to say, first buy or sell according to the present stock price $S(t)$ and right after the portfolio value changes with the variation of prices dS and dC.

[10] The number of shares per call, which we have called n_s, is usually denoted by δ and, since Eq. (9.80) covers from risk, hence the name "delta hedging".

We equate (9.81) with (9.82)

$$rM = \frac{\partial C}{\partial t} + \frac{1}{2}S^2 g^2 \frac{\partial^2 C}{\partial S^2}. \qquad (9.83)$$

We can turn this expression into a closed equation for the call price. In effect, from Eqs. (9.77) and (9.80) we have

$$M = C - S\frac{\partial C}{\partial S},$$

which after substituting into (9.83) finally yields

$$\frac{\partial C}{\partial t} + rS\frac{\partial C}{\partial S} + \frac{1}{2}S^2 g^2(t, S)\frac{\partial^2 C}{\partial S^2} - rC = 0. \qquad (9.84)$$

This is the celebrated Back-Scholes equation for the call price $C(t, S)$. It is a backward equation which has to be solved under the final condition at maturity [cf. Eq. (9.72)]

$$C(T, S) = \max[S(T) - K, 0]. \qquad (9.85)$$

We also need to know two boundary conditions with respect to the state variable S. Notice that the equation is defined on the semi-infinite interval $0 \le S \le \infty$. In such a case, since $C(t, S)$ is assumed to be sufficiently well behaved for all S, we only need to specify one boundary condition at $S = 0$ (see, for instance, Carslaw and Jaeger, 1990). Despite this we will specify below the boundary condition at $S = \infty$ as well (Wilmott *et al.*, 1995). Let us note that all financial derivatives (options of any kind, forwards, futures, swaps, etc.) have the same boundary conditions but different initial or final condition (Wilmott, 1998). We will now specify the boundary conditions. We see from the multiplicative character of (9.73) that if at some time t_0 the price drops to zero then $dS(t) = 0$ for all $t \ge t_0$. In other words, $S(t) = 0$ forever. It such a case, the call option is obviously worthless:

$$C(t, 0) = 0. \qquad (9.86)$$

A second boundary condition appears as $S \to \infty$. Indeed, if the stock price increases without bound the difference between share price and option price tends to vanish, because in such a case the option becomes more and more likely to be exercised and the value of the option will agree with the stock price. We, therefore, have

$$\lim_{S \to \infty} \frac{C(t, S)}{S} = 1. \qquad (9.87)$$

9.4.2   *The Black-Scholes formula*

To obtain the price of the European call we have to solve Eq. (9.84) with final and boundary conditions given by Eqs. (9.85)–(9.87). In order to achieve this we need to specify the market model governing the stock price evolution [cf. Eq. (9.73)]. Although the Black-Sholes equation (9.84) does not depend on the function $f(t, S)$, it depends on the function $g^2(t, S)$. Unfortunately, we cannot solve Eq. (9.84) for any functional form of g.

In the original formulation of the Black-Sholes theory, the market model used was the log-normal model discussed in Sect. 9.1. For such a model the evolution equation for the stock price is given (9.3) which means $f(t, S) = \mu$ (here an irrelevant constant) and

$$g(t, S) = \sigma, \tag{9.88}$$

also constant. The Black-Scholes equation then reads

$$\frac{\partial C}{\partial t} + rS\frac{\partial C}{\partial S} + \frac{1}{2}(\sigma S)^2\frac{\partial^2 C}{\partial S^2} - rC = 0. \tag{9.89}$$

The change of variables $(t, S) \to (\tau, z)$:

$$\tau = T - t, \qquad z = \ln(S/K) + (r - \sigma^2/2)\tau, \tag{9.90}$$

(K is the exercise price and T is the maturity time) turns the backward equation into the non homogeneous diffusion equation

$$\frac{\partial C}{\partial \tau} = \frac{1}{2}\sigma^2\frac{\partial^2 C}{\partial z^2} - rC. \tag{9.91}$$

The final condition (9.85) now becomes the following initial condition at $\tau = 0$:

$$C(0, z) = K\max(e^z - 1, 0),$$

which, taking into account that $e^z - 1 > 0$ is equivalent to $z > 0$, can be written as

$$C(0, z) = K(e^z - 1)\Theta(z), \tag{9.92}$$

where $\Theta(z)$ is the Heaviside step function.

Defining the new dependent variable

$$u(\tau, z) = e^{r\tau}C(\tau, z) \tag{9.93}$$

transforms (9.91) into the homogeneous diffusion equation

$$\frac{\partial u}{\partial \tau} = \frac{1}{2}\sigma^2\frac{\partial^2 u}{\partial z^2}, \tag{9.94}$$

with the initial condition

$$u(0, z) = K(e^z - 1)\Theta(z). \tag{9.95}$$

Solving the initial problem (9.94)–(9.95) is standard. We Fourier transform Eq. (9.94) with the result

$$\frac{\partial \tilde{u}}{\partial \tau} = -\frac{1}{2}\sigma^2 \omega^2 \tilde{p},$$

where $\tilde{p}(\tau, \omega) = \mathcal{F}\{p(\tau, z)\}$ is the Fourier transform. The solution to this equation with initial condition $\tilde{u}(0, \omega)$ reads

$$\tilde{u}(\tau, \omega) = \tilde{u}(0, \omega)e^{-\sigma^2 \omega^2 \tau/2},$$

or by Fourier inverting

$$u(\tau, z) = \frac{1}{\sigma\sqrt{2\pi\tau}} \int_{-\infty}^{\infty} u(0, z - z')e^{-z'^2/2\sigma^2\tau}dz',$$

which after taking into account the initial condition (9.95) yields

$$u(\tau, z) = \frac{K}{\sigma\sqrt{2\pi\tau}} \int_{-\infty}^{z} \left(e^{z-z'} - 1\right)e^{-z'^2/2\sigma^2\tau}dz'. \tag{9.96}$$

In the original variables S and t and for the original function $C(t, S)$ [cf. Eqs. (9.90) and (9.93)], the Black-Scholes formula for the price of the European call is thus given by

$$C(t, S) = SN(d_1) - Ke^{-r(T-t)}N(d_2), \tag{9.97}$$

$(0 \leq t \leq T)$, where

$$N(z) = \frac{1}{\sqrt{2\pi}} \int_{-\infty}^{z} e^{-z'^2/2}dz'$$

is the probability integral, and

$$d_1 = \frac{1}{\sigma\sqrt{T-t}}[\ln(S/K) + (r + \sigma^2/2)(T - t)], \qquad d_2 = d_1 - \sigma\sqrt{T-t}.$$

Finally, recalling that

$$\lim_{z \to \infty} N(z) = 1, \qquad \lim_{z \to -\infty} N(z) = 0,$$

we see that the Black-Scholes formula (9.97) satisfies the boundary conditions (9.86) and (9.87):

$$C(t, 0) = 0, \qquad \lim_{S \to \infty} \frac{C(t, S)}{S} = 1.$$

PART 3

The level crossing problem

Chapter 10

First-passage, escape and extremes.
General settings

First-passage and extreme values problems are often referred to under the
collective name of *level-crossing problems* and have a long and standing
tradition in many branches of science and engineering. They are a significant aspect of stochastic methods, both theoretically and, very specially,
for their practical applications.

In the third and last part of this book we present a relatively complete
approach to the problem. In this chapter we address the general settings,
not only on first passage and escape problems but also on the extreme values
(maxima and minima) that any random process may attain. In the next
chapter we apply the general formalism to diffusion processes and discuss
several relevant examples. In the last chapter we consider some selected
applications to socio-economic systems.[1]

10.1 Introductory remarks

From a general point of view, the level-crossing problem covers several
related issues such as first-passage and escape problems as well as the theory
of extreme values. One of its most relevant applications is the noise-induced
escape problem in which a physical system leaves, due to noise, a given
stable state to either go to a different stable state (this would be the case
of a phase transition) or else running away without being entangled in
any other stable state. The noise-induced escape problem is ubiquitous in
countless fields ranging from physics, chemistry, engineering and biology
and it has been the object of intense research since the last quarter of the

---

[1]From a theoretical point of view, some basic references are: Gardiner's book (Gardiner,
1986) and the review articles (Masoliver and Perelló, 2014b; Masoliver, 2014b). From
the point of view of applications to physical sciences, see Redner's book (Redner, 2007)
and also the review volume (Metzler *et al.*, 2014) for recent developments in this field.

nineteen century when Arrhenius law was published and specially after the
works of Kramers in the 1940's.

A classical example of level-crossing in engineering consists in knowing
the time required by a mechanical structure to first reach a critical amplitude exceeding its stability threshold and collapse due to external random
vibrations such as wind, ocean waves, earthquakes, etc. Another classical
example in communication theory is the so-called "false alarm problem"
where fluctuations cause the current of an electric circuit to attain a critical value for which an alarm is triggered. Other natural phenomena also
need similar approaches as, for instance, floods, very high temperatures,
solar flares and earthquakes, just to name a few.

In this chapter we will review the general aspects of the problem. In
the next chapter we apply the formalism to diffusion processes, and in
the last chapter we will report some selected applications to a number of
problems of natural and socio-economic sciences. We will mainly focus on
one-dimensional random processes with brief excursions to higher dimensional systems. The extension to higher dimensions is a rather difficult task
for which few general results can be obtained and it is still object of current
research.

In its most comprehensive definition, the level-crossing problem consists
in the statistical characterization for a random process to reach, for the first
time, the boundary of a given region of the phase space. Let us suppose a
N-dimensional process $\mathbf{X}(t) = (X_1(t), \ldots, X_N(t))$ and let $\mathbf{R} \subset \mathbb{R}^N$ be some
region of the phase space with boundary $\mathbf{S}$. If $\mathbf{R}$ is a bounded region with
a closed boundary surface $\mathbf{S}$ and the process starts initially inside $\mathbf{R}$, we
will have an *exit or escape problem* when $\mathbf{X}(t)$ first crosses the boundary $\mathbf{S}$
and leaves the region $\mathbf{R}$. If $\mathbf{R}$ is unbounded and $\mathbf{S}$ is an open surface we
will have a *first-passage or hitting problem* when the process reaches $\mathbf{S}$ for
the first time.

Limiting ourselves to one-dimensional processes we can make these definitions more specific. Thus, if the boundary consists of one point x_c we
have a first-passage or hitting problem when $X(t) = x_c$ for the first time
and the value x_c is called a critical value or threshold. If the boundary consists of two points a and b and initially $a < X(t_0) < b$ we will have an exit
or escape problem out of the interval (a, b) when $X(t) \notin (a, b)$ for the first
time. In the one dimensional case both problems are related because exit
problems out of semi-infinite intervals $(-\infty, b)$ or (a, ∞) are first-passage
problems to thresholds b or a respectively. In the first case the process $X(t)$
starts below the critical value ($x_c = b$) while in the second it does above

threshold ($x_c = a$).

In what follows we will restrict ourselves to one dimensional processes, although many definitions and results can be straightforwardly extended to an arbitrary number of dimensions.

10.2 The escape problem

The problem consists in knowing whether or not the process $X(t)$, starting at some point inside the interval (a, b), has exited this interval for the first time. The problem is solved when one knows the *survival probability* denoted by $S_{ab}(t|x)$ and defined as the probability that, starting initially at $x \in (a, b)$, the process at time t has not left the interval at that time or during any previous time. Formally

$$S_{ab}(t|x) = \text{Prob}\Big\{ X(t') \in (a, b) \text{ for all } t' \in (t_0, t) \big| X(t_0) = x \in (a, b) \Big\}.$$

The complementary of $S_{ab}(t|x)$ is the *escape or exit probability*,

$$W_{ab}(t|x) = 1 - S_{ab}(t|x), \tag{10.1}$$

which gives the probability that, starting initially at $x \in (a, b)$, the process has first left the interval at a latter time t.

Very closely connected with these probabilities is *the escape or exit time* which is defined as the minimum time required by the process to leave the interval. It is a random variable denoted by $\tau_{ab}(x)$ (indicating its dependence on the initial position x) and formally defined as

$$\tau_{ab}(x) = \inf\Big\{ t \big| X(t) \notin (a, b); X(t_0) = x \in (a, b) \Big\}. \tag{10.2}$$

Notice that the event $\{\tau_{ab}(x) < t\}$ is exactly the event $\{X(t) \notin (a, b)\}$ because if the observation time t is greater than the exit time τ_{ab} this is equivalent to say that, at time t, the process has left the interval. Consequently, both events have the same probability [2]

$$\text{Prob}\big\{ \tau_{ab}(x) < t | X(0) = x \big\} = \text{Prob}\big\{ X(t) \notin (a, b) | X(0) = x \in (a, b) \big\}.$$

However, the probability on the left is by definition the distribution function of the exit time, which we denote by $F_{ab}(t|x)$, while the probability on the right is the escape probability $W_{ab}(t|x)$. Therefore, the escape probability is precisely the distribution function of the escape time,

$$F_{ab}(t|x) = W_{ab}(t|x). \tag{10.3}$$

---

[2]We will deal exclusively with time-homogeneous processes which are invariant under time translations, so that the initial time $t_0 = 0$ can be set equal to zero without loss of generality.

Since the probability density function (PDF) –denoted by $f_{ab}(t|x)$– is the derivative of the distribution function, we conclude that $f_{ab}(t|x)$ is the time derivative of the escape probability

$$f_{ab}(t|x) = \frac{\partial W_{ab}(t|x)}{\partial t}. \tag{10.4}$$

Note that in terms of the survival probability we can also write (cf. Eq. (10.1))

$$f_{ab}(t|x) = -\frac{\partial S_{ab}(t|x)}{\partial t}. \tag{10.5}$$

Other statistics of great importance in level crossing problems are the moments of the exit time. In terms of the exit PDF they are defined by

$$T_{ab}^{(n)}(x) = \int_0^\infty t^n f_{ab}(t|x)dt, \tag{10.6}$$

$(n = 1, 2, 3, \dots)$ and the mean escape time is the first moment:

$$T_{ab}^{(1)}(x) \equiv T_{ab}(x).$$

Mean exit time moments can be written in terms of the escape probability. Indeed, by combining Eqs. (10.5) and (10.6) we have

$$T_{ab}^{(n)}(x) = -\int_0^\infty t^n \frac{\partial S_{ab}(t|x)}{\partial t}dt.$$

Integration by parts yields

$$T_{ab}^{(n)}(x) = -\lim_{t\to\infty}\left[t^n S_{ab}(t|x)\right] + n\int_0^\infty t^{n-1} S_{ab}(t|x)dt,$$

$(n = 1, 2, 3, \dots)$. Assuming that survival becomes more and more improbable as time increases; specifically, assuming that $S_{ab}(t|x)$ decreases faster than t^{-n}, that is, $t^n S_{ab}(t|x) \to 0$ as $t \to \infty$, we get

$$T_{ab}^{(n)}(x) = n\int_0^\infty t^{n-1} S_{ab}(t|x)dt, \tag{10.7}$$

$(n = 1, 2, 3, \dots)$ and the mean exit time is simply the time integral of the survival probability

$$T_{ab}(x) = \int_0^\infty S_{ab}(t|x)dt. \tag{10.8}$$

Let us bear in mind that the mean escape time is a meaningful statistic as long as all higher-order moments remain relatively small. Indeed if, for instance, the second moment is large this would imply that the escape time $\tau_{a,b}(x)$ has a broad distribution around its mean value which renders the mean escape time rather useless from a practical point of view. The same applies to the first-passage time that we will discuss below.[3]

---

[3] For a thorough discussion on this rather forgotten but relevant aspect of level crossings see (Mattos *et al.*, 2012).

10.2.1 *Laplace transform analysis*

Many aspects regarding level crossings and their relations with escape times become simpler by using Laplace transform methods which, as we will soon see, provide some simple and useful asymptotic expressions for the escape probability as $t \to \infty$.

Let us thus denote by $\hat{f}_{ab}(s|x) = \mathcal{L}\{f_{ab}(t|x)\}$ the Laplace transform of the escape time density,

$$\hat{f}_{ab}(s|x) = \int_0^\infty e^{-st} f_{ab}(t|x)dt.$$

The nth derivative with respect to s of this expression yields

$$\frac{\partial^n \hat{f}_{ab}(s|x)}{\partial s^n} = (-1)^n \int_0^\infty e^{-st} t^n f_{ab}(t|x)dt.$$

Setting $s = 0$ and recalling Eq. (10.6) we see that the escape time moments are

$$T_{ab}^{(n)}(x) = (-1)^n \left. \frac{\partial^n \hat{f}_{ab}(s|x)}{\partial s^n} \right|_{s=0}. \tag{10.9}$$

Moments can thus be obtained as derivatives of the transformed density $\hat{f}_{ab}(s|x)$ instead of integrals involving $f_{ab}(t|x)$.

Expanding the transformed PDF in powers of s

$$\hat{f}_{ab}(a,b) = \sum_{n=0}^\infty \frac{s^n}{n!} \left. \frac{\partial^n \hat{f}_{ab}(s|x)}{\partial s^n} \right|_{s=0}$$

and using Eq. (10.9) we see that the Laplace transform of the escape time PDF has the following expansion in powers of s [4]

$$\hat{f}_{ab}(s|x) = \sum_{n=0}^\infty \frac{(-1)^n}{n!} T_{ab}^{(n)}(x) s^n. \tag{10.10}$$

Note that in the first term of this expansion we must set $T_{ab}^{(0)}(x) = 1$. This is a direct consequence of the normalization of the density $f_{ab}(t|x)$. Indeed, setting $n = 0$ in Eq. (10.6) we see that

$$T_{ab}^{(0)}(x) = \int_0^\infty f_{ab}(t|x)dt = 1$$

because of the normalization of $f_{ab}(t|x)$.

---

[4]This is true provided that all moments $T_{ab}^{(n)}(x)$ exist ($n = 1, 2, 3, \dots$) which may not always be the case.

We can obtain similar expansions for the Laplace transform of the escape and survival probabilities. Let us start with the escape probability. Using the well known property of the Laplace transform

$$\mathcal{L}\left\{\frac{\partial W_{ab}(t|x)}{\partial t}\right\} = s\hat{W}_{ab}(s|x) - W_{ab}(t = 0|x),$$

we see that the Laplace transform of Eq. (10.4) yields

$$\hat{f}_{ab}(s|x) = s\hat{W}_{ab}(s|x) - W_{ab}(t = 0|x). \tag{10.11}$$

However, since initially escape is impossible, we have $W_{ab}(t = 0|x) = 0$ and

$$\hat{W}_{ab}(s|x) = \frac{1}{s}\hat{f}_{ab}(s|x). \tag{10.12}$$

Combining this equation with Eq. (10.10) we get the following expansion for the transformed escape probability:

$$\hat{W}_{ab}(s|x) = \sum_{n=0}^{\infty} \frac{(-1)^n}{n!} T_{ab}^{(n)}(x) s^{n-1}. \tag{10.13}$$

Let us next address the expansion of the transformed survival probability $\hat{S}_{ab}(s|x) = \mathcal{L}\{S_{ab}(t|x)\}$. Taking the Laplace transform of Eq. (10.1) we have

$$\hat{S}_{ab}(s|x) = \frac{1}{s} - \hat{W}_{ab}(s|x), \tag{10.14}$$

where we have taken into account that $\mathcal{L}\{1\} = 1/s$. Substituting into Eq. (10.13) and the cancellation of the first term of the expansion yields

$$\hat{S}_{ab}(s|x) = \sum_{n=1}^{\infty} \frac{(-1)^{n-1}}{n!} T_{ab}^{(n)}(x) s^{n-1}, \tag{10.15}$$

from which we see that

$$T_{ab}(x) = \hat{S}_{ab}(0|x), \tag{10.16}$$

or, equivalently (cf. Eq. (10.14))

$$T_{ab}(x) = \lim_{s\to 0}\left[\frac{1}{s} - \hat{W}(s|x)\right]. \tag{10.17}$$

From Eq. (10.15) we also see that the moments of the mean exit time are given by

$$T_{ab}^{(n)}(x) = (-1)^{n-1} n \left.\frac{\partial^{n-1}\hat{S}_{ab}(s|x)}{\partial s^{n-1}}\right|_{s=0}, \tag{10.18}$$

$(n = 2, 3, 4, \ldots)$.

10.2.2 *Asymptotic analysis*

The expansions given by Eqs. (10.13) and (10.15) allow for an asymptotic analysis of the escape and survival probabilities for large values of time. We first focus on the escape probability.

The asymptotic analysis is based on the so-called *Tauberian theorems*. These are a collection of results which determine that the long time behavior of a function $\phi(t)$ is determined by the small s behavior of its Laplace transform $\hat{\phi}(s)$. Thus, if as $s \to 0$ $\hat{\phi}(s)$ is approximated by the function $\hat{\phi}_0(s)$, i.e.,

$$\hat{\phi}(s) \sim \hat{\phi}_0(s), \qquad (s \to 0),$$

then, under rather general circumstances, we have the following long-time approximation for the inverse transform $\phi(t)$

$$\phi(t) \sim \phi_0(t), \qquad (t \to \infty),$$

where $\phi_0(t)$ is the inverse transform of $\hat{\phi}_0(s)$. For more information see, for instance, Feller (1991), Handelsman and Lew (1974) and Pitt (1958).

Returning to the escape problem we see from Eq. (10.13) that the small s behavior of $\hat{W}_{ab}$ is

$$\hat{W}_{ab}(s|x) = \frac{1}{s} - T_{ab}(x) + O(s) = \frac{1}{s}\left[1 - sT_{ab}(x) + O(s^2)\right], \qquad (10.19)$$

where $T_{ab}(x)$ is the mean exit time. Using the Taylor expansion $1/(1+x) = 1 - x + O(x^2)$ we write

$$1 - sT_{ab}(x) + O(s^2) = \frac{1}{1 + sT_{ab}(x) + O(s^2)},$$

and Eq. (10.19) can be written as

$$\hat{W}_{ab}(s|x) = \frac{1}{s\left[1 + sT_{ab}(x) + O(s^2)\right]}.$$

But Tauberian theorems imply that the long-time behavior of the escape probability will be given by the Laplace inversion of this approximation. Taking into account the inverse transformation (Roberts and Kaufman, 1966)

$$\mathcal{L}^{-1}\left\{\frac{1}{s(1 + \tau s)}\right\} = 1 - e^{-t/\tau}$$

we get

$$W_{ab}(t|x) \simeq 1 - e^{-t/T_{ab}(x)}, \qquad (t \to \infty), \qquad (10.20)$$

which shows that the long-time behavior of the exit problem is determined by the mean exit time. According to Eq. (10.20) the long-time survival probability is given by (see Eq. (10.1))

$$S_{ab}(t|x) \simeq e^{-t/T_{ab}(x)}, \qquad (t \to \infty), \qquad (10.21)$$

and survival decreases exponentially at a rate given by the inverse of the mean exit time.

10.3 The first-passage problem

We now address the general background of the first-passage problem –also called hitting problem– to some threshold x_c. The main objective consists in knowing the first-passage probability, denoted by $W_c(t|x)$, for the process to reach x_c *for the first time* at the instant t having started at some initial value $x = X(0)$. As before, we denote by $S_c(t|x)$ survival probability, that is the probability that the process starting at x has not reached x_c at time t,

$$S_c(t|x) = \text{Prob}\Big\{ X(t') \neq x_c,\ 0 \leq t' \leq t \big| X(0) = x \Big\}.$$

We obviously have

$$W_c(t|x) = 1 - S_c(t|x). \qquad (10.22)$$

Let us recall that the first-passage problem is equivalent to an escape problem out of a semi infinite interval. Indeed, if the initial point x is above threshold, $x > x_c$, first-passage to x_c is equivalent to escape out of (x_c, ∞),

$$W_c(t|x) = W_{x_c\infty}(t|x), \qquad (x_c < x), \qquad (10.23)$$

while if the initial point is below threshold, first-passage is equivalent to escape out of $(-\infty, x_c)$,

$$W_c(t|x) = W_{-\infty x_c}(t|x), \qquad (x_c > x). \qquad (10.24)$$

Therefore, all definitions and expressions obtained above for the escape problem are valid for the first-passage problem. Thus, for instance, and having in mind Eq. (10.2), we see that *the first-passage time $\tau_c(x)$* (i.e., the time to reach a given level x_c for the first time) is the random variable formally defined by

$$\tau_c(x) = \inf\big\{ t \big| X(t) \notin (x_c, \infty); x_c < x = X(0) \big\},$$

if the process starts above threshold; or by

$$\tau_c(x) = \inf\big\{ t \big| X(t) \notin (-\infty, x_c); x_c > x = X(0) \big\},$$

if the process starts below threshold.

Suppose that the first-passage time is less than the observation time, $\tau_c(x) < t$, this means that at time t the process has already crossed threshold x_c with certainty. Therefore, the distribution function of the first-passage time –which is by definition the probability of the event $\{\tau_c(x) < t\}$– coincides with the hitting probability $W_c(t|x)$. Consequently, the PDF of the first-passage time, $f_c(t|x)$, will be the time derivative of W_c

$$f_c(t|x) = \frac{\partial W_c(t|x)}{\partial t}, \qquad (10.25)$$

or, equivalently,

$$f_c(t|x) = -\frac{\partial S_c(t|x)}{\partial t}. \tag{10.26}$$

The moments of the first-passage time $T_c^{(n)}(x)$ are defined by

$$T_c^{(n)}(x) = \int_0^\infty t^n f_c(t|x)dt, \tag{10.27}$$

$(n = 1, 2, 3, \dots)$ and the first moment, $T_c^{(1)}(x) \equiv T_c(x)$, is the mean first-passage time. Similarly to the escape problem, the first-passage time moments can be written in terms of the survival probability $S_c(t|x)$. Thus introducing Eq. (10.26) into Eq. (10.27) and integrating by parts we get (see Eqs. (10.7) and (10.8))

$$T_c^{(n)}(x) = n \int_0^\infty t^{n-1} S_c(t|x)dt, \tag{10.28}$$

$(n = 1, 2, 3, \dots)$ and the mean first-passage time is the time integral of the survival probability

$$T_c(x) = \int_0^\infty S_c(t|x)dt. \tag{10.29}$$

In terms of the Laplace transform

$$\hat{f}_c(s|x) = \int_0^\infty e^{-st} f_c(t|x)dt,$$

the first-passage moments can be written as [see Eq. (10.9)]

$$T_c^{(n)}(x) = (-1)^n \left. \frac{\partial^n \hat{f}_c(s|x)}{\partial s^n} \right|_{s=0}. \tag{10.30}$$

Expanding the transformed PDF $\hat{f}_c(s|x)$ in powers of s and using Eq. (10.30) we write

$$\hat{f}_c(s|x) = \sum_{n=0}^\infty \frac{(-1)^n}{n!} T_c^{(n)}(x) s^n,$$

where $T_c^0(x) = 1$ due to the normalization of $f_c(t|x)$.

Laplace transforming Eq. (10.25) and bearing in mind that hitting the threshold x_c is impossible at $t = 0$, that is, $W_c(t = 0|x) = 0$, we have (see Eqs. (10.11)-(10.12))

$$\hat{W}_c(s|x) = \frac{1}{s}\hat{f}_c(s|x),$$

which results in the following power series for the Laplace transform of the hitting probability

$$\hat{W}_c(s|x) = \sum_{n=0}^{\infty} \frac{(-1)^n}{n!} T_c^{(n)}(x) s^{n-1}, \qquad (10.31)$$

from which we see that the mean first-passage time is given by the limit

$$T_c(x) = \lim_{s \to 0} \left[\frac{1}{s} - \hat{W}_c(s|x) \right], \qquad (10.32)$$

as long as the limit exists.

A similar analysis performed on the survival probability will result in the following expansion (cf. Eq. (10.15))

$$\hat{S}_c(s|x) = \sum_{n=1}^{\infty} \frac{(-1)^{n-1}}{n!} T_c^{(n)}(x) s^{n-1}, \qquad (10.33)$$

which implies

$$T_c(x) = \hat{S}_c(0|x) \qquad (10.34)$$

and

$$T_c^{(n)}(x) = (-1)^{n-1} n \left. \frac{\partial^{n-1} \hat{S}_c(s|x)}{\partial s^{n-1}} \right|_{s=0}, \qquad (10.35)$$

$(n = 2, 3, 4, \dots)$.

Finally, assuming $s \to 0$ in Eq. (10.31) and following an identical reasoning as that of the previous section, which involves the use of Tauberian Theorems, we can see that the long-time behavior of the first-passage and survival probabilities is given by

$$W_c(t|x) \simeq 1 - e^{-t/T_c(x)} \quad \text{and} \quad S_c(t|x) \simeq e^{-t/T_c(x)}, \qquad (10.36)$$

$(t \to \infty)$, showing that, similarly to the escape problem (cf. Eqs. (10.20) and (10.21)), the long-time behavior of the hitting problem is determined by the mean first-passage time.

10.4   Escape and first-passage problems in N dimensions

We next address first-passage and escape problems for N dimensional random processes $\mathbf{X}(t) = \left(X_1(t), \dots, X_N(t) \right)$. We, therefore, generalize to several dimensions the general settings of the first-passage and escape problems that we have presented in one dimension. The general expressions for the

N dimensional case are basically the same than those of the one dimensional case.

Suppose that at some initial time t_0 the random process is inside some N dimensional region $\mathbf{R} \subset \mathbb{R}^N$. The problem consists in knowing the probability that at some later time $t > t_0$ the process has crossed the boundary $\mathbf{S}$ of $\mathbf{R}$ for the first time. As we mentioned at the beginning of this chapter, if $\mathbf{R}$ is bounded (that is, $\mathbf{S}$ is a closed surface) we have an exit or escape problem while if $\mathbf{R}$ is unbounded ($\mathbf{S}$ is open) we have a first-passage or hitting problem.

The main objective is thus characterizing the survival probability $S(t|\mathbf{x})$ of the process which, like the one dimensional case, is the probability that the process, initially at $\mathbf{x} \in \mathbf{R}$, has not left $\mathbf{R}$ during any later time $t > t_0$. The complementary of the survival probability is the *escape probability* (or exit or first-passage probability) defined as

$$W(t|\mathbf{x}) = 1 - S(t|\mathbf{x}). \tag{10.37}$$

That is, W is the probability that $\mathbf{X}(t)$ has left the region $\mathbf{R}$ for the first time at the instant t or before.

Following the one dimensional development we define the first passage time (or escape or exit time) $\tau = \tau(\mathbf{x})$ as the earliest time at which the process initially at $\mathbf{x} \in \mathbf{R}$ leaves the region $\mathbf{R}$:

$$\tau(\mathbf{x}) = \inf\big\{t\big|\mathbf{X}(t) \notin \mathbf{R}; \mathbf{X}(t_0) = \mathbf{x} \in \mathbf{R}\big\}.$$

As a result of the above terminology, if $\mathbf{R}$ is bounded τ is an exit or escape time out of $\mathbf{R}$, while if $\mathbf{R}$ is unbounded, τ is a first-passage time or hitting time to the (open) boundary $\mathbf{S}$ of $\mathbf{R}$.

Note that τ is a random variable (depending on any particular realization of the process $\mathbf{X}(t)$) and that if $\tau(\mathbf{x}) < t$ then at time t the process has left $\mathbf{R}$ at some previous time. In other words, the event $\{\tau(\mathbf{x}) < t\}$ coincides with the event $\{\mathbf{X}(t) \notin \mathbf{R}\}$ and the probabilities of both events are equal

$$\text{Prob}\big\{\tau(\mathbf{x}) < t\big|\mathbf{X}(t_0) = \mathbf{x}\big\} = \text{Prob}\big\{\mathbf{X}(t) \notin \mathbf{R}\big|\mathbf{X}(t_0) = \mathbf{x} \in \mathbf{R}\big\}.$$

Since the right hand side is the distribution function $F(t|\mathbf{x})$ of the exit time while the right hand side is the exit probability $W(t|\mathbf{x})$ defined above. We thus have $F(t|\mathbf{x}) = W(t|\mathbf{x})$ and the PDF of the exit time, $f(t|\mathbf{x})$, is therefore given by

$$f(t|\mathbf{x}) = \frac{\partial W(t|\mathbf{x})}{\partial t}. \tag{10.38}$$

In terms of the exit PDF the moments of the exit time are

$$T_n(\mathbf{x}) = \int_0^\infty t^n f(t|\mathbf{x})dt, \qquad (10.39)$$

($n = 1, 2, 3, \dots$) and the mean exit (or escape) time (MET) is the first moment $T(\mathbf{x}) = T_1(\mathbf{x})$. For unbounded regions the mean exit time is usually called the mean first-passage time (MFPT).

MET moments can be also written as integrals of the survival probability S. Indeed, from Eqs. (10.37) and (10.38) we can write

$$f(t|\mathbf{x}) = -\frac{\partial S(t|\mathbf{x})}{\partial t},$$

and from Eq. (10.39) we have

$$T_n(\mathbf{x}) = -\int_0^\infty t^n \frac{\partial S(t|\mathbf{x})}{\partial t}dt, \qquad (10.40)$$

which, integrating by parts and assuming that $t^n S(t|\mathbf{x}) \to 0$ as $t \to \infty$ (meaning that survival is impossible as $t \to \infty$), yields

$$T_n(\mathbf{x}) = n \int_0^\infty t^{n-1} S(t|\mathbf{x})dt. \qquad (10.41)$$

For $n = 1$, we see that the MET is the time integral of the survival probability

$$T(\mathbf{x}) = n \int_0^\infty S(t|\mathbf{x})dt. \qquad (10.42)$$

Before proceeding further, let us note that the general results on the exit time discussed in previous sections for one dimensional problems keep their validity in higher dimensions. We summarize some of the main results and refer the reader to Sections 10.2 and 10.3 for more results and details. Thus, in terms of the Laplace transform of the escape time PDF,

$$\hat{f}(s|\mathbf{x}) = \int_0^\infty e^{-st} f(t|\mathbf{x})dt,$$

the exit moments can be written as

$$T_n(\mathbf{x}) = (-1)^n \left.\frac{\partial^n \hat{f}(s|\mathbf{x})}{\partial s^n}\right|_{s=0}. \qquad (10.43)$$

Like in one dimension, the MET is given in terms of the Laplace transform of the survival probability $\hat{S}(s\mathbf{x})$ evaluated at $s = 0$:

$$T(\mathbf{x}) = \hat{S}(0|\mathbf{x}). \qquad (10.44)$$

Alternatively, in terms of the escape probability the MET is

$$T(\mathbf{x}) = \lim_{s \to 0} \left[\frac{1}{s} - \hat{W}(s|\mathbf{x}) \right], \qquad (10.45)$$

provided that the limit exists.

Finally the MET, when it exists, also determines the long-time behavior of the survival and escape probability by means of the following asymptotic expressions as $t \to \infty$:

$$S(t|\mathbf{x}) \simeq e^{-t/T(\mathbf{x})}, \qquad W(t|\mathbf{x}) \simeq 1 - e^{-t/T(\mathbf{x})}. \qquad (10.46)$$

10.5 Extremes and the first-passage problem

First-passage problems are closely related to the theory of extremes which was initiated in the late nineteen twenties by the works of Frechet, Fisher and Tippet and subsequently developed by Gnedenko and Gumbel during late forties and early fifties.[5] The classical theory applies to series of independent random variables and the central result is the Fisher-Tippet theorem which states that under suitable conditions the asymptotic probability distributions of extremes are restricted to be of three types, called Gumbel, Frechet and Weibull (Gumbel, 2004). We remark that all of this applies only to series of independent random variables and not to random processes defined in continuous time. Let us note, however, that when extreme events are rare (which is often the case) they can be approximately treated as independent variables for which the Fisher-Tippet theorem holds. This approximation, however, reduces the question to a problem of statistics and time series analysis and neglects the underlying dynamics and the correlations induced by it. We shall, therefore, present an essentially dynamical approach based on random processes (in continuous time) and not only on sequences of independent random variables.

As remarked in the introduction to this chapter, the statistics of extremes is important for a wide variety of problems in physics, biology, economics and engineering, specially for the understanding of critical values and failure modes in complex systems. A classical case is that of the noise-induced extreme values attained by a given system at a given time. For example, the distribution of the maximum vibrations that can reach critical breaking amplitudes. Other examples, among many, are critical populations in ecology, the "false alarm" in communication theory (where

---

[5]For more information see, for instance, Masoliver (2014a).

fluctuations cause the current of an electric circuit to attain an extreme value), or the distribution of market crashes in finance.

The extreme-value problem includes the maximum and minimum values attained by a given random process during a certain time interval. It also embraces the maximum absolute value and the range or span, the latter defined as the difference between the maximum and the minimum. In physics this problem has been traditionally related to level crossings and first-passage times and it has been mostly restricted to diffusion processes (see next chapter).[6]

The problem of extreme values is essentially one dimensional and we will develop the theory for one dimensional random processes. As we will see next the maximum and minimum are extremes related to the first-passage problem. In the next section we will study the maximum absolute value and the span which are both related to the exit problem.

10.5.1 *The maximum*

We denote by $M(t)$ the maximum value reached by the random process $X(t)$ over the time span $(0, t)$. Formally,

$$M(t) = \max\{X(t'); 0 \leq t' \leq t\}.$$

Clearly $M(t)$ is a random quantity whose value depends on the particular trajectory of the process $X(t)$. Let us denote by $\Phi_{\max}(\xi, t|x)$ the distribution function of the maximum $M(t)$, that is, the probability of the event $\{M(t) < \xi\}$:

$$\Phi_{\max}(\xi, t|x) = \mathrm{Prob}\{M(t) < \xi | X(0) = x\}. \qquad (10.47)$$

We will now see that $\Phi_{\max}(\xi, t|x)$ is related to the first-passage probability. To this end we must distinguish the cases $\xi > x$ and $\xi < x$. Suppose first that the value of the maximum ξ is greater than the initial value, $\xi > x$, in this case the process $X(t)$ has not crossed threshold ξ at time t, and the probability of the event $\{M(t) < \xi | X(0) = x\}$ equals the survival probability $S_\xi(t|x)$,

$$\Phi_{\max}(\xi, t|x) = S_\xi(t|x), \qquad (\xi > x).$$

If, on the other hand, the value of the maximum is lower than the initial point, $\xi < x$, the event $\{M(t) < \xi | X(0) = x\}$ is impossible and has zero probability. Indeed, in such a case the probability that the maximum $M(t)$

---

[6]See also Berman (1992) for similar developments aimed also to diffusion processes but oriented to the pure mathematician.

of $X(t)$ be lower than ξ is zero if ξ is lower than the initial value of the process. Therefore,

$$\Phi_{\max}(\xi, t|x) = 0, \qquad (\xi < x).$$

We summarize both cases into the single expression:

$$\Phi_{\max}(\xi, t|x) = S_\xi(t|x)\Theta(\xi - x), \qquad (10.48)$$

where $\Theta(\cdot)$ is the Heaviside step function.

We denote by $\varphi_{\max}(\xi, t|x)$ probability density function of $M(t)$. In terms of the distribution function, the PDF is given by

$$\varphi_{\max}(\xi, t|x) = \frac{\partial}{\partial \xi}\Phi_{\max}(\xi, t|x).$$

Substituting for (10.48) and bearing in mind that the derivative of the Heaviside step function is Dirac delta function,

$$\frac{\partial}{\partial \xi}\Theta(\xi - x) = \delta(\xi - x),$$

we have

$$\varphi_{\max}(\xi, t|x) = \frac{\partial S_\xi(t|x)}{\partial \xi}\Theta(\xi - x) + S_\xi(t|x)\delta(\xi - x).$$

However,

$$S_\xi(t|x)\delta(\xi - x) = S_x(t|x)\delta(\xi - x) = 0,$$

because $S_x(t|x) = 0$, in other words, survival is impossible starting at the boundary. Therefore, the PDF for the maximum value of $X(t)$ is related to the survival probability of the hitting problem by

$$\varphi_{\max}(\xi, t|x) = \frac{\partial S_\xi(t|x)}{\partial \xi}\Theta(\xi - x). \qquad (10.49)$$

We next address the question of obtaining the moments of the maximum value. Let us begin with the first moment, that is, the mean maximum value which we denote by $\left\langle M(t)|x \right\rangle$ and define as

$$\left\langle M(t)|x \right\rangle = \int_{-\infty}^{\infty} \xi\varphi_{\max}(\xi, t|x)d\xi. \qquad (10.50)$$

Substituting for Eq. (10.49) we have

$$\left\langle M(t)|x \right\rangle = \int_{x}^{\infty} \xi\frac{\partial S_\xi(t|x)}{\partial \xi}d\xi. \qquad (10.51)$$

At first sight this expression can be simplified by an integration by parts. This is, however, not possible because $S_\xi \to 1$ as $\xi \to \infty$ which leads to a divergent result. The situation can be amended using the hitting probability W_ξ instead of the survival probability S_ξ. As seen in the previous section both probabilities are related by

$$S_\xi(t|x) = 1 - W_\xi(t|x) \qquad (10.52)$$

and Eq. (10.51) can be written as

$$\Big\langle M(t)\big|x\Big\rangle = -\int_x^\infty \xi\frac{\partial W_\xi(t|x)}{\partial\xi}d\xi.$$

Integrating by parts now yields

$$\Big\langle M(t)\big|x\Big\rangle = -\lim_{\xi\to\infty}\big[\xi W_\xi(t|\xi)\big] + xW_x(t|x) + \int_x^\infty W_\xi(t|x)d\xi.$$

Note that now $W_\xi \to 0$ as $\xi \to \infty$ but, in order to obtain a finite result, we have to be more precise and assume that the hitting probability decreases faster than $1/\xi$. That is, $\xi W_\xi \to 0$ as $\xi \to \infty$. Finally, and recalling also that $W_x(t|x) = 1$ (crossing is a sure event starting at the threshold), we get

$$\Big\langle M(t)\big|x\Big\rangle = x + \int_x^\infty W_\xi(t|x)d\xi. \qquad (10.53)$$

Attending that W_ξ is always positive this equation shows, the otherwise obvious result, that the mean maximum value attained by the process is greater than its initial value x.

Following an analogous reasoning we can easily see that the moments of the maximum, defined by

$$\Big\langle M^n(t)\big|x\Big\rangle = \int_{-\infty}^\infty \xi^n\varphi_{\max}(\xi,t|x)d\xi, \qquad (10.54)$$

are given by

$$\Big\langle M^n(t)\big|x\Big\rangle = x^n + n\int_x^\infty \xi^{n-1}W_\xi(t|x)d\xi, \qquad (10.55)$$

($n = 1, 2, 3, \dots$). In writing this equation we have assumed that $\xi^n W_\xi \to 0$ as $\xi \to \infty$ which is the condition imposed on W_ξ for moments to exist.

10.5.2   *The minimum*

We denote by

$$m(t) = \min\{X(t'); 0 \le t' \le t\}$$

the minimum value attained by $X(t)$ during the time interval $(0,t)$, and let

$$\Phi_{\min}(\xi,t|x) = \text{Prob}\{m(t) < \xi | X(0) = x\}$$

be its distribution function. Note that if $\xi < x$, the event $\{m(t) < \xi | X(0) = x\}$ implies that the process has crossed threshold ξ at time t or before. Hence the distribution function agrees with the hitting probability to level ξ, *i.e.*

$$\Phi_{\min}(\xi, t|x) = W_\xi(t|x), \qquad (\xi < x).$$

On the other hand, when $\xi > x$ the event $\{m(t) < \xi | X(0) = x\}$ is a sure event. Indeed, in this case the probability that the minimum $m(t)$ of the process be lower than ξ is equal to 1 if the initial value x is greater than ξ. That is,

$$\Phi_{\min}(\xi, t|x) = 1, \qquad (\xi > x).$$

Summing up

$$\Phi_{\min}(\xi, t|x) = \Theta(\xi - x) + W_\xi(t|x)\Theta(x - \xi). \qquad (10.56)$$

Let us denote by $\varphi_{\min}(\xi, t|x)$ the PDF of the minimum $m(t)$,

$$\varphi_{\min}(\xi, t|x) = \frac{\partial}{\partial \xi}\Phi_{\min}(\xi, t|x).$$

From Eq. (10.56) we have

$$\varphi_{\min}(\xi, t|x) = \delta(\xi - x) + \frac{\partial W_\xi(t|x)}{\partial \xi}\Theta(x - \xi) - W_\xi(t|x)\delta(x - \xi).$$

But

$$W_\xi(t|x)\delta(x - \xi) = W_\xi(t|x)\delta(x - \xi) = \delta(\xi - x)$$

[recall that $W_\xi(t|\xi) = 1$ and $\delta(x - \xi) = \delta(\xi - x)$]. Hence,

$$\varphi_{\min}(\xi, t|x) = \frac{\partial W_\xi(t|x)}{\partial \xi}\Theta(x - \xi). \qquad (10.57)$$

The mean minimum value, defined as

$$\left\langle m(t)|x \right\rangle = \int_{-\infty}^{\infty} \xi\varphi_{\min}(\xi, t|x)d\xi, \qquad (10.58)$$

is then given by

$$\left\langle m(t)|x \right\rangle = \int_{-\infty}^{x} \xi\frac{\partial W_\xi(t|x)}{\partial \xi}d\xi. \qquad (10.59)$$

An integration by parts yields

$$\left\langle m(t)|x \right\rangle = x - \lim_{\xi \to -\infty}[\xi W_\xi(t|x)] - \int_{-\infty}^{x} W_\xi(t|x)d\xi.$$

Because $W_{-\infty}(t|x) = 0$ (*i.e.*, hitting an infinite threshold is impossible) then, if we also assume that W_ξ decreases faster than $1/|\xi|$, we have $\xi W_\xi \to 0$ as $\xi \to -\infty$. Hence

$$\left\langle m(t)|x \right\rangle = x - \int_{-\infty}^{x} W_\xi(t|x)d\xi, \tag{10.60}$$

which shows that the mean minimum value is indeed lower than the initial value.

Analogously to the maximum value, the moments of the minimum are given by

$$\left\langle m^n(t)|x \right\rangle = x^n - n\int_{-\infty}^{x} \xi^{n-1}W_\xi(t|x)d\xi, \tag{10.61}$$

as long as W_ξ decreases faster than $|\xi|^{-n}$ as $\xi \to -\infty$.

10.6   Other extremes and the escape problem *

Having seen that the maximum and minimum values are determined by the first-passage problem, we will next address the maximum absolute value and the range or span and see that both statistics are related to the exit problem.

10.6.1   *The maximum absolute value*

For processes taking on either positive or negative values it is also interesting to know the statistical distribution of the maximum absolute value. We thus consider the maximum value attained by the absolute value of the process, $|X(t)|$, during the time span $(0,t)$. We denote by $G_{\max}(\xi,t|x)$ its distribution function,

$$G_{\max}(\xi,t|x) = \text{Prob}\left\{\max|X(t')| < \xi, 0 \le t' \le t | X(0) = x\right\}, \tag{10.62}$$

where $0 \le t' \le t$ and $\xi > 0$. Obviously ξ cannot be negative and hence

$$G_{\max}(\xi,t|x) = 0, \qquad (\xi < 0).$$

In order to connect this distribution function with the escape problem we must distinguish two cases according to which the initial point is inside or outside the interval $(-\xi,\xi)$ spanned by the level $\xi > 0$ of the absolute maximum. For the first case where $-\xi < x < \xi$, we have

$$\left\{\max|X(t')| < \xi, 0 \le t' \le t | X(0) = x\right\}$$

$$= \left\{-\xi < X(t') < \xi, 0 \le t' \le t | X(0) = x\right\},$$

meaning that during the time span $(0, t)$ the process $X(t)$ has not left the interval $(-\xi, \xi)$. Hence, the distribution function (10.62) coincides with the survival probability of the interval $(-\xi, \xi)$

$$G_{\max}(\xi, t|x) = S_{-\xi,\xi}(t|x), \qquad (|x| < \xi).$$

Note that when the initial value is outside the interval $(-\xi, \xi)$, the event $\{\max|X(t')| < \xi | X(0) = x\}$ $(0 \le t' \le t)$ is impossible and

$$G_{\max}(\xi, t|x) = 0, \qquad (|x| > \xi).$$

We, therefore, conclude

$$G_{\max}(\xi, t|x) = S_{-\xi,\xi}(t|x)\Theta(\xi - |x|), \tag{10.63}$$

$(\xi > 0)$ where $\Theta(\cdot)$ is the Heaviside step function.

The PDF of the absolute maximum is defined by

$$g_{\max}(\xi, t|x) = \frac{\partial}{\partial \xi} G_{\max}(\xi, t|x).$$

Substituting for Eq. (10.63)

$$g_{\max}(\xi, t|x) = \frac{\partial S_{-\xi,\xi}(t|x)}{\partial \xi}\Theta(\xi - |x|) + S_{-\xi,\xi}(t|x)\delta(\xi - |x|).$$

But

$$S_{-\xi,\xi}(t|x)\delta(\xi - |x|) = S_{-|x|,|x|}(t|x)\delta(\xi - |x|) = 0,$$

because $S_{-|x|,|x|}(t|x) = 0$ (survival out of an interval is impossible if the process starts at one end of that interval). Therefore,

$$g_{\max}(\xi, t|x) = \frac{\partial S_{-\xi,\xi}(t|x)}{\partial \xi}\Theta(\xi - |x|), \tag{10.64}$$

$(\xi > 0)$. In terms of the escape probability $W_{-\xi,\xi}$ this PDF can also be written as

$$g_{\max}(\xi, t|x) = -\frac{\partial W_{-\xi,\xi}(t|x)}{\partial \xi}\Theta(\xi - |x|). \tag{10.65}$$

Let us next evaluate the mean value of the absolute maximum defined by

$$\left\langle \max|X(t)| \, \big| x \right\rangle = \int_0^\infty \xi g_{\max}(\xi, t|x)d\xi.$$

From Eq. (10.65) we have

$$\left\langle \max|X(t)| \, \big| x \right\rangle = -\int_{|x|}^\infty \xi \frac{\partial W_{-\xi,\xi}(t|x)}{\partial \xi}d\xi.$$

Integration by parts yields

$$\Big\langle \max |X(t)| \big| x \Big\rangle = |x| + \int_{|x|}^{\infty} W_{-\xi,\xi}(t|x) d\xi, \tag{10.66}$$

where we have taken into account that $W_{-|x|,|x|}(t|x) = 1$ and made the assumption that the escape probability $W_{-\xi,\xi}$ decreases faster than $1/\xi$, so that $\xi W_{-\xi,\xi} \to 0$ as $\xi \to \infty$.

Following an analogous reasoning one can easily show that the moments of the maximum absolute value can be written as

$$\Big\langle \big(\max |X(t)|\big)^n \big| x \Big\rangle = |x|^n + n \int_{|x|}^{\infty} \xi^{n-1} W_{-\xi,\xi}(t|x) d\xi, \tag{10.67}$$

$(n = 1, 2, 3, \dots)$. These moments exist as long as $W_{-\xi,\xi}$ decreases faster than $|\xi|^{-n}$ as $|\xi| \to \infty$.

We finally remark that obtaining the minimum absolute value is of little use and in those processes where the origin is attainable the minimum absolute value is zero.

10.6.2 *The range or span*

The range or span (also termed as "the oscillation") of a random process $X(t)$ over the time interval $(0, t)$ is defined as the distance between the maximum and the minimum:

$$R(t) = M(t) - m(t). \tag{10.68}$$

This random quantity is either characterized by the distribution function,

$$F_R(r, t|x) = \text{Prob}\{R(t) < r | X(0) = x\},$$

or by the PDF,

$$f_R(r, t|x) dr = \text{Prob}\{r < R(t) < r + dr | X(0) = x\},$$

which is related to the distribution function by

$$f_R(r, t|x) = \frac{\partial}{\partial r} F_R(r, t|x). \tag{10.69}$$

We can associate the span distribution with the escape problem out of a variable interval. This connection, which is a bit convoluted, is given by the following result

Theorem

The probability density function of the span of a random process is given by

$$f_R(r, t|x) = \int_{x-r}^{x} \frac{\partial^2 S_{v,r+v}(t|x)}{\partial r^2} dv, \tag{10.70}$$

$(r > 0)$, *where $S_{v,r+v}(t|x)$ is the survival probability of the (variable) interval $(v, r+v)$.*

Proof. Let us denote by $F_2(\xi, \eta, t|x)$ the joint distribution function of the maximum and the minimum:

$$F_2(\xi, \eta, t|x) = \text{Prob}\{M(t) < \xi, m(t) < \eta | X(0) = x\}.$$

Note that the event $\{M(t) < \xi\}$ is the union of two disjoint events:

$$\{M(t) < \xi\} = \{M(t) < \xi, m(t) < \eta\}$$
$$\cup \{M(t) < \xi, m(t) > \eta\},$$

where we have dropped the dependence on the initial value x which is, nonetheless, implied in all what follows. We thus have

$$\text{Prob}\{M(t) < \xi, m(t) < \eta\}$$
$$= \text{Prob}\{M(t) < \xi\} - \text{Prob}\{M(t) < \xi, m(t) > \eta\},$$

but (see Eqs. (10.47) and (10.48))

$$\text{Prob}\{M(t) < \xi\} = S_\xi(t|x)\Theta(\xi - x),$$

where $S_\xi(t|x)$ is the survival probability up to the single threshold ξ. If, on the other hand, $S_{\eta,\xi}(t|x)$ is the survival probability of the interval (η, ξ) one easily realizes that

$$\text{Prob}\{M(t) < \xi, m(t) > \eta\} = S_{\eta,\xi}(t|x)\Theta(\xi - x)\Theta(x - \eta).$$

Collecting results we write

$$F_2(\xi, \eta, t|x) = S_\xi(t|x)\Theta(\xi - x) - S_{\eta,\xi}(t|x)\Theta(\xi - x)\Theta(x - \eta).$$

The joint PDF of the maximum and the minimum, defined as the second derivative of the joint distribution function

$$f_2(\xi, \eta, t|x) = \frac{\partial^2}{\partial\xi\partial\eta} F_2(\xi, \eta, t|x),$$

is then given by

$$f_2(\xi, \eta, t|x) = -\frac{\partial}{\partial\xi}\left[\frac{\partial S_{\eta,\xi}}{\partial\eta}\Theta(\xi - x)\Theta(x - \eta) - S_{\eta,\xi}(t|x)\delta(x - \eta)\Theta(\xi - x)\right].$$

Recalling that starting at any boundary point renders survival impossible we see that

$$S_{\eta,\xi}(t|x)\delta(x - \eta) = S_{x,\xi}(t|x)\delta(x - \eta) = 0.$$

Hence

$$f_2(\xi, \eta, t|x) = -\frac{\partial^2 S_{\eta, \xi}}{\partial \xi \partial \eta} \Theta(\xi - x)\Theta(x - \eta) - \frac{\partial S_{\eta, \xi}}{\partial \eta} \delta(\xi - x)\Theta(x - \eta),$$

but again $S_{\eta, x}(t|x) = 0$, so that

$$\frac{\partial S_{\eta, \xi}}{\partial \eta} \delta(\xi - x) = \frac{\partial}{\partial \eta} \big[S_{\eta, x}(t|x)\delta(\xi - x) \big] = 0.$$

Therefore

$$f_2(\xi, \eta, t|x) = -\frac{\partial^2 S_{\eta, \xi}}{\partial \xi \partial \eta} \Theta(\xi - x)\Theta(x - \eta). \qquad (10.71)$$

In terms of the joint density the PDF of the span, Eq. (10.69), is given by

$$f_R(r, t|x) = \int_{-\infty}^{\infty} d\xi \int_{-\infty}^{\infty} \delta[r - (\xi - \eta)] f_2(\xi, \eta, t|x) d\eta, \qquad (10.72)$$

which, after substituting for Eq. (10.71) and integrating the delta function, yields

$$f_R(r, t|x) = -\int_{x-r}^{x} \frac{\partial^2 S_{\eta, \xi}(t|x)}{\partial \eta \partial \xi} \bigg|_{\xi = r + \eta} d\eta, \qquad (10.73)$$

where $r > 0$ (recall that, by definition, $R(t)$ is always positive). This expression for f_R is more conveniently written by making the change of variables

$$r = \xi - \eta, \quad v = \eta.$$

Indeed, $d\eta = dv$ and

$$\frac{\partial^2 S_{\eta, \xi}}{\partial \eta \partial \xi} \bigg|_{\xi = r + \eta} = -\frac{\partial^2 S_{v, r+v}}{\partial r^2} + \frac{\partial^2 S_{v, r+v}}{\partial v \partial r}.$$

Substituting into Eq. (10.73) we have

$$f_R(r, t|x) = \int_{x-r}^{x} \frac{\partial^2 S_{v, r+v}}{\partial r^2} dv - \int_{x-r}^{x} \frac{\partial^2 S_{v, r+v}}{\partial v \partial r} dv,$$

and taking into account (recall that $S_{x, x+r}(t|x) = S_{x-r, x}(t|x) = 0$)

$$\int_{x-r}^{x} \frac{\partial^2 S_{v, r+v}(t|x)}{\partial r \partial v} dv = \frac{\partial}{\partial r} \int_{x-r}^{x} \frac{\partial S_{v, r+v}(t|x)}{\partial v} dv$$

$$= \frac{\partial}{\partial r} \big[S_{x, x+r}(t|x) - S_{x-r, x}(t|x) \big] = 0,$$

we finally get

$$f_R(r, t|x) = \int_{x-r}^{x} \frac{\partial^2 S_{v, r+v}(t|x)}{\partial r^2} dv, \qquad (10.74)$$

$(r > 0)$, which proves Eq. (10.70). $\qquad\qquad\qquad\qquad\qquad\qquad$ $\square$

Having the expression for the span PDF, we next address the issue of the mean span defined by

$$\Big\langle R(t)\big|x\Big\rangle = \int_0^\infty r f_R(r,t|x)dr. \tag{10.75}$$

Unfortunately the introduction of Eq. (10.70) into this definition followed by an integration by parts results in indeterminate boundary terms as the reader may easily check. The correct expression for the mean span which is free of these inconsistencies is given by the following result

Theorem

The mean span of a random process $X(t)$ is given by

$$\Big\langle R(t)\big|x\Big\rangle = \int_{-\infty}^\infty \xi \frac{\partial S_\xi(t|x)}{\partial \xi} d\xi, \tag{10.76}$$

where $S_\xi(t|x)$ if the survival probability of $X(t)$ up to a varying threshold ξ $(-\infty < \xi < \infty)$.

Proof. In order to prove Eq. (10.76) –which avoids the divergencies appearing in the evaluation of the mean span– we proceed as follows. Instead of using Eq. (10.70) as the expression for the span PDF we will use the following expression of f_R which is the result of combining Eqs. (10.71) and (10.72):

$$f_R(r,t|x) = -\int_{-\infty}^\infty d\xi \int_{-\infty}^\infty d\eta \frac{\partial^2 S_{\eta,\xi}}{\partial\xi\partial\eta} \delta[r - (\xi - \eta)]\Theta(\xi - x)\Theta(x - \eta).$$

Plugging into

$$\Big\langle R(t)\big|x\Big\rangle = \int_0^\infty r f_R(r,t|x)dr,$$

and performing the integration over r using the delta function, we obtain

$$\Big\langle R(t)\big|x\Big\rangle = -\int_{-\infty}^\infty d\xi \int_{-\infty}^\infty d\eta (\xi - \eta) \frac{\partial^2 S_{\eta,\xi}}{\partial\xi\partial\eta} \Theta(\xi - x)\Theta(x - \eta). \tag{10.77}$$

We rewrite this equation as

$$\Big\langle R(t)\big|x\Big\rangle = -\int_{-\infty}^\infty d\xi \Theta(\xi - x)\xi \frac{\partial}{\partial\xi} \int_{-\infty}^\infty d\eta \Theta(x - \eta) \frac{\partial S_{\eta,\xi}}{\partial\eta}$$
$$+ \int_{-\infty}^\infty d\eta \Theta(x - \eta)\eta \frac{\partial}{\partial\eta} \int_{-\infty}^\infty d\xi \Theta(\xi - x) \frac{\partial S_{\eta,\xi}}{\partial\xi}, \tag{10.78}$$

but

$$\int_{-\infty}^{\infty} \Theta(x - \eta) \frac{\partial S_{\eta,\xi}}{\partial \eta} d\eta = \int_{-\infty}^{x} \frac{\partial S_{\eta,\xi}}{\partial \eta} d\eta = S_{\xi,x}(t|x) - S_{-\infty,\xi}(t|x).$$

However, $S_{\xi,x}(t|x) = 0$ and

$$S_{-\infty,\xi}(t|x) = S_\xi(t|x),$$

because the escape problem out of the semi-infinite interval $(-\infty, \xi)$ coincides with the first-passage problem to threshold ξ. Hence

$$\int_{-\infty}^{\infty} \Theta(x - \eta) \frac{\partial S_{\eta,\xi}}{\partial \eta} d\eta = -S_\xi(t|x). \qquad (10.79)$$

Proceeding similarly we get

$$\int_{-\infty}^{\infty} \Theta(\xi - x) \frac{\partial S_{\eta,\xi}}{\partial \xi} d\xi = S_\eta(t|x). \qquad (10.80)$$

Plugging Eqs. (10.79)-(10.80) into Eq. (10.78) and applying the Heaviside functions $\Theta(\xi - x)$ and $\Theta(x - \eta)$ we get

$$\Big\langle R(t)|x \Big\rangle = \int_x^{\infty} \xi \frac{\partial S_\xi(t|x)}{\partial \xi} d\xi + \int_{-\infty}^{x} \eta \frac{\partial S_\eta(t|x)}{\partial \eta} d\eta.$$

That is

$$\Big\langle R(t)|x \Big\rangle = \int_{-\infty}^{\infty} \xi \frac{\partial S_\xi(t|x)}{\partial \xi} d\xi,$$

which is Eq. (10.76).

□

In terms of the hitting probability $W_\xi(t|x)$ the expression (10.76) for the mean span is greatly simplified. In effect, substituting $S_\xi = 1 - W_\xi$ into Eq. (10.76), followed by an integration by parts, yield

$$\Big\langle R(t)|x \Big\rangle = -\int_{-\infty}^{\infty} \xi \frac{\partial W_\xi(t|x)}{\partial \xi} d\xi$$

$$= -\xi W_\xi(t|x) \Big|_{\xi=-\infty}^{\xi=+\infty} + \int_{-\infty}^{\infty} W_\xi(t|x) d\xi.$$

However, $W_\xi \to 0$ as $\xi \to \pm\infty$ since crossing becomes impossible as threshold grows. If, in addition, we assume that this decay is faster than $1/|\xi|$, i.e., $\xi W_\xi \to 0$ ($\xi \to \pm\infty$), we have

$$\Big\langle R(t)|x \Big\rangle = \int_{-\infty}^{\infty} W_\xi(t|x) d\xi. \qquad (10.81)$$

It is worth noticing that one can arrive at this expression in a more direct way. In effect, recalling the definition of the range as the difference between the maximum and the minimum, we have

$$\langle R(t)|x \rangle = \langle M(t)|x \rangle - \langle m(t)|x \rangle, \qquad (10.82)$$

and substituting for Eqs. (10.53) and (10.60) of the average maximum and minimum respectively, we get

$$\langle R(t)|x \rangle = \int_x^\infty W_\xi(t|x)d\xi + \int_{-\infty}^x W_\xi(t|x)d\xi$$

$$= \int_{-\infty}^\infty W_\xi(t|x)d\xi,$$

which is precisely Eq. (10.81).

There are no simple expressions, beside that of Eq. (10.81), for higher moments of the span as there are for the other extremes treated above. In the present case moments have to be evaluated through their definition and the use of Eq. (10.70)

$$\langle R^n(t)|x \rangle = \int_0^\infty r^n f_R(r,t|x)dr$$

$$= \int_0^\infty r^n dr \int_{x-r}^x \frac{\partial^2 S_{v,r+v}(t|x)}{\partial r^2} dv.$$

This is quite unfortunate because the evaluation of span moments becomes a complicated business even numerically. The reason for not having a more convenient expression lies essentially in the fact that maxima and minima are generally correlated quantities and these correlations appear in all moments greater than the first one.

Chapter 11

The level crossing problem for diffusion processes

Having introduced the general theory of the level-crossing problem, we will now apply it to diffusion processes. We have extensively studied such kind of processes in Chapter 4 for one-dimensional diffusions, Chapter 5 for multidimensional diffusions and also Chapter 7 from the point of view of stochastic differential equations. We first address the problem in one dimension and for time homogeneous processes where drift $f(x)$ and diffusion coefficient $D(x)$ do not depend explicitly of time. In the last part of this chapter we will address the problem for multidimensional diffusions.

Assuming that the process starts at the initial value $X(0) = x$, we know that the transition density (or propagator)

$$p(x', t|x)dx' = \text{Prob}\{X(t) = x'|X(0) = x\}$$

obeys the Fokker-Planck equation which in its backward version reads[1]

$$\frac{\partial p}{\partial t} = f(x)\frac{\partial p}{\partial x} + \frac{1}{2}D(x)\frac{\partial^2 p}{\partial x^2}, \qquad (11.1)$$

with the initial condition

$$p(x', 0|x) = \delta(x' - x). \qquad (11.2)$$

We will now show that the knowledge of the transition density, with appropriate boundary conditions, is equivalent to knowing the escape and first-passage time probabilities. We begin with the escape problem.

11.1 The escape problem

Suppose a time-homogeneous diffusion process $X(t)$ defined in an interval (a, b) with transition density $p(x', t|x)$, where $x = X(0) \in (a, b)$ is the initial

---

[1]See Eq. (4.38) of Chapter 4 with the replacements $x \to x'$ and $x_0 \to x$.

value of the process. The probability that $X(t)$ has not left (a,b) at time t, that is, the survival probability, is thus given by the integral

$$S_{ab}(t|x) = \int_a^b p(x',t|x)dx'. \qquad (11.3)$$

Integrating the backward Fokker-Planck equation (11.1) with respect to final value x', commuting derivatives with integrals and taking into account Eqs. (11.2) and (11.3), we see that the survival probability obeys the equation

$$\frac{\partial S_{ab}}{\partial t} = f(x)\frac{\partial S_{ab}}{\partial x} + \frac{1}{2}D(x)\frac{\partial^2 S_{ab}}{\partial x^2}, \qquad (11.4)$$

with initial condition

$$S_{ab}(0|x) = 1, \qquad (11.5)$$

where $x \in (a,b)$. Realizing that for any diffusion process survival is impossible if the process starts at each end of the interval (a,b), we see that $S_{ab}(t|x)$ must obey the following boundary conditions

$$S_{ab}(t|a) = S_{ab}(t|b) = 0. \qquad (11.6)$$

Let us incidentally note that the boundary conditions on $S_{ab}(t|x)$ along with Eq. (11.3) imply that the propagator $p(x',t|x)$ also obeys the boundary conditions [2]

$$p(x',t|a) = p(x',t|b) = 0. \qquad (11.7)$$

Therefore, the propagator is the solution of the initial and boundary-value problem (11.1), (11.2) and (11.7).

Recall that the escape, or exit, probability out of (a,b) is given by

$$W_{ab}(t|x) = 1 - S_{ab}(t|x). \qquad (11.8)$$

Hence, W_{ab} follows the same Fokker-Planck equation than S_{ab},

$$\frac{\partial W_{ab}}{\partial t} = f(x)\frac{\partial W_{ab}}{\partial x} + \frac{1}{2}D(x)\frac{\partial^2 W_{ab}}{\partial x^2}, \qquad (11.9)$$

but with initial and boundary conditions reversed:

$$W_{ab}(0|x) = 0, \qquad W_{ab}(t|a) = W_{ab}(t|b) = 1. \qquad (11.10)$$

---

[2]In effect, remember that $p(x',t|x)$ is a non-negative function, so that the condition

$$S(a|t) = \int_a^b p(x',t|a)dx' = 0$$

necessarily implies that $p(x',t|a) = 0$ and similarly when $x = b$.

Using Laplace transform methods the initial and boundary value problem (11.4)-(11.6) becomes simpler. Indeed, the Laplace transform

$$\hat{S}_{ab}(s|x) = \int_0^\infty e^{-st} S_{ab}(t|x) dt,$$

and the standard property

$$\mathcal{L}\left\{ \frac{\partial S_{ab}(t|x)}{\partial t} \right\} = s\hat{S}_{ab}(s|x) - S_{ab}(0|x)$$

turn Eq. (11.4) into the ordinary differential equation (recall the initial condition $S_{ab}(0|x) = 1$)

$$\frac{1}{2}D(x)\frac{d^2\hat{S}_{ab}}{dx^2} + f(x)\frac{d\hat{S}_{ab}}{dx} - s\hat{S}_{ab} = -1, \tag{11.11}$$

with boundary conditions

$$\hat{S}_{ab}(s|a) = \hat{S}_{ab}(s|b) = 0. \tag{11.12}$$

Since the Laplace transform of the survival probability $\hat{S}_{ab}(s|x)$ is related to the exit probability $\hat{W}_{ab}(s|x)$ by [3]

$$\hat{S}_{ab}(s|x) = \frac{1}{s} - \hat{W}_{ab}(s|x), \tag{11.13}$$

we see from Eqs. (11.11) and (11.12) that the transformed exit probability $\hat{W}_{ab}$ is the solution to the homogeneous equation

$$\frac{1}{2}D(x)\frac{d^2\hat{W}_{ab}}{dx^2} + f(x)\frac{d\hat{W}_{ab}}{dx} - s\hat{W}_{ab} = 0, \tag{11.14}$$

with inhomogeneous boundary conditions

$$\hat{W}_{ab}(s|a) = \hat{W}_{ab}(s|b) = \frac{1}{s}. \tag{11.15}$$

11.1.1 *The mean exit time*

Let us now show that the mean exit time, $T_{ab}(x)$, obeys an even simpler equation than Eq. (11.11). In effect, recalling that (cf. Eq. (10.16), Chapter 10)

$$T_{ab}(x) = \hat{S}_{ab}(0|x), \tag{11.16}$$

and setting $s = 0$ in Eqs. (11.11) and (11.12) we see that the mean exit time obeys the ordinary differential equation

$$\frac{1}{2}D(x)\frac{d^2T_{ab}}{dx^2} + f(x)\frac{dT_{ab}}{dx} = -1, \tag{11.17}$$

---

[3]Take the Laplace transform of Eq. (11.8) and recall that $\mathcal{L}\{1\} = 1/s$.

with boundary conditions

$$T_{ab}(a) = T_{ab}(b) = 0. \tag{11.18}$$

Note that Eq. (11.17) is a first-order linear equation for the derivative $dT_{ab}(x)/dx$, hence the first step toward the solution is straightforward and reads

$$\frac{dT_{ab}}{dx} = e^{-\psi(x)}\left[C - \Psi(x)\right], \tag{11.19}$$

where C is an integration constant and $\psi(x)$ and $\Psi(x)$ are the indefinite integrals:

$$\psi(x) = \exp\left\{2\int \frac{f(x)}{D(x)}dx\right\}, \qquad \Psi(x) = 2\int \frac{e^{\psi(x)}}{D(x)}dx. \tag{11.20}$$

Integrating Eq. (11.19) yields

$$T_{ab}(x) = C\int_a^x e^{-\psi(x')}dx' - \int_a^x e^{-\psi(x')}\Psi(x')dx'. \tag{11.21}$$

Let us note that this expression satisfies the boundary condition $T_{ab}(a) = 0$. We finally obtain constant C by using the second boundary condition $T_{ab}(b) = 0$. The result is

$$C = \frac{\int_a^b e^{-\psi(x')}\Psi(x')dx'}{\int_a^b e^{-\psi(x')}dx'}. \tag{11.22}$$

This development can be extended without any difficulty to higher-order moments. In effect, taking successive derivatives with respect to s in Eq. (11.11) and recalling that (cf. Eq. (10.18))

$$T_{ab}^{(n)}(x) = (-1)^{n-1}n\left.\frac{\partial^{n-1}\hat{S}_{ab}}{\partial s^{n-1}}\right|_{s=0}$$

we easily see that the mean exit time moments obey the following set of recursive equations:

$$\frac{1}{2}D\frac{d^2 T_{ab}^{(n)}}{dx^2} + f(x)\frac{dT_{ab}^{(n)}}{dx} = -nT_{ab}^{(n-1)}, \tag{11.23}$$

$(n = 2, 3, \dots)$. The boundary conditions attached to these equations are easily obtained by combining Eqs. (10.18) and (11.12) and read

$$T_{ab}^{(n)}(a) = T_{ab}^{(n)}(b) = 0. \tag{11.24}$$

11.2 The first-passage problem

We next address the hitting problem. As shown in Sect. 10.3, the problem is equivalent to the exit problem out of a semi-infinite interval. Thus, if the initial value of the process is above threshold, first-passage to x_c is equivalent to escape out of (x_c, ∞), while if the initial value is below threshold, first-passage is equivalent to escape out of $(-\infty, x_c)$. That is,

$$S_c(t|x) = S_{x_c,\infty}(t|x) \ (x > x_c), \quad S_c(t|x) = S_{-\infty,x_c}(t|x) \ (x < x_c), \quad (11.25)$$

where $S_c(t|x)$ is the survival probability of not having reached threshold x_c at time t starting initially at x and $S_{x_c,\infty}(t|x)$ and $S_{-\infty,x_c}(t|x)$ are the survival probabilities of the semi-infinite intervals (x_c, ∞) and $(-\infty, x_c)$ respectively.

We, therefore, see that all definitions and expressions obtained above for the exit problem are valid for the first-passage problem as well. In particular $S_c = S_c(t|x)$ obeys the Fokker-Planck equation

$$\frac{\partial S_c}{\partial t} = f(x)\frac{\partial S_c}{\partial x} + \frac{1}{2}D(x)\frac{\partial^2 S_c}{\partial x^2}, \tag{11.26}$$

with initial and boundary conditions

$$S_c(0|x) = 1, \qquad S_c(t|x_c) = 0. \tag{11.27}$$

Before proceeding further we note that the equation satisfied by the hitting, or first-passage, probability,

$$W_c(t|x) = 1 - S_c(t|x),$$

is given by

$$\frac{\partial W_c}{\partial t} = f(x)\frac{\partial W_c}{\partial x} + \frac{1}{2}D(x)\frac{\partial^2 W_c}{\partial x^2}, \tag{11.28}$$

with initial and boundary conditions

$$W_c(0|x) = 0, \qquad W_c(t|x_c) = 1. \tag{11.29}$$

There are other boundary conditions which are singular because they refer to the behavior of $S_c(t|x)$ and $W_c(t|x)$ as $x \to \pm\infty$. Thus, for instance, in the case $x > x_c$ (starting above threshold) the initial value x is bounded from below and we must assume that the survival probability behaves properly as $x \to \infty$. That is,

$$\lim_{x\to\infty} S_c(t|x) = \text{finite}, \qquad (x > x_c). \tag{11.30}$$

On the other hand, when the process starts below threshold, x is bounded from above $(x < x_c)$ and we must add the following singular boundary condition

$$\lim_{x \to -\infty} S_c(t|x) = \text{finite}, \qquad (x < x_c). \qquad (11.31)$$

Following previous steps we see that the Laplace transform, $\hat{S}_c(s|x)$, of the survival probability obeys the ordinary differential equation

$$\frac{1}{2} D(x) \frac{d^2 \hat{S}_c}{dx^2} + f(x) \frac{d \hat{S}_c}{dx} - s\hat{S}_c(s|x) = -1, \qquad (11.32)$$

with boundary condition

$$\hat{S}_c(s|x_c) = 0. \qquad (11.33)$$

Equation (11.32) is a linear ordinary differential equation of second order. The only boundary condition (11.33) does not suffice to obtain a complete solution to the problem. A second boundary condition –in this case a singular one– is provided by one of the additional conditions (11.30) or (11.31),

$$\lim_{x \to \infty} \hat{S}_c(s|x) = \text{finite}, \quad (x > x_c); \quad \lim_{x \to -\infty} \hat{S}_c(s|x) = \text{finite}, \quad (x < x_c).$$
$$(11.34)$$

From Eqs. (11.32)-(11.33) we see that the mean first-passage time, given as in Eq. (11.16) by

$$T_c(x) = \hat{S}_c(0|x) \qquad (11.35)$$

obeys the equation [see also Eqs. (11.17)-(11.18)]

$$\frac{1}{2} D(x) \frac{d^2 T_c}{dx^2} + f(x) \frac{dT_c}{dx} = -1, \qquad (11.36)$$

with boundary condition

$$T_c(x_c) = 0. \qquad (11.37)$$

Note that obtaining the mean first-passage time through Eqs. (11.36)-(11.37) leads to indeterminate results because the singular conditions given in Eq. (11.34) become meaningless. Indeed, starting at infinite distance from threshold results in an infinite mean first-passage time:

$$\lim_{x \to \pm\infty} T_c(x) = \infty.$$

In many cases this condition is of no help for solving Eq. (11.36) and, therefore, obtaining the mean first-passage time has to be done through Eq. (11.35) after solving problem (11.32)-(11.34) for the survival probability.

Pursuing the analogy of first-passage as an escape problem out of a semi-infinite interval, we easily see that the first-passage time moments obey the recursive equation

$$\frac{1}{2} D \frac{d^2 T_c^{(n)}}{dx^2} + f(x) \frac{dT_c^{(n)}}{dx} = -n T_c^{(n-1)}, \qquad (11.38)$$

$(n = 2, 3, \dots)$. The boundary condition attached to this equation is easily obtained by combining Eq. (10.35) of Chapter 10 with Eq. (11.12). It reads

$$T_c^{(n)}(c) = 0. \qquad (11.39)$$

As before, the solution of the problem (11.38)-(11.39) leads to indeterminate results and obtaining the mean first-passage time moments has to be done through Eq. (10.35):

$$T_c^{(n)}(x) = (-1)^{n-1} n \left. \frac{\partial^{n-1} \hat{S}_c(s|x)}{\partial s^{n-1}} \right|_{s=0}$$

after solving Eqs. (11.32)-(11.34) for $\hat{S}_c(s|x)$.

In the next two sections we review the level crossing problem for two relevant diffusion processes which still allow for an exact treatment of the problem. These are the Brownian motion and the Ornstein-Uhlenbeck processes. There is a third diffusion process, called the Feller process, that is also solvable.[4] The process has the remarkable property that it never attains negative values, a feature that makes the Feller process a good candidate for modeling many applications particularly in socio-economic systems and population dynamics. We will treat the Feller processes in the next chapter.

11.3   The Wiener process

As we have seen in Chapter 4, the Wiener process or Brownian motion is the simplest but most fundamental diffusion process. The simplicity of the process arises from the fact that it has no drift and the diffusion coefficient is constant. We next study the first-passage escape problems as well as the extremes of the process.

---

[4]See, for instance, Masoliver (2014a) for a complete treatment of the Wiener and Feller processes. Also, Masoliver (2014b) for the Ornstein-Uhlenbeck process.

11.3.1 *Hitting and escape problems*

We begin with the first-passage problem. We know that the Laplace transform of the survival probability $\hat{S}_c(s|x)$ to threshold x_c satisfies the boundary-value problem (11.32)-(11.33) which in this case, where $f(x) = 0$ and $D(x) = D,$[5] reduces to

$$\frac{1}{2}D\frac{d^2\hat{S}_c}{dx^2} - s\hat{S}_c = -1, \qquad \hat{S}_c(s|x_c) = 0. \qquad (11.40)$$

As we can easily check by direct substitution, the solution to this problem that is finite for both $x > x_c$ and $x < x_c$ is simple and reads

$$\hat{S}_c(s|x) = \frac{1}{s}\left[1 - e^{-\sqrt{2s/D}|x-x_c|}\right]. \qquad (11.41)$$

Recall that in terms of $\hat{S}_c(s|x)$ the mean first-passage time is given by $T_c(x) = \hat{S}_c(0|x)$. We thus expand Eq. (11.41) in powers of s and see that

$$\hat{S}_c(s|x) = \sqrt{\frac{2}{sD}}|x - x_c| + O(1),$$

from which we conclude that the mean first-passage time is infinite,

$$T_c(x) = \infty,$$

and the same applies to all moments. Therefore, the Wiener process possesses no first-passage time moments.

Bearing in mind the following inversion formula (Roberts and Kaufman, 1966)

$$\mathcal{L}^{-1}\left\{\left(1 - e^{-a\sqrt{s}}\right)/s\right\} = \text{erf}\left[a/\sqrt{t}\right],$$

where $\text{erf}(z)$ is the error function defined as

$$\text{erf}(z) = \frac{2}{\sqrt{\pi}}\int_0^z e^{-\xi^2}d\xi, \qquad (11.42)$$

we see that the Laplace inversion of Eq. (11.41) leads to the survival probability in real time

$$S_c(t|x) = \text{erf}\left[\frac{|x - x_c|}{\sqrt{2Dt}}\right]. \qquad (11.43)$$

We next consider the escape problem out of the interval (a, b). The Laplace transform of the survival probability, $\hat{S}_{ab}(s|x)$, obeys the same equation than that of $\hat{S}_c$ [cf. Eq. (11.40)],

$$\frac{1}{2}D\frac{d^2\hat{S}_{ab}}{dx^2} - s\hat{S}_{ab} = -1,$$

---

[5]In Chapter 4 we have assumed that $D = 1$.

but with two boundary conditions instead of one

$$\hat{S}_{ab}(s|a) = \hat{S}_{ab}(s|b) = 0.$$

The solution to this problem is readily found to be

$$\hat{S}_{ab}(s|x) = \frac{1}{s}\left[1 - \frac{\cosh\sqrt{2s/D}[x - (a+b)/2]}{\cosh\sqrt{2s/D}[(a-b)/2]}\right]. \tag{11.44}$$

The mean exit time is given by

$$T_{ab}(x) = \hat{S}_{ab}(0|x)$$

and, since as $s \to 0$ a Taylor expansion yields

$$1 - \frac{\cosh\sqrt{2s/D}[x - (a+b)/2]}{\cosh\sqrt{2s/D}[(a-b)/2]} = \frac{s}{D}(x-a)(b-x) + O(s^2),$$

we see that the mean exit time follows the quadratic law:

$$T_{ab}(x) = \frac{1}{D}(x-a)(b-x), \tag{11.45}$$

which for a symmetric interval around the origin ($a = -\xi$ and $b = \xi$, where $\xi > 0$) yields the inverted parabola

$$T_{-\xi\xi}(x) = (\xi^2 - x^2)/D.$$

The inverse Laplace transform of Eq. (11.44) may be found in standard tables (Roberts and Kaufman, 1966) but we will not write here for any interval (a, b). However, for a symmetric interval $(-\xi, \xi)$ the inverse transform is somewhat simpler and yields

$$S_{-\xi\xi}(t|x) = \frac{2}{\pi}\sum_{n=0}^{\infty}\frac{(-1)^n}{n+1/2}e^{-D(n+1/2)^2\pi^2 t/\xi^2}\cos\left[(n+1/2)\pi x/\xi\right].$$

11.3.2 *Extremes*

We know from Chapter 10 that the probability density function of the maximum value is related to the survival probability of the first-passage problem by

$$\varphi_{\max}(\xi, t|x) = \frac{\partial S_\xi(t|x)}{\partial \xi}\Theta(\xi - x), \tag{11.46}$$

which, after substituting for Eq. (11.43) yields the following truncated Gaussian density

$$\varphi_{\max}(\xi, t|x) = \left(\frac{2}{\pi Dt}\right)^{1/2}e^{-(\xi-x)^2/2Dt}\Theta(\xi - x). \tag{11.47}$$

The mean maximum defined by (cf. Eqs. (10.50))

$$\left\langle M(t)|x \right\rangle = \int_{-\infty}^{\infty} \xi \varphi_{\max}(\xi, t|x) d\xi,$$

is then given by

$$\left\langle M(t)|x \right\rangle = \left(\frac{2}{\pi Dt}\right)^{1/2} \int_{x}^{\infty} \xi e^{-(\xi-x)^2/2Dt} d\xi.$$

That is,

$$\left\langle M(t)|x \right\rangle = x + \left(\frac{2Dt}{\pi}\right)^{1/2}. \tag{11.48}$$

Likewise, the probability density of the minimum value is given by (cf. Eqs. (10.52), (10.57) and (11.43))

$$\varphi_{\min}(\xi, t|x) = \left(\frac{2}{\pi Dt}\right)^{1/2} e^{-(x-\xi)^2/2Dt} \Theta(x - \xi), \tag{11.49}$$

and the mean minimum reads

$$\left\langle m(t)|x \right\rangle = x - \left(\frac{2Dt}{\pi}\right)^{1/2}. \tag{11.50}$$

Notice that both extreme values grow like $t^{1/2}$ as $t \to \infty$, which is the typical behavior of the Wiener process.

These results can be generalized to include any moment of the maximum and the minimum. By combining Eqs. (10.54) and (11.47) we easily see that

$$\left\langle M^n(t)|x \right\rangle = \frac{1}{\sqrt{\pi}} \sum_{k=0}^{n} \binom{n}{k} \Gamma\left(\frac{k+1}{2}\right) (2Dt)^{k/2} x^{n-k} \tag{11.51}$$

($n = 1, 2, 3, \dots$). Following an analogous reasoning we show that the moments of the minimum are

$$\left\langle m^n(t)|x \right\rangle = \frac{1}{\sqrt{\pi}} \sum_{k=0}^{n} (-1)^k \binom{n}{k} \Gamma\left(\frac{k+1}{2}\right) (2Dt)^{k/2} x^{n-k} \tag{11.52}$$

($n = 1, 2, 3, \dots$). With increasing n these expressions become rather clumsy. We can get, however, simpler expressions if instead of the maximum or the minimum we consider their "distance" from the initial position. This is defined by $M(t) - x$ in the case of the maximum or by $x - m(t)$ for the minimum. We have

$$\left\langle [M(t) - x]^n|x \right\rangle = \left\langle [x - m(t)]^n|x \right\rangle = \frac{1}{\sqrt{\pi}} \Gamma\left(\frac{n+1}{2}\right) (2Dt)^{n/2}. \tag{11.53}$$

Both distances are equal showing the otherwise obvious symmetry of the process.

11.4 The Ornstein-Uhlenbeck process *

The Ornstein-Uhlenbeck (OU) process has also been treated in Chapter 4. It is a one-dimensional process with linear drift and constant diffusion coefficient. In its most general form the process is governed by the stochastic differential equation

$$dX(t) = -\alpha[X(t) - \mu]dt + \sigma dW(t),$$

where $\alpha > 0$, μ and $\sigma > 0$ are constant parameters and μ is the stationary mean value to which the process reverts as $t \to \infty$ (sometimes called "normal level"). In order to study first-passage properties, it is convenient to scale time and shift and scale the process in the following way:

$$t' = \alpha t, \qquad X' = \frac{\alpha^{1/2}}{\sigma}(X - \mu), \tag{11.54}$$

and the Langevin equation above reads

$$dX'(t') = -X'(t')dt' + dW'(t'),$$

where $W'(t') = \alpha^{-1/2}W(t'/\alpha)$ (in what follows we will drop the primes). This equation corresponds to a diffusion process with $f(x) = -x$ and $D = 1$.

11.4.1 *First-passage probability*

We address the level-crossing problem for the OU problem. Focussing on the hitting problem to some threshold x_c, we have seen that the first-passage probability $W_c(t|x)$ is the solution to the initial and boundary-value problem given in Eqs. (11.28) and (11.29). In this case, where $f(x) = -x$ and $g(x) = 1$, we have the following partial differential equation

$$\frac{\partial W_c}{\partial t} = -x\frac{\partial W_c}{\partial x} + \frac{1}{2}\frac{\partial^2 W_c}{\partial x^2}, \tag{11.55}$$

with initial and boundary conditions

$$W_c(0|x) = 0, \qquad W_c(t|x_c) = 1. \tag{11.56}$$

The Laplace transform

$$\hat{W}_c(s|x) = \int_0^\infty e^{-st}W_c(t|x)dt$$

yields the ordinary differential equation

$$\frac{d^2\hat{W}_c}{dx^2} - 2x\frac{d\hat{W}_c}{dx} - 2s\hat{W}_c = 0, \tag{11.57}$$

with the boundary condition

$$\hat{W}_c(s|x_c) = \frac{1}{s}. \tag{11.58}$$

In order to solve this problem we define the new variable $z = x^2$ and see that $\hat{W}_c$ obeys the Kummer equation

$$\frac{d^2\hat{W}_c}{dz^2} + \left(\frac{1}{2} - z\right)\frac{d\hat{W}_c}{dz} - \frac{s}{2}\hat{W}_c = 0. \tag{11.59}$$

The general solution to this equation is given by a linear combination of the Kummer functions $F(s/2, 1/2, z)$ and $U(s/2, 1/2, z)$ (Magnus *et al.*, 1966). In the original variable x we have

$$\hat{W}_c(s|x) = AF(s/2, 1/2, x^2) + BU(s/2, 1/2, x^2),$$

where A and B are arbitrary constants which have to be found not only through the boundary condition given by Eq. (11.56) –which is insufficient to get both A and B– but also by imposing the finiteness of W_c for any value of the initial position x. To this end we must distinguish two situations: (i) $|x| \leq |x_c|$, and (ii) $|x| \geq |x_c|$. Taking into account the following asymptotic limits of the Kummer functions (Magnus *et al.*, 1966)

$$F(s/2, 1/2, x^2) \to \infty, \ |x| \to \infty \quad \text{and} \quad U(s/2, 1/2, x^2) \to 0, \ |x| \to \infty,$$

we see that for case (i) when x^2 remains bounded ($0 \leq x^2 \leq x_c^2$), a finite hitting probability is given by $W_c(t|x) = AF(s/2, 1/2, x^2)$ and the boundary condition (11.58) fixes A to be $A = 1/sF(s/2, 1/2, x_c^2)$. Hence,

$$\hat{W}_c(s|x) = \frac{F(s/2, 1/2, x^2)}{sF(s/2, 1/2, x_c^2)}, \qquad |x| \leq |x_c|. \tag{11.60}$$

In the case (ii) when x^2 is unbounded from above ($x_c^2 \leq x^2 \leq \infty$), the finite hitting probability must be of the form $W_c(t|x) = BU(s/2, 1/2, x^2)$ and the boundary condition (11.58) yields $B = 1/sU(s/2, 1/2, x_c^2)$. That is,

$$\hat{W}_c(s|x) = \frac{U(s/2, 1/2, x^2)}{sU(s/2, 1/2, x_c^2)}, \qquad |x| \geq |x_c|. \tag{11.61}$$

Reaching the origin

In general, the analytical inversion of Eqs. (11.60) and (11.61) seems to be beyond reach, even though numerical inversions are perfectly feasible. There is, however, a special case in which the exact analytical inversion is still possible. This is the case of the first-passage probability to threshold $x_c = 0$. This probability is more relevant than it might seem at first sight, because in the original variables (cf. Eq. (11.54)) hitting the origin for $X(t)$ implies reaching the stationary mean value μ of the original OU process which constitutes a very relevant and rather practical information in many applications.

Let us denote by $W_0(t|x)$ the first-passage probability to the origin starting at x. Since now $x_c = 0$ and obviously $|x| \geq 0$, the Laplace transform of this probability will be given by Eq. (11.61)

$$\hat{W}_0(s|x) = \frac{U(s/2, 1/2, x^2)}{sU(s/2, 1/2, 0)}.$$

We have shown elsewhere (Masoliver, 2014b) that the Laplace inversion of this equation results in the following Gaussian probability

$$W_0(t|x) = \mathrm{Erfc}\left[\frac{|x|e^{-t}}{\sqrt{1 - e^{-2t}}}\right], \qquad (11.62)$$

where $\mathrm{Erfc}(\cdot)$ is the complementary error function.

The mean first-passage time

We have also shown in Chapter 10 (cf. Eq. (10.32)) that in terms of $\hat{W}_c(s|x)$ the mean first-passage time to x_c is given by

$$T_c(x) = \lim_{s \to 0}\left[\frac{1}{s} - \hat{W}_c(s|x)\right].$$

Hence when $|x| \leq |x_c|$ we have

$$T_c(x) = \lim_{s \to 0}\left[\frac{F(s/2, 1/2, x_c^2) - F(s/2, 1/2, x^2)}{sF(s/2, 1/2, x_c^2)}\right], \qquad (11.63)$$

while when $|x| \geq x_c$ we have

$$T_c(x) = \lim_{s \to 0}\left[\frac{U(s/2, 1/2, x_c^2) - U(s/2, 1/2, x^2)}{sU(s/2, 1/2, x_c^2)}\right]. \qquad (11.64)$$

On the other hand, we have proved (Masoliver, 2014b) that as $s \to 0$

$$F(s/2, 1/2, x^2) = 1 + s\int_0^{x^2} F(1, 3/2, z)dz + O(s^2), \qquad (11.65)$$

and

$$U(s/2, 1/2, x^2) = 1 - \frac{s}{2}\left[\psi(1/2) + \int_0^{x^2} U(1, 3/2, z)dz\right] + O(s^2), \quad (11.66)$$

where $\psi(z) = \Gamma'(z)/\Gamma(z)$ is the digamma function. Hence,

$$sF(s/2, 1/2, x_c^2) = s + O(s^2) \quad \text{and} \quad sU(s/2, 1/2, x_c^2) = s + O(s^2).$$

Therefore, plugging Eqs. (11.65) and (11.66) into Eqs. (11.63) and (11.64) yields

$$T_c(x) = \int_{x^2}^{x_c^2} F(1, 3/2, z)dz, \qquad (|x| \leq |x_c|), \qquad (11.67)$$

and

$$T_c(x) = \frac{1}{2} \int_{x_c^2}^{x^2} U(1, 3/2, z)dz, \qquad (|x| \geq |x_c|). \qquad (11.68)$$

11.4.2 *The exit problem*

We now address the exit problem out of some interval (a, b) for the OU process. As we know, the Laplace transform of the exit probability $\hat{W}_{ab}(s|x)$ obeys the same equation than the hitting probability (see Eq. (11.57))

$$\frac{d^2\hat{W}_{ab}}{dx^2} - 2x\frac{d\hat{W}_{ab}}{dx} - 2s\hat{W}_{ab} = 0, \qquad (11.69)$$

but with two boundary conditions

$$\hat{W}_{ab}(s|a) = \hat{W}_{ab}(s|b) = \frac{1}{s}. \qquad (11.70)$$

In order to solve this problem it turns out to be more convenient to follow a different way than that we have taken for the first-passage problem treated above. Thus, instead of making the change of variable $z = x^2$ and transform Eq. (11.69) into the Kummer equation (11.59), we recognize Eq. (11.69) as the Weber differential equation for which the general solution is given by (Magnus *et al.*, 1966)

$$\hat{W}_{ab}(s|x) = e^{x^2/2}\big[AD_{-s}(\sqrt{2}x) + BD_{-s}(-\sqrt{2}x)\big],$$

where $D_\nu(\pm z)$ are Weber functions and A and B arbitrary constants to be determined by the boundary conditions (11.70). For a symmetric interval $(-\xi, \xi)$ the solution has a simpler form and reads

$$\hat{W}_{-\xi\xi}(s|x) = e^{(x^2-\xi^2)/2}\frac{D_{-s}(\sqrt{2}x) + D_{-s}(-\sqrt{2}x)}{s[D_{-s}(\sqrt{2}\xi) + D_{-s}(-\sqrt{2}\xi)]}, \qquad (11.71)$$

as the reader can easily check. This expression can be written in a simpler form after using the following relationship between the Weber function D and the Kummer functions F and U (Magnus *et al.*, 1966)

$$D_{-s}(\pm\sqrt{2}x) = \frac{e^{-x^2/2}}{2^{s/2}}\left[\frac{\Gamma(1/2)}{\Gamma(s/2+1/2)}F\left(\frac{s}{2}, \frac{1}{2}, x^2\right) \right.$$

$$\left. \pm\frac{\Gamma(-1/2)}{\Gamma(s/2)}F\left(\frac{1+s}{2}, \frac{3}{2}, x^2\right) \right].$$

Substituting into Eq. (11.71) and some cancellation of terms yields

$$\hat{W}_{-\xi\xi}(s|x) = \frac{F(s/2, 1/2, x^2)}{sF(s/2, 1/2, \xi^2)}, \qquad (11.72)$$

$(-\xi < x < \xi)$. Let us incidentally note that this expression for the escape probability out of the symmetric interval $(-\xi, \xi)$ is formally equal to the first-passage probability to threshold ξ (cf. Eq. (11.60)).

In terms of $\hat{W}_{ab}$ the mean exit time is given by

$$T_{ab}(x) = \lim_{s\to 0}\left[\frac{1}{s} - \hat{W}_{ab}(s|x)\right]$$

(cf. Eq. (10.17) of Chapter 10). In the case of a symmetric interval we then have

$$T_{-\xi\xi}(x) = \lim_{s\to 0}\frac{F(s/2, 1/2, \xi^2) - F(s/2, 1/2, x^2)}{sF(s/2, 1/2, \xi^2)},$$

which, after using the expansion given in Eq. (11.65), reads

$$T_{-\xi\xi}(x) = \int_{x^2}^{\xi^2} F(1, 3/2, z)dz. \qquad (11.73)$$

Again, the mean exit time out of $(-\xi, \xi)$ is equal to the mean first-passage time to threshold ξ (cf. Eq. (11.67)).

11.4.3 *Extremes*

Let us now report on the extreme values attained by the OU process. We recall from Chapter 10 that the distribution functions of the maximum and minimum values that a random process attains at time t are respectively given by (cf. Eqs. (10.48), (10.52) and (10.56))

$$\Phi_{\max}(\xi, t|x) = \left[1 - W_\xi(t|x)\right]\Theta(\xi - x), \qquad (11.74)$$

and

$$\Phi_{\min}(\xi, t|x) = \Theta(\xi - x) + W_\xi(t|x)\Theta(x - \xi), \qquad (11.75)$$

where $W_\xi(t|x)$ is the first-passage probability to level ξ and $\Theta(\cdot)$ is the Heaviside step function.

Since, as shown in Eqs. (11.60) and (11.61), the value of $\hat{W}_\xi(s|t)$ depends on whether $|x| > |\xi|$ or $|x| < |\xi|$ we must thus relate these two conditions with $\Theta(\xi - x)$ and $\Theta(x - \xi)$.

Let us observe that if $\xi > 0$ condition $\xi > x$ implies $|\xi| > |x|$, while if $x < 0$ the condition $\xi > x$ implies $|\xi| < |x|$. Therefore,

$$\Theta(\xi - x) = \Theta(\xi)\Theta(|\xi| - |x|) + \Theta(-x)\Theta(|x| - |\xi|). \qquad (11.76)$$

Exchanging ξ and x we also have

$$\Theta(x - \xi) = \Theta(x)\Theta(|x| - |\xi|) + \Theta(-\xi)\Theta(|\xi| - |x|). \tag{11.77}$$

Collecting results we see that for the OU process the Laplace transform of the maximum and minimum distribution functions respectively read

$$\hat{\Phi}_{\max}(\xi, s|x) = \frac{1}{s}\left[1 - \frac{F(s/2, 1/2, x^2)}{F(s/2, 1/2, \xi^2)}\right]\Theta(\xi)\Theta(|\xi| - |x|) \tag{11.78}$$

$$+ \frac{1}{s}\left[1 - \frac{U(s/2, 1/2, x^2)}{U(s/2, 1/2, \xi^2)}\right]\Theta(-x)\Theta(|x| - |\xi|),$$

and

$$\hat{\Phi}_{\min}(\xi, s|x) = \frac{1}{s}\left[\Theta(\xi) + \theta(-\xi)\frac{F(s/2, 1/2, x^2)}{F(s/2, 1/2, \xi^2)}\right]\Theta(|\xi| - |x|) \tag{11.79}$$

$$+ \frac{1}{s}\left[\Theta(-x) + \Theta(x)\frac{U(s/2, 1/2, x^2)}{U(s/2, 1/2, \xi^2)}\right]\Theta(|x| - |\xi|).$$

The average maximum and minimum

The average maximum $M(t)$ and minimum $m(t)$ of the process at time t are given by (cf. Eqs. (10.53) and (10.60) of Chapter 10)

$$\left\langle M(t)|x \right\rangle = x + \int_x^\infty W_\xi(t|x)d\xi, \tag{11.80}$$

and

$$\left\langle m(t)|x \right\rangle = x - \int_{-\infty}^x W_\xi(t|x)d\xi, \tag{11.81}$$

where $W_\xi(t|x)$ is the first-passage probability to the level ξ. For the OU process we know the Laplace transform of this probability. We, therefore, transform these equations:

$$\left\langle \hat{M}(s)|x \right\rangle = \frac{x}{s} + \int_x^\infty \hat{W}_\xi(s|x)d\xi \tag{11.82}$$

and

$$\left\langle \hat{m}(s)|x \right\rangle = \frac{x}{s} - \int_{-\infty}^x \hat{W}_\xi(s|x)d\xi, \tag{11.83}$$

where

$$\left\langle \hat{M}(s)|x \right\rangle \equiv \int_0^\infty e^{-st}\left\langle M(t)|x \right\rangle dt,$$

is the Laplace transform of the maximum and with a similar expression for the minimum.

For the OU process we have (cf. Eqs. (11.82) and (11.76))

$$\left\langle \hat{M}(s)|x\right\rangle = \frac{x}{s} + \int_{-\infty}^{\infty} \Theta(\xi - x)\hat{W}_\xi(s|x)d\xi$$

$$= \frac{x}{s} + \int_{-\infty}^{\infty} \left[\Theta(\xi)\Theta(|\xi| - |x|) + \Theta(-x)\Theta(|x| - |\xi|)\right]\hat{W}_\xi(s|x)d\xi,$$

which, after substituting for Eqs. (11.60) and (11.61), yields

$$\left\langle \hat{M}(s)|x\right\rangle = \frac{1}{s}\left[x + F(s/2, 1/2, x^2)\int_{-\infty}^{\infty} \Theta(\xi)\Theta(|\xi| - |x|)\frac{d\xi}{F(s/2, 1/1, \xi^2)}\right.$$

$$\left. + \Theta(-x)U(s/2, 1/2, x^2)\int_{-\infty}^{\infty} \Theta(|x| - |\xi|)\frac{d\xi}{U(s/2, 1/1, \xi^2)}\right],$$

but

$$\int_{-\infty}^{\infty} \Theta(\xi)\Theta(|\xi| - |x|)\frac{d\xi}{F(s/2, 1/1, \xi^2)} = \int_{|x|}^{\infty} \Theta(\xi)\frac{d\xi}{F(s/2, 1/1, \xi^2)}$$

$$= \int_{|x|}^{\infty} \frac{d\xi}{F(s/2, 1/1, \xi^2)},$$

and

$$\int_{-\infty}^{\infty} \Theta(|x| - |\xi|)\frac{d\xi}{U(s/2, 1/1, \xi^2)} = 2\int_{0}^{\infty} \Theta(|x| - |\xi|)\frac{d\xi}{U(s/2, 1/1, \xi^2)}$$

$$= 2\int_{0}^{|x|} \frac{d\xi}{U(s/2, 1/1, \xi^2)},$$

and we finally write

$$\left\langle \hat{M}(s)|x\right\rangle = \frac{1}{s}\left[x + F(s/2, 1/2, x^2)\int_{|x|}^{\infty} \frac{d\xi}{F(s/2, 1/2, \xi^2)}\right. \tag{11.84}$$

$$\left. + 2\Theta(-x)U(s/2, 1/2, x^2)\int_{0}^{|x|} \frac{d\xi}{U(s/2, 1/2, \xi^2)}\right].$$

We incidentally note that when the process starts at the origin (i.e., $x = 0$), and after using the property $F(a, b, 0) = 1$ (Magnus *et al.*, 1966), this expression reduces to[6]

$$\left\langle \hat{M}(s)|0\right\rangle = \frac{1}{s}\int_{0}^{\infty} \frac{d\xi}{F(s/2, 1/2, \xi^2)}. \tag{11.85}$$

The exact analytical inversion of these expressions to get the mean maximum in real time seems to be beyond reach. In the case $x = 0$ we can,

---

[6]Let us recall that in the original variables given in Eq. (11.54) starting at the origin implies starting at the normal level $x = \mu$.

nonetheless, obtain an approximate expression. In effect, since the Kummer function $F(a, b, z)$ grows exponentially as $|z| \to \infty$ (Magnus et al., 1966), we see that $1/F(s/2, 1/2, \xi^2)$ decays exponentially as $\xi \to \infty$. Hence, the major contribution to the integral in Eq. (11.85) comes from the neighborhood of $\xi = 0$ (recall again that $F(s/2, 1/2, 0) = 1$). We, thus, expand to the lower order in ξ^2 the Kummer function $F(s/2, 1/2, \xi^2)$ (Magnus et al., 1966)

$$F(s/2, 1/2, \xi^2) = 1 + s\xi^2 + O(\xi^4)$$

and approximate Eq. (11.85) as

$$\langle \hat{M}(s)|0 \rangle \simeq \frac{1}{s} \int_0^\infty \frac{d\xi}{1 + s\xi^2} = \frac{\pi}{2} s^{-3/2}.$$

Laplace inverting this approximation is straightforward, yielding

$$\langle M(t)|0 \rangle \simeq (\pi t)^{1/2}. \tag{11.86}$$

Note that the mean maximum grows like $t^{1/2}$ as in Brownian motion (see Sect. 11.3).

Proceeding in an analogous way with the transformed minimum, Eq. (11.83), it is rather straightforward to show that [Masoliver (2014b)]

$$\langle \hat{m}(s)|x \rangle = \frac{1}{s} \left[x - F(s/2, 1/2, x^2) \int_{|x|}^\infty \frac{d\xi}{F(s/2, 1/2, \xi^2)} \right. \tag{11.87}$$

$$\left. - 2\Theta(x) U(s/2, 1/2, x^2) \int_0^{|x|} \frac{d\xi}{U(s/2, 1/2, \xi^2)} \right].$$

When $x = 0$ we have

$$\langle \hat{m}(s)|0 \rangle = -\frac{1}{s} \int_0^\infty \frac{d\xi}{F(s/2, 1/2, \xi^2)}. \tag{11.88}$$

Showing the anti-symmetry between the maximum and the minimum with respect to the origin –compare with Eq. (11.85)–, a sort of reflection principle:

$$\langle \hat{m}(s)|0 \rangle = -\langle \hat{M}(s)|0 \rangle,$$

which obviously implies $\langle m(t)|0 \rangle = -\langle M(t)|0 \rangle$. Hence [cf. Eq. (11.86)]

$$\langle m(t)|0 \rangle \simeq -(\pi t)^{1/2}, \tag{11.89}$$

which exhibits the same growth as the Brownian motion.

Maximum absolute value and span

In Chapter 10 we saw that the distribution function for the maximum absolute value is [cf. Eq. (10.63)]

$$G_{\max}(\xi, t|x) = \left[1 - W_{-\xi,\xi}(t|x)\right]\Theta(\xi - |x|),$$

$(\xi > 0)$, where $W_{-\xi,\xi}(t|x)$ is the exit probability out of a symmetric interval. For the OU process the Laplace transform of this probability is given in Eq. (11.72) and, therefore, the Laplace transform of $G_{\max}$ reads

$$\hat{G}_{\max}(\xi, s|x) = \left[1 - \frac{F(s/2, 1/2, x^2)}{sF(s/2, 1/2, \xi^2)}\right]\Theta(\xi - |x|). \tag{11.90}$$

The Laplace transform of the maximum absolute value is [cf. Eq. (10.67) of Chapter 10]

$$\mathcal{L}\left\{\left\langle[\max|X(t)|\,|x]\right\rangle\right\} = \frac{1}{s}\left[|x| + F(s/2, 1/2, x^2)\int_{|x|}^{\infty}\frac{d\xi}{F(s/2, 1/2, x^2)}\right].$$

When $x = 0$ and recalling that $F(s/2, 1/2, 0) = 1$ we see that this transform coincides with that of the average maximum, Eq. (11.85):

$$\mathcal{L}\left\{\left\langle[\max|X(t)|\,|0]\right\rangle\right\} = \frac{1}{s}\int_{0}^{\infty}\frac{d\xi}{F(s/2, 1/2, x^2)},$$

whence

$$\left\langle[\max|X(t)|]\,|x\right\rangle = \left\langle M(t)|0\right\rangle \simeq (\pi t)^{1/2}. \tag{11.91}$$

Let us finally recall from Chapter 10 that we can evaluate the average span of any process by subtracting the mean maximum from the mean minimum:

$$\left\langle R(t)|x\right\rangle = \left\langle M(t)|x\right\rangle - \left\langle m(t)|x\right\rangle.$$

Thus, for the OU process we have the following expression for the Laplace transform of the average span (cf. Eqs. (11.84) and (11.87))

$$\left\langle \hat{R}(s)|x\right\rangle = \frac{2}{s}\left[F(s/2, 1/2, x^2)\int_{|x|}^{\infty}\frac{d\xi}{F(s/2, 1/2, \xi^2)}\right.$$
$$\left. + U(s/2, 1/2, x^2)\int_{0}^{|x|}\frac{d\xi}{U(s/2, 1/2, \xi^2)}\right]. \tag{11.92}$$

From which we see

$$\left\langle \hat{R}(s)|0\right\rangle = \frac{2}{s}\int_{0}^{\infty}\frac{d\xi}{F(s/2, 1/2, \xi^2)},$$

that is (cf. Eqs. (11.85) and (11.86))

$$\left\langle R(t)|0\right\rangle = 2\left\langle M(t)|0\right\rangle \simeq 2(\pi t)^{1/2}.$$

11.5   Higher dimensional diffusions *

In Sect. 10.4 of the previous chapter we have outlined the general settings of the level-crossing problem for N dimensional random processes. We shall now apply the general formalism to multidimensional diffusion processes.

Suppose $\mathbf{X}(t) = (X_1(t), \ldots, X_N(t))$ is a time-homogeneous diffusion processes defined in some N-dimensional region $\mathbf{R}$ with boundary $\mathbf{B}$. Initially, at $t = 0$, the process is at $\mathbf{x} \in \mathbf{R}$ and at time $t > 0$ is at $\mathbf{x}'$ and the transition density $p(\mathbf{x}', t|\mathbf{x})$ satisfies the backward Fokker-Planck equation[7]

$$\frac{\partial p}{\partial t} = \sum_{j=1}^{N} f_j(\mathbf{x}) \frac{\partial}{\partial x_j} p(\mathbf{x}', t|\mathbf{x}) + \frac{1}{2} \sum_{j,k=1}^{N} D_{jk}(\mathbf{x}) \frac{\partial^2}{\partial x_j \partial x_k} p(\mathbf{x}', t|\mathbf{x}), \quad (11.93)$$

with initial condition

$$p(\mathbf{x}', 0|\mathbf{x}) = \delta(\mathbf{x}' - \mathbf{x}). \quad (11.94)$$

As in the one-dimensional case, the survival probability is given by

$$S(t|\mathbf{x}) = \int_{\mathbf{R}} p(\mathbf{x}', t|\mathbf{x}) d^N \mathbf{x}', \quad (11.95)$$

where the N-fold integral extends over region $\mathbf{R}$. Recall from Sect. 10.2 that if τ is the first-passage (or exit) time, then

$$S(t|\mathbf{x}) = \text{Prob}\{\tau \geq t | \mathbf{X}(0) = \mathbf{x} \in \mathbf{R}\} = 1 - \text{Prob}\{\tau < t | \mathbf{X}(0) = \mathbf{x} \in \mathbf{R}\},$$

which implies that the probability density of the first-passage time, $f(t|\mathbf{x})$, is related to the survival probability by (cf. Sect. 10.2)

$$f(t|\mathbf{x}) = -\frac{\partial S(t|\mathbf{x})}{\partial t}. \quad (11.96)$$

The mean first-passage time, defined as

$$T(\mathbf{x}) = \int_0^\infty t f(t|\mathbf{x}) dt,$$

will thus be given by

$$T(\mathbf{x}) = \int_0^\infty S(t|\mathbf{x}) dt. \quad (11.97)$$

By combining Eq. (11.95) with Eqs. (11.93)-(11.94) we find that the survival probability obeys the equation

$$\frac{\partial S}{\partial t} = \sum_{j=1}^{N} f_j(\mathbf{x}) \frac{\partial S}{\partial x_j} + \frac{1}{2} \sum_{j,k=1}^{N} D_{jk}(\mathbf{x}) \frac{\partial^2 S}{\partial x_j \partial x_k}. \quad (11.98)$$

---

[7]See Eq. (5.40) of Chapter 5 with the replacements $\mathbf{x} \to \mathbf{x}'$ and $\mathbf{x}_0 \to \mathbf{x}$.

From Eqs. (11.94) and (11.95) we see that at $t = 0$

$$S(0|\mathbf{x}) = \int_{\mathbf{R}} \delta(\mathbf{x}' - \mathbf{x})d^N\mathbf{x}' = 1,$$

and the initial condition for Eq. (11.98) is

$$S(0|\mathbf{x}) = 1. \tag{11.99}$$

Recall that starting at the boundary $\mathbf{B}$ of the region $\mathbf{R}$ implies no survival, which means the following boundary condition

$$S(t|\mathbf{x}) = 0, \qquad (\mathbf{x} \in \mathbf{B}). \tag{11.100}$$

Let us incidentally note that this indicates that the transition density also obeys the absorbing boundary condition

$$p(\mathbf{x}', t|\mathbf{x}) = 0, \qquad (\mathbf{x} \in \mathbf{B}). \tag{11.101}$$

Before proceeding further let us recall that the first-passage probability

$$W(t|\mathbf{x}) = 1 - S(t|\mathbf{x}),$$

obeys the same Fokker-Planck equation than $S(t|\mathbf{x})$,

$$\frac{\partial W}{\partial t} = \sum_{j=1}^N f_j(\mathbf{x})\frac{\partial W}{\partial x_j} + \frac{1}{2}\sum_{j,k=1}^N D_{jk}(\mathbf{x})\frac{\partial^2 W}{\partial x_j \partial x_k}, \tag{11.102}$$

but with initial and boundary conditions reversed

$$W(0|\mathbf{x}) = 0, \qquad W(t|\mathbf{x}) = 1 \quad (\mathbf{x} \in \mathbf{B}). \tag{11.103}$$

Let us finally obtain the equation satisfied by the mean exit time given in Eq. (11.97) as the time integral of the survival probability. Thus the time integration of Eq. (11.98) yields the following equation for the mean escape time out of region $\mathbf{R}$:

$$\sum_{j=1}^N f_j(\mathbf{x})\frac{\partial T}{\partial x_j} + \frac{1}{2}\sum_{j,k=1}^N D_{jk}(\mathbf{x})\frac{\partial^2 T}{\partial x_j \partial x_k} = -1, \tag{11.104}$$

with the boundary condition

$$T(\mathbf{x}) = 0, \qquad (\mathbf{x} \in \mathbf{B}). \tag{11.105}$$

In writing Eq. (11.104) we taken into account [cf. Eq. (11.99) and $S(t|\mathbf{x}) \to 0$ as $t \to \infty$]

$$\int_0^\infty \frac{\partial S(t|\mathbf{x})}{\partial t}dt = \lim_{t\to\infty} S(t|\mathbf{x}) - S(0|\mathbf{x}) = -1.$$

One can also easily show that the nth moments of the first-passage time,

$$T_n(\mathbf{x}) = \int_0^\infty t^{n-1} S(t|\mathbf{x})dt,$$

satisfy the recursive equations

$$\sum_{j=1}^N f_j(\mathbf{x}) \frac{\partial T_n}{\partial x_j} + \frac{1}{2} \sum_{j,k=1}^N D_{jk}(\mathbf{x}) \frac{\partial^2 T_n}{\partial x_j \partial x_k} = -n T_{n-1}(\mathbf{x}) \qquad (11.106)$$

($n = 2, 3, 4, \dots$) with the boundary condition

$$T_n(\mathbf{x}) = 0, \qquad (\mathbf{x} \in \mathbf{B}). \qquad (11.107)$$

In some practical settings the surface boundary $\mathbf{B}$ of the region $\mathbf{R}$ is divided into two surfaces $\mathbf{B}_a$ and $\mathbf{B}_r$, such that the process is absorbed when hitting $\mathbf{B}_a$ and reflected when it reaches $\mathbf{B}_r$. We have shown in Sect. 1.5 of Chapter 5 that the boundary conditions on the survival probability $S(t|\mathbf{x})$ now are

$$S(t|\mathbf{x}) = 0, \quad \text{for} \quad \mathbf{x} \in \mathbf{B}_a, \qquad (11.108)$$

and

$$\sum_{j,k=1}^N n_j D_{jk}(\mathbf{x}) \frac{\partial}{\partial x_k} S(t|\mathbf{x}) = 0, \quad \text{for} \quad \mathbf{x} \in \mathbf{B}_r, \qquad (11.109)$$

where n_j is the jth component of the unitary normal vector pointing outside surface $\mathbf{B}_r$.

11.6 Free diffusions in two and three dimensions *

As an example of level crossings in several dimensions we will solve the first-passage and exit problems of free, isotropic and homogeneous diffusion processes in two and three dimensions.

By free diffusion we mean that the diffusive process moves freely without the influence of any external field, that is, with no drift. Isotropy implies that all directions are equivalent which amounts to having a diagonal diffusion matrix that we also assume to be constant (homogeneous diffusion). We thus have

$$\mathbf{f}(\mathbf{x}) = 0, \qquad D_{jk}(\mathbf{x}) = D\delta_{jk}, \qquad (11.110)$$

where D is the (constant) diffusion coefficient, δ_{ij} is the Kronecker symbol and $j, k = 1, \dots, N$ ($N = 2, 3$). Let us incidentally note that the process defined by Eq. (11.110) is also termed as the N dimensional Brownian motion which has countless applications in many branches of physics, chemistry and biology.

We want to address the exit and first-passage problem for the free Brownian motion out of some N dimensional region **R** with boundary **B**. The first-passage probability $W(t|\mathbf{x})$ obeys the Fokker-Planck equation (11.102) which, after using (11.110), simplifies into the diffusion equation

$$\frac{\partial W}{\partial t} = \frac{1}{2}D\nabla^2 W,$$

where

$$\nabla^2 = \sum_{j=1}^{N} \frac{\partial^2}{\partial x_j^2},$$

is the Laplacian. The initial and boundary conditions are given by Eq. (11.103).

Since we will finally deal with Brownian motion in two and three dimensions, in what follows we will use the more standard notation

$$\mathbf{r} = \mathbf{x} = (x_1, \ldots, , x_N) \qquad (N = 2, 3, \ldots),$$

and $|\mathbf{r}| = (x_1^2 + \cdots + x_N^2)^{1/2}$. In order to proceed further we need to specify the geometry of the region **R** and its boundary **B**. We will next study the escape and first-passage problems when **R** is a spherical region of radius a that, for the sake of simplicity, we assume centered at the origin of coordinates. In this case region **R** is determined by the inequality $|\mathbf{r}| < a$ and the boundary **B** by $|\mathbf{r}| = a$. The problem to solve is thus given by

$$\frac{\partial W}{\partial t} = \frac{1}{2}D\nabla^2 p, \tag{11.111}$$

and

$$W(0|\mathbf{r}) = 0 \quad (|\mathbf{r}| < a), \qquad W(t|\mathbf{r})\Big|_{|\mathbf{r}|=a} = 1. \tag{11.112}$$

We observe that this problem has spherical symmetry and $W = W(t|r)$, where $r = |\mathbf{r}|$. Since the radial part of the Laplacian in N dimensions is

$$\nabla^2 W(t|r) = \frac{1}{r^{N-1}} \frac{\partial}{\partial r}\left(r^{N-1}\frac{\partial W}{\partial r}\right) = \frac{(N-1)}{r}\frac{\partial W}{\partial r} + \frac{\partial^2 W}{\partial r^2},$$

the problem (11.111) and (11.112) reads

$$\frac{\partial W}{\partial t} = \frac{D}{2}\frac{(N-1)}{r}\frac{\partial W}{\partial r} + \frac{D}{2}\frac{\partial^2 W}{\partial r^2}, \tag{11.113}$$

with

$$W(0|r) = 0, \qquad W(t|a) = 1. \tag{11.114}$$

Solving this problem becomes much simpler if we take the time Laplace transform. Thus, defining

$$\hat{W}(s|r) = \mathcal{L}\{W(t|r)\} = \int_0^\infty e^{-st} W(t|r)\, dt,$$

as the Laplace transform of the exit probability and taking into account the well known properties:

$$\mathcal{L}\left\{\frac{\partial W}{\partial t}\right\} = s\hat{W} - W(0|r), \qquad \mathcal{L}\{1\} = \frac{1}{s},$$

we turn the initial and boundary value problem (11.113)-(11.114) which involves a partial differential equation into a simpler boundary value problem involving only an ordinary differential equation

$$\frac{d^2\hat{W}}{dr^2} + \frac{(N-1)}{r}\frac{d\hat{W}}{dr} - \left(\frac{2s}{D}\right)\hat{W} = 0, \tag{11.115}$$

with boundary condition

$$\hat{W}(s|a) = \frac{1}{s}. \tag{11.116}$$

Defining the new function $\hat{q}(s;r)$,

$$\hat{W}(s|r) = r^\nu \hat{q}(s;r), \qquad \nu = 1 - N/2, \tag{11.117}$$

and the new independent variable

$$z = r\left(\frac{2s}{D}\right)^{1/2}, \tag{11.118}$$

turn Eq. (11.115) into the modified Bessel equation

$$z^2 \frac{d^2\hat{q}}{dz^2} + z\frac{d\hat{q}}{dz} - (\nu^2 + z^2)\hat{q} = 0, \tag{11.119}$$

with boundary condition

$$\hat{q}\left[s; (2s/D)^{1/2}a\right] = \frac{a^{-\nu}}{s}. \tag{11.120}$$

The solution to Eq. (11.119) is given by modified Bessel functions of order $\nu = 1 - N/2$. For any value of ν the general solution is given by a linear combination of modified Bessel functions of first kind $I_\nu(z)$ and second kind, $K_\nu(z)$ (Magnus *et al.*, 1966). If, however, ν is not an integer the solution is also given by a combination of Bessel functions of first kind $I_{\pm\nu}(z)$ ($\nu \neq 0, \pm 1, \pm 2, \ldots$). We will next specialize in two and three dimensions.

11.6.1 *Two dimensions*

In two dimensions $N = 2$ and $\nu = 0$. The solution to Eq. (11.119) is $\hat{q}(s,z) = AI_0(z) + BK_0(z)$, where A and B are arbitrary constants. From Eq. (11.117) we see that now $\hat{W} = \hat{q}$ and, in the original variable r, the general solution reads

$$\hat{W}(s|r) = AI_0\big[(2s/D)^{1/2}r\big] + BK_0\big[(2s/D)^{1/2}r\big].\qquad(11.121)$$

Before proceeding further we should bear in mind that $W(t|r)$ is a probability which means that it must be finite for all values of $t \geq 0$ and $r \geq 0$ implying that its Laplace transform $\hat{W}(s|r)$ exists and it is finite for $s \geq 0$ and all possible initial values $r \geq 0$.

In order to obtain the unknown constants A and B from the boundary condition (11.120) we thus have to distinguish two regions (i) $r < a$ and (ii) $r > a$, and the exit problem and the first-passage problem, which up to now we have treated jointly, depart from each other. Indeed, if the Brownian walker starts inside the circle $r = a$, the region **R** is given by $r < a$ which is a bounded region implying that we are dealing with an exit problem out of the circular region $r < a$. On the other hand, when $r > a$, the region **R** is unbounded and we have a first-passage problem when the process crosses the circle $r = a$.

(i) *The exit probability*

In this case $r < a$ and taking into account that $K_0(0) = \infty$ while $I_0(0)$ is finite [Magnus *et al.* (1966)] we see that the solution to Eq. (11.119) should be of the form (recall that now $\hat{q} = \hat{W}$)

$$\hat{W}(s|r) = AI_0\big[(2s/D)^{1/2}r\big].$$

From the boundary condition (11.120) we obtain the constant A and finally get

$$\hat{W}(s|r) = \frac{I_0\big[(2s/D)^{1/2}r\big]}{sI_0\big[(2s/D)^{1/2}a\big]},\qquad (r < a).\qquad(11.122)$$

(ii) *The hitting probability*

In this case $r > a$ and the region **R** is now unbounded. Since $I_0(\infty) = \infty$ and $K_0(\infty) = 0$ the solution of Eq. (11.119) is of the form

$$\hat{W}(s|r) = BK_0\big[(2s/D)^{1/2}r\big],$$

and from the boundary condition we have

$$\hat{W}(s|r) = \frac{K_0\big[(2s/D)^{1/2}r\big]}{sK_0\big[(2s/D)^{1/2}a\big]},\qquad (r > a).\qquad(11.123)$$

Unfortunately the analytical Laplace inversion of Eqs. (11.122) and (11.123) seems to be beyond reach and the expression for $W(t|r)$ in real time can be only achieved by the numerical inversion of these expressions. We will see next that in three dimensions it is possible to obtain the exact expressions for the exit probability.

11.6.2 *Three dimensions*

In this case $N = 3$ and $\nu = -1/2$ and the solution to Eq. (11.119) can be written not only as a linear combination of $I_{-1/2}(z)$ and $K_{-1/2}(z)$ but also by a linear combination of $I_{-1/2}(z)$ and $I_{1/2}(z)$, where (Magnus *et al.*, 1966)

$$I_{-1/2}(z) = \left(\frac{2}{\pi z}\right)^{1/2} \cosh(z), \qquad I_{1/2}(z) = \left(\frac{2}{\pi z}\right)^{1/2} \sinh(z), \quad (11.124)$$

and

$$K_{\pm 1/2}(z) = \left(\frac{\pi}{2z}\right)^{1/2} e^{-z}. \qquad (11.125)$$

As in two dimensions we will distinguish two regions: $r < a$ (bounded) and $r > a$ (unbounded) which, in turn, determines the escape and first-passage probabilities.

(i) *The exit probability*

Inside the sphere $r < a$ we write the solution to Eq. (11.119) in the form

$$\hat{q}(s, z) = A I_{-1/2}(z) + B I_{1/2}(z).$$

Since $I_{-1/2}(0) = \infty$, a finite solution implies $A = 0$ and the unknown constant B is determined from the boundary condition (11.120). Taking into account that now $\hat{W}(s|r) = r^{-1/2}\hat{q}(s|r)$ (cf. Eq. (11.117)) we see that the solution for $\hat{W}$ in the original variable reads

$$\hat{W}(s|r) = \left(\frac{a}{r}\right)^{1/2} \frac{I_{1/2}\left[(2s/D)^{1/2}r\right]}{s I_{1/2}\left[(2s/D)^{1/2}a\right]},$$

that is

$$\hat{W}(s|r) = \left(\frac{a}{r}\right) \frac{\sinh\left[(2s/D)^{1/2}r\right]}{s \sinh\left[(2s/D)^{1/2}a\right]}, \qquad (11.126)$$

$(r < a)$. The Laplace inversion of this expression is readily obtained by using the following inversion formula (Roberts and Kaufman, 1966)

$$\mathcal{L}^{-1}\left\{\frac{\sinh x\sqrt{s}}{s \sinh \beta\sqrt{s}}\right\} = \frac{x}{\beta} - \frac{2}{\pi}\sum_{n=1}^{\infty}\frac{(-1)^{n-1}}{n}e^{-n^2\pi^2 t/\beta^2}\sin(n\pi x/\beta).$$

Thus

$$W(t|r) = 1 - \frac{2a}{\pi r} \sum_{n=1}^{\infty} \frac{(-1)^{n-1}}{n} e^{-n^2\pi^2 Dt/2a^2} \sin\left(n\pi\sqrt{D/2}r/a\right), \quad (r < a).$$

$$(11.127)$$

(ii) *The hitting probability*

Outside the sphere, $r > a$, we write the solution to Eq. (11.119) as

$$\hat{q}(s,z) = AI_{-1/2}(z) + BK_{-1/2}(z)$$

and since, when $z \to \infty$, $I_{-1/2}(z) \to \infty$ while $K_{-1/2}(z) \to 0$, we must take $A = 0$ and B is determined from Eq. (11.120). The solution for $\hat{W}$ thus reads

$$\hat{W}(s|r) = \left(\frac{a}{r}\right)^{1/2} \frac{K_{-1/2}\left[(2s/D)^{1/2}r\right]}{sK_{-1/2}\left[(2s/D)^{1/2}a\right]},$$

or, equivalently (see Eq. (11.125))

$$\hat{W}(s|r) = \left(\frac{a}{r}\right) \frac{1}{s} e^{-(2s/D)^{1/2}(r-a)}, \quad (r > a). \qquad (11.128)$$

Taking into account the Laplace inversion formula

$$\mathcal{L}^{-1}\left\{\frac{e^{-\beta\sqrt{s}}}{s}\right\} = \text{Erfc}\left[\frac{\beta}{2\sqrt{t}}\right],$$

where $\text{Erfc}(\cdot)$ is the complementary error function, we finally get

$$W(t|r) = \left(\frac{a}{r}\right) \text{Erfc}\left[\frac{r-a}{\sqrt{2Dt}}\right], \quad (r > a). \qquad (11.129)$$

11.6.3 *A comment on the one dimensional problem*

We have focussed on two and three dimensional diffusions with spherical symmetry. However, the one dimensional case, which we have treated in Sect. 11.3, can also be recovered from the above formalism. Indeed, for this case $\mathbf{r} = (x,0,0)$ ($-\infty < x < \infty$) and the "spherical region" $r < a$ is now the interval $-a < x < a$. Moreover, $N = 1$ and the problem given by Eqs. (11.113) and (11.114) reads

$$\frac{\partial W}{\partial t} = \frac{1}{2}D\frac{\partial^2 W}{\partial x^2},$$

with

$$W(0|x) = 0, \qquad W(t|\pm a) = 1.$$

As we have seen in Sect. 11.3, the solution of this problem is rather straight-forward after using the time Laplace transform [see Eqs. (11.115) and (11.116) with $N = 1$]. Nevertheless, the solution can also be obtained from the above development for 2 and 3 dimensions. Indeed, in this case $\nu = 1/2$ which results in an equivalent solution to that of the three dimensional case where $\nu = -1/2$ [Bessel functions are the same, cf. Eqs. (11.124) and (11.125)]. As a result we have [see Eqs. (11.127) and (11.129)]

$$W(t|x) = 1 - \frac{2a}{\pi r} \sum_{n=1}^{\infty} \frac{(-1)^{n-1}}{n} e^{-n^2 \pi^2 Dt/2a^2} \sin\left(n\pi \sqrt{D/2} x/a\right), \quad (11.130)$$

for $x \in (-a, a)$, and

$$W(t|x) = \left(\frac{a}{r}\right) \operatorname{Erfc}\left[\frac{x-a}{\sqrt{2Dt}}\right], \quad (11.131)$$

for $x \notin (-a, a)$.

11.6.4   *Mean exit and mean first-passage times*

We finally discuss the MET and MFPT for the spherical region $r < a$. We know that the MET in terms of the exit probability is given by (cf. Eq. (10.45) of Chapter 10)

$$T(\mathbf{x}) = \lim_{s \to 0} \left[\frac{1}{s} - \hat{W}(s|\mathbf{x})\right]. \quad (11.132)$$

(a) Two dimensions

(i) *The mean exit time.*

    Let us assume that $r < a$ which corresponds to the exit problem out of the circle $r = a$. In this case $\hat{W}(s|\mathbf{r})$ is given by Eq. (11.122) and, since as $z \to 0$

$$I_0(z) = 1 + \frac{z^2}{4} + O(z^4),$$

we have

$$\hat{W}(s|r) = \frac{1 + (s/2D)r^2 + O(s^2)}{s[1 + (s/2D)a^2 + O(s^2)]} = \frac{1}{s}\left[1 + (s/2D)(r^2 - a^2) + O(s^2)\right],$$

and the MET reads

$$T(\mathbf{r}) = \frac{1}{2D}(a^2 - r^2), \quad (r < a). \quad (11.133)$$

(ii) *The mean first-passage time.*

For the unbounded region $r > a$, we have a first-passage problem to the circle $r = a$ and the hitting probability is given by Eq. (11.123). Since (Magnus *et al.*, 1966)

$$K_0(z) = \left[\ln(z/2) + \gamma\right]I_0(z) + z^2/4 + O(z^4),$$

where γ is Euler's constant. Substituting into Eq. (11.123) and after some simple algebra we get

$$\hat{W}(s|r) = \frac{1}{s}\frac{\ln\left[(s/2D)^{1/2}r\right] + \gamma + O[(s\ln s)]}{\ln\left[(s/2D)^{1/2}a\right] + \gamma + O(s\ln s)}$$

$$= \frac{1}{s}\frac{\ln\left[(s/2D)^{1/2}r\right]}{\ln\left[(s/2D)^{1/2}a\right]}\left[1 + O(1/\ln s)\right].$$

From Eq. (11.132) we see that

$$T(\mathbf{r}) = \lim_{s \to 0}\left\{\frac{\ln(a/r)}{s\ln\left[(s/2D)^{1/2}a\right]} + O(1/s\ln^2 s)\right\},$$

and, since $s\ln s \to 0$ as $s \to 0$ we conclude that the MFPT is infinite:

$$T(\mathbf{r}) = \infty, \qquad (r > a). \tag{11.134}$$

We can interpret these results as follows. For unbounded regions in the plane the probability that a given trajectory never crosses a finite boundary is different from zero. However for such trajectories the first-passage time is infinite and, since they have a non vanishing weight in the average of the hitting time, this results in an infinite mean first-passage time. On the other hand, for bounded regions the escape time is always finite and we have a finite mean exit time.

(a) Three dimensions

(i) *The mean exit time.*

When $r < a$ we have an exit problem out of the sphere $r = a$. Now $\hat{W}(s|r)$ is given by Eq. (11.126) and, bearing in mind that

$$\sinh z = z + \frac{1}{3!}z^3 + O(z^4),$$

we get

$$\hat{W}(s|r) = \frac{1 + (s/2D)r^2/3! + O(s^2)}{s[1 + (s/2D)a^2/3! + O(s^2)]}$$

$$= \frac{1}{s}\left[1 + \frac{1}{3!}\left(\frac{2s}{D}\right)(r^2 - a^2) + O(s^2)\right],$$

and the MET reads

$$T(\mathbf{r}) = \frac{1}{3D}(a^2 - r^2), \qquad (r < a). \qquad (11.135)$$

(ii) *The mean first-passage time.*

For the unbounded region $r > a$, the first-passage probability to the sphere $r = a$ is given by Eq. (11.128):

$$\hat{W}(s|r) = \left(\frac{a}{r}\right)\frac{1}{s}e^{-(2s/D)^{1/2}(r-a)}, \qquad (r > a).$$

Hence

$$T(\mathbf{r}) = \lim_{s \to 0}\left\{\frac{1}{s}\left[1 - \left(\frac{a}{r}\right)e^{-(2s/D)^{1/2}(r-a)}\right]\right\} = \infty,$$

and, as in two dimensions, the MFPT is infinite

$$T(\mathbf{r}) = \infty, \qquad (r > a). \qquad (11.136)$$

The interpretation of these results is the same than in two dimensions, as we have discussed above.

11.7 Inertial processes *

In mechanics, as well as in many other parts of physics, one often encounters random processes whose time evolution is not given by a first-order stochastic differential equation (cf. Eq. (8.96) of chapter 8) but by a second order equation involving first and second derivatives. The general form of this kind of equations is

$$\frac{d^2 X(t)}{dt^2} = F\left(t, X(t), \frac{dX(t)}{dt}, Z(t)\right), \qquad (11.137)$$

where $Z(t)$ is a given random process (the input noise) which is often modeled as Gaussian white noise. The origin of such equations usually arises from Newton second law of motion, where $X(t)$ is the position of a particle moving under the influence of deterministic and random forces altogether represented by the function F. A paradigmatic example is the so-called "noisy oscillator", a linear (or non-linear) oscillator disturbed by random influences, either in the frequency (the Kubo oscillator) or with an external random force or even with random damping (Gitterman, 2012).

Although the analysis that follows is valid for any kind of random process obeying equations like (11.137), for a more intuitive approach we will suppose that $X(t)$ is the position of a particle of unit mass moving under the influence of a total force represented by F. In many practical application the force acting on the particle has a deterministic part and a random part. Thus, in what follows we will suppose that F does not depend explicitly of time and is linear in the input noise which is taken to be Gaussian white noise. We thus have the second-order equation (dots represent derivatives with respect to time)

$$\ddot{X}(t) = f(X, \dot{X}) + g(X, \dot{X})\xi(t), \qquad (11.138)$$

where $\xi(t)$ is zero-centered Gaussian white noise,

$$\langle \xi(t) \rangle = 0, \qquad \langle \xi(t)\xi(t') \rangle = \delta(t - t').$$

$f(x, v)$ and $g(x, v)$ are given functions of the position and velocity of the particle, f represents the deterministic force and g modulates the input noise.[8]

Defining $V(t) = \dot{X}(t)$ as the random velocity of the particle, the second order process (11.138) is equivalent to the bidimensional process[9] $(X(t), V(t))$ and the stochastic differential equation (11.138) corresponds to the bidimensional system

$$\dot{X}(t) = V$$
$$\dot{V}(t) = f(X, V) + g(X, V)\xi(t). \qquad (11.139)$$

We know (cf. Chapter 7) that the process (X, V), defined by Eq. (11.139), is a bidimensional diffusion process with vector drift and (singular) diffusion matrix given by

$$\mathbf{f}(x, v) = \big(v, f(x, v)\big), \qquad \mathbf{D} = \begin{pmatrix} 0 & 0 \\ 0 & g^2(x, v) \end{pmatrix}. \qquad (11.140)$$

We want to address the level crossing problem for inertial processes described by Eq. (11.138). The problem consists in the escape of the particle out of some interval (a, b) and the first-passage when position $X(t)$ reaches a given value x_c for the first time. In what follows we focus on the exit problem since, as we have seen the first-passage problem is included in the escape problem by setting $a \to -\infty$ and $x_c = b$ or $b \to \infty$ and $x_c = a$.

Let us denote by $S(t|x, v)$ the survival probability, that is to say, the probability that the position $X(t)$ of the particle has not left the interval

---

[8]The Kramers problem discussed at the end of chapter 5 is a particular case of Eq. (11.138) (cf. Eq. (5.80))

[9]Let us recall from Chapter 3 (cf. Sect. 3.2) that $X(t)$ is not Markovian while (X, V) it is.

(a, b) at time t or during any previous time. From Eqs. (11.98) and (11.140) we see that $S(t|x, v)$ obeys the equation

$$\frac{\partial S}{\partial t} = v\frac{\partial S}{\partial x} + f(x, v)\frac{\partial S}{\partial v} + \frac{1}{2}g^2(x, v)\frac{\partial^2 S}{\partial v^2}, \qquad (11.141)$$

with initial condition (cf. Eq. (11.99))

$$S(0|x, v) = 1. \qquad (11.142)$$

The boundary condition –which is in general given by Eq. (11.100), that is, $S(t|x, v) = 0$ for $(x, v) \in \mathbf{B}$, where $\mathbf{B}$ is the boundary of the escape region $\mathbf{R}$– takes now a more intricate form. Indeed, if the particle starts at the lower boundary, $x = a$, it won't leave the interval unless the initial velocity v is negative (i.e., pointing to the left) because if the velocity points to the right ($v > 0$) the particle moves to the right and remains inside (a, b). On the other hand, if initially the particle is at the upper boundary, $x = b$, it will leave (a, b) only if the initial velocity $v > 0$ is positive, otherwise it will remain inside (a, b). The boundary conditions for the problem are thus given by

$$S(t|x = a, v < 0) = 0, \qquad S(t|x = b, v > 0) = 0. \qquad (11.143)$$

From a geometrical point of view this implies that the bidimensional boundary $\mathbf{B}$ has a discontinuity at $v = 0$. Indeed, in the half plane $v < 0$ the boundary is the vertical line $x = a$ whereas in the half plane $v > 0$ the boundary is the vertical line $x = b$.

Recalling that the mean exit time is the time integral of the survival probability,

$$T(x, y) = \int_0^\infty S(t|x, y)dt,$$

we see from Eqs. (11.141), (11.142) and (11.143) that the mean exit time obeys the equation (see also Eq. (11.104))

$$\frac{1}{2}g^2(x, y)\frac{\partial^2 T}{\partial v^2} + f(x, v)\frac{\partial T}{\partial v} + v\frac{\partial T}{\partial x} = -1, \qquad (11.144)$$

with boundary conditions

$$T(a, v) = 0 \quad \text{if} \quad v < 0, \qquad T(b, v) = 0 \quad \text{if} \quad v > 0. \qquad (11.145)$$

The special form of the boundary conditions, Eqs. (11.143) or (11.145), with data on a non-smooth boundary at $v = 0$ makes very difficult obtaining the analytical solution to these problems even in the simplest cases as we will see next.[10]

---

[10]Boundary value problems like these are known in the mathematics literature as a "problem of Fichera" and it was shown in the late nineteen fifties that they are well-posed problems (Langer, 1960)

11.7.1 *Mean exit time for free inertial processes*

We present an example in which has been possible to obtain an analytical solution to the boundary value problem (11.144)–(11.145) (Masoliver and Porrà, 1995, 1996). We consider the exact analytical solution to the mean exit time out of an interval $(0, L)$ for the displacement of an unbounded and undamped particle under the influence of a random acceleration

$$\ddot{X}(t) = D^{1/2}\xi(t), \qquad (11.146)$$

where $\xi(t)$ is Gaussian white noise and $D > 0$ is the diffusion coefficient. Note that this is the simplest inertial process corresponding to Eq. (11.138) with $f(x, v) = 0$ and $g(x, y) = D$.

From Eqs. (11.144)–(11.145) we see that the mean exit time for process (11.146) out of the interval $(0, L)$ is the solution to

$$\frac{1}{2}D\frac{\partial^2 T}{\partial v^2} + v\frac{\partial T}{\partial x} = -1, \qquad (11.147)$$

with boundary conditions

$$T(0, v) = 0 \quad \text{if} \quad v < 0, \qquad T(L, v) = 0 \quad \text{if} \quad v > 0. \qquad (11.148)$$

In order to solve this problem we first observe that the problem is invariant under the changes $x \to L - x$ and $v \to -v$. In other words, $T(x, v)$ satisfies the symmetry relation

$$T(x, v) = T(L - x, -v). \qquad (11.149)$$

This symmetry implies the following matching conditions at $v = 0$

$$T(x, 0) = T(L - x, 0) \qquad (11.150)$$

$$\left.\frac{\partial T(x, v)}{\partial v}\right|_{v=0} = -\left.\frac{\partial T(L - x, v)}{\partial v}\right|_{v=0}. \qquad (11.151)$$

We note that Eq. (11.149) allows us to write the solution $T(x, v)$ for all v once we know the solution of (11.147) and (11.148) for, say, $v \le 0$. We thus assume that $v \le 0$ and define

$$T_-(x, v) \equiv T(x, v) \quad \text{if} \quad v \le 0. \qquad (11.152)$$

In dimensionless units defined by

$$y = x/L, \qquad u = -(2/LD)^{1/3}v, \qquad (11.153)$$

Eq. (11.147) reads

$$\frac{\partial^2 \overline{T}_-}{\partial u^2} - u\frac{\partial \overline{T}_-}{\partial y} = -1, \qquad (11.154)$$

306 Random processes: First-passage and escape

where $\overline{T}_-(y,u)$ is the scaled, and dimensionless, mean exit time defined as

$$\overline{T}_-(y,u) = \left(\frac{D}{2L^2}\right)^{1/3} T_-(x,v). \qquad (11.155)$$

We see from (11.153) that the dimensionless velocity u has opposite sign than v. That is to say, and regarding $\overline{T}_-(y,u)$, the scaled velocity u is positive. Therefore, from boundary conditions (11.148) we get the following one for the scaled mean exit time:

$$\overline{T}_-(0,u) = 0 \qquad (u > 0). \qquad (11.156)$$

Since (11.154) is a second order differential equation another boundary condition is needed and it is provided by the behavior of the exit time at infinite speed. Indeed, if the initial velocity of the particle tends to infinity the exit time tends to zero. We thus have[11]

$$\lim_{u\to\infty} \overline{T}_-(y,u) = 0 \qquad (0 \le y \le 1). \qquad (11.157)$$

In spite of the fact that the range of the scaled position y is bounded by an upper bound, it is still permissible to define the Laplace transform of $\overline{T}_-(y,u)$ with respect to y (LePage, 1990)

$$\hat{\overline{T}}_-(\sigma,u) = \int_0^\infty e^{-\sigma y}\overline{T}_-(y,u)dy.$$

Thus, taking into account the properties

$$\mathcal{L}\left\{\frac{\partial \overline{T}_-(y,u)}{\partial y}\right\} = \sigma\hat{\overline{T}}_-(\sigma,u) - \overline{T}_-(y=0,u)., \qquad \mathcal{L}\{1\} = \frac{1}{\sigma},$$

and the boundary condition (11.156), the transformation of Eq. (11.154) leads to the following ordinary differential equation

$$\frac{d^2\hat{\overline{T}}_-}{du^2} - \sigma u\hat{\overline{T}}_- = -\frac{1}{\sigma},$$

which, after defining the new variable

$$\eta = \sigma^{1/3}u, \qquad (11.158)$$

results in the following inhomogeneous linear equation

$$\frac{d^2\hat{\overline{T}}_-}{d\eta^2} - \eta\hat{\overline{T}}_- = -\sigma^{-5/3}. \qquad (11.159)$$

---

[11]This boundary condition on the velocity is implicit in the original problem given in Eqs. (11.147) and (11.148) (and also for the survival probability) since $T(x,v) \to 0$ as $v \to \pm\infty$.

The solution to Eq. (11.159) can be written in terms of the Airy functions $Ai(\eta)$ and $Bi(\eta)$ which are two independent solutions of the Airy equation (Abramowitz and Stegun, 1965). That is,

$$Ai''(\eta) - \eta Ai = 0, \qquad Bi''(\eta) - \eta Bi = 0.$$

As the reader may check by direct substitution, the general solution of Eq. (11.159) is

$$\hat{\bar{T}}_-(\sigma, u) = \frac{\alpha(\sigma)}{\sigma^{5/3}} Ai(\eta) + \frac{\pi}{\sigma^{5/3}} \left[Bi(\eta) \int_\eta^\infty Ai(t)dt + Ai(\eta) \int_0^\eta Bi(t)dt \right],$$

$$(11.160)$$

where $\alpha(\sigma)$ is an unknown quantity independent of η, that is to say, independent of velocity u.

We shall relate $\alpha(\sigma)$ with the derivative of $\hat{\bar{T}}_-(\sigma, u)$ with respect to u evaluated at $u = 0$. In effect, let us denote by $\hat{\phi}(\sigma)$ this derivative

$$\hat{\phi}(\sigma) = \left. \frac{\partial \hat{\bar{T}}_-(\sigma, u)}{\partial u} \right|_{u=0}. \tag{11.161}$$

Taking into account that $\partial/\partial u = \sigma^{1/3} \partial/\partial \eta$ (cf. Eq. (11.158)) the derivative of Eq. (11.160) with respect to u yields

$$\hat{\phi}(\sigma) = \frac{\alpha(\sigma)}{\sigma^{4/3}} Ai'(0) + \frac{\pi}{\sigma^{4/3}} Bi'(0) \int_0^\infty Ai(t)dt.$$

Bearing in mind (Abramowitz and Stegun, 1965)

$$Ai'(0) = -\frac{3^{-1/3}}{\Gamma(1/3)}, \quad Bi'(0) = \frac{3^{-1/3}\sqrt{3}}{\Gamma(1/3)}, \quad \int_0^\infty Ai(t)dt = \frac{1}{3}$$

we get

$$\alpha(\sigma) = \frac{\pi}{\sqrt{3}} - 3^{1/3}\Gamma(1/3)\sigma^{4/3}\hat{\phi}(\sigma).$$

We substitute this expression into Eq. (11.160) and recalling (11.158) we write

$$\hat{\bar{T}}_-(\sigma, u) = -3^{1/3}\Gamma(1/3)\sigma^{-1/3}Ai(u\sigma^{1/3})\hat{\phi}(\sigma) + \hat{R}(\sigma, u), \tag{11.162}$$

where

$$\hat{R}(\sigma, u) = \frac{\pi}{\sigma^{5/3}} \left[\frac{1}{\sqrt{3}} Ai\left(u\sigma^{1/3}\right) \right. \tag{11.163}$$

$$\left. + Ai\left(u\sigma^{1/3}\right) \int_0^{u\sigma^{1/\sigma^3}} Bi(t)dt + Bi\left(u\sigma^{1/3}\right) \int_{u\sigma^{1/\sigma^3}}^\infty Ai(t)dt \right]$$

Equation (11.162) apparently provides the solution to the problem in the Laplace space of the position. However, Eq. (11.162) is only a formal solution because it depends on the unknown function $\phi(\sigma)$. We will find this function using the symmetry relation (11.149). In what follows, and for the sake of simplicity and clarity, we will develop the procedure when $u = 0$, that is to say, we will obtain the mean exit time when the initial velocity of the particle is zero. We address the interested reader to Masoliver and Porrà (1996) for the complete solution of the problem for any value of the velocity.

When $u = 0$ Eq. (11.162) reduces to

$$\hat{\overline{T}}_-(\sigma,0) = -3^{1/3}\Gamma(1/3)\sigma^{-1/3}Ai(0)\hat{\phi}(\sigma)$$
$$+ \frac{\pi}{\sigma^{5/3}}\left[\frac{1}{\sqrt{3}}Ai(0) + Bi(0)\int_0^\infty Ai(t)dt\right],$$

but (Abramowitz and Stegun, 1965)

$$Ai(0) = \frac{3^{-2/3}}{\Gamma(2/3)}, \quad Bi(0) = \frac{3^{-2/3}\sqrt{3}}{\Gamma(2/3)}, \quad \int_0^\infty Ai(t)dt = \frac{1}{3},$$

and we get

$$\hat{\overline{T}}_-(\sigma,0) = -\frac{\Gamma(1/3)}{3^{1/3}\Gamma(2/3)}\sigma^{-1/3}\hat{\phi}(\sigma) + \frac{2\pi 3^{-2/3}}{\sqrt{3}\Gamma(2/3)}\sigma^{-5/3}. \qquad (11.164)$$

We proceed to Laplace inversion. Taking into account $\mathcal{L}^{-1}\{\sigma^{-\beta}\} = y^{\beta-1}/\Gamma(\beta)$ and the convolution theorem

$$\mathcal{L}^{-1}\left\{\sigma^{-1/3}\hat{\phi}(\sigma)\right\} = \frac{1}{\Gamma(1/3)}\int_0^y \frac{\phi(z)}{(y-z)^{2/3}}dz,$$

the inverse transform of Eq. (11.164) yields

$$\overline{T}_-(y,0) = -\frac{3^{-1/3}}{\Gamma(2/3)}\int_0^y \frac{\phi(z)}{(y-z)^{2/3}}dz + \frac{\pi 3^{-1/6}}{\Gamma^2(2/3)}y^{2/3}, \qquad (11.165)$$

where function $\phi(z)$ is related to the derivative of the mean exit time by (cf. Eq. (11.161))

$$\phi(z) = \left.\frac{\partial\overline{T}_-(z,u)}{\partial u}\right|_{u=0}, \qquad (11.166)$$

which is unknown unless we have an expression for the mean exit time. Fortunately we can get the expression for $\phi(z)$ by the symmetry of the problem as expressed in Eq. (11.149). Thus, we see from the first matching condition given in Eq. (11.150) that

$$\overline{T}_-(y,0) = \overline{T}_-(1-y,0).$$

Substituting Eq. (11.165) into this condition and reorganizing terms we write

$$\int_0^y \frac{\phi(z)}{(y-z)^{2/3}}dz - \int_0^{1-y} \frac{\phi(z)}{(1-y-z)^{2/3}}dz = \frac{\pi 3^{1/6}}{\Gamma(2/3)}\left[y^{2/3} - (1-y)^{2/3}\right].$$

(11.167)

In the second integral we make the change of variable $z \to 1 - z$ and use the second matching condition (11.151), that is $\phi(1-z) = -\phi(z)$, we have

$$\int_0^{1-y} \frac{\phi(z)}{(1-y-z)^{2/3}}dz = \int_y^1 \frac{\phi(1-z)}{(z-y)^{2/3}}dz = -\int_y^1 \frac{\phi(z)}{(z-y)^{2/3}}dz.$$

This allows us to write

$$\int_0^y \frac{\phi(z)}{(y-z)^{2/3}}dz - \int_0^{1-y} \frac{\phi(z)}{(1-y-z)^{2/3}}dz$$

$$= \int_0^y \frac{\phi(z)}{(y-z)^{2/3}}dz + \int_y^1 \frac{\phi(z)}{(z-y)^{2/3}}dz = \int_0^1 \frac{\phi(z)}{|y-z|^{2/3}}dz,$$

and Eq. (11.167) reads

$$\int_0^1 \frac{\phi(z)}{|y-z|^{2/3}}dz = \frac{3^{2/3}\Gamma(1/3)}{2}\left[y^{2/3} - (1-y)^{2/3}\right],$$

(11.168)

where we have used the reflection formula of the gamma function $\Gamma(\nu)\Gamma(1-\nu) = \pi/\sin(\pi\nu)$ (Abramowitz and Stegun, 1965).

Equation (11.168) is a weakly singular Fredholm integral equation of first kind for the unknown function $\phi(y)$. This equation is a particular case of more general integral equations of the form

$$\int_0^1 \frac{\phi(z)}{|y-z|^\beta}dz = f(y), \qquad (0 \le y \le 1),$$

(11.169)

where $0 < \beta < 1$. The solution of this equation is given by (Porter and Stirling, 1990) (see, also, Masoliver and Porrà, 1996)

$$\phi(y) = -\frac{\Gamma(\beta)\Gamma[(1-\beta)/2]}{\Gamma[(1+\beta)/2]}\frac{\cos^2 \pi\beta/2}{\pi^2}$$

$$\times y^{(\beta-1)/2}\frac{d}{dy}\int_y^1 dt \frac{t^{1-\beta}}{(t-y)^{(1-\beta)/2}}\frac{d}{dt}\int_0^t \frac{z^{(\beta-1)/2}}{(t-z)^{(1-\beta)/2}}f(z)dz.$$

In our case $\beta = 2/3$ and

$$f(y) = \frac{3^{1/6}}{2\Gamma^2(5/6)}\left[y^{2/3} - (1-y)^{2/3}\right].$$

After some amount of algebra we have (see Masoliver and Porrà, 1996, Appendix A)

$$\phi(y) = My^{-1/6}(1-y)^{-1/6}\left[F\left(1,-\frac{2}{3};\frac{5}{6};1-y\right) - F\left(1,-\frac{2}{3};\frac{5}{6};y\right)\right],$$
(11.170)

where $F(a,b;c;y)$ is the Gauss hypergeometric function (Abramowitz and Stegun, 1965) and

$$M = \frac{3^{1/6}\Gamma(3/2)}{2\Gamma(5/6)\Gamma(4/3)}.$$

Having obtained the explicit expression of $\phi(y)$ we are now in the position to evaluate $T(x, v = 0)$. In effect, the substitution of Eq. (11.170) into Eq. (11.165) yields, after lengthy calculations (see Masoliver and Porrà, 1996, Appendix B) the exact expression of $T(x,0)$ [12]. In the original units [cf. Eqs. (11.153) and (11.155)] this expression reads

$$T(x,0) = Nx^{1/6}(L-x)^{1/6}\left[F\left(1,-\frac{1}{3};\frac{7}{6};\frac{x}{L}\right) + F\left(1,-\frac{1}{3};\frac{7}{6};1-\frac{x}{L}\right)\right],$$
(11.171)

where

$$N = \frac{2^{2/3}}{3^{1/6}\Gamma(7/3)}\left(\frac{L}{D}\right)^{1/3}.$$

Another interested quantity, closely related to the exit time problem, is the averaged mean exit time over all possible initial positions x. We denote by $T_L(v)$ this averaged time and if we assume that x is uniformly distributed in the interval $(0, L)$ then

$$T_L(v) = \frac{1}{L}\int_0^L T(x,v)dx.$$

When $v = 0$ this averaged time reads (Masoliver and Porrà, 1995)

$$T_L(0) = \frac{3^{7/3}\Gamma(1/6)}{40\sqrt{\pi}}\left(\frac{2L^2}{D}\right)^{1/3},$$

which shows that the average time $T_L(0)$ grows, with the size L of the escape interval, as $L^{2/3}$. It is interesting to compare this growth with that of the Brownian motion. If the motion of the particle is described by the Wiener process, the velocity (instead of the acceleration) is driven by Gaussian white noise, i.e., $\dot{X}(t) = D\xi(t)$, and the mean exit time out of the interval

---

[12]From Eq. (11.152) we see that at zero velocity we have $T(x,0) = T_-(x,0)$.

$(0, L)$ is given by $T(x) = x(L - x)/D$ (cf. Eq. (11.45)) and the averaged time over a uniform interval reads

$$T_L = L^2/6D$$

which grows as L^2, faster than the inertial process.

Two final remarks. Although for the sake of simplicity we have only presented the case of zero initial velocity, the reader can find the exact results of the mean exit time for any value of the initial velocity in (Masoliver and Porrà, 1996). A second remark is that a number of practical applications of the formalism have been developed, specially in the field of polymers, and we refer the interested reader to the literature for more information (see, for instance, the review Burkhardt, 2014).

11.8 Anomalous and fractional diffusion *

In Chapter 4 we have outlined the basic elements of the theory of anomalous diffusion processes for which their basic characteristic is that the mean square deviation follows the asymptotic law

$$\langle \Delta X(t)^2 \rangle \sim t^\alpha,$$

where $\alpha > 0$ and $t \to \infty$. For free processes in the isotropic case (as, for instance, unbounded particles moving in isotropic media) the probability density function obeys the fractional diffusion equation

$$\frac{\partial^\alpha p}{\partial t^\alpha} = \frac{1}{2} D \frac{\partial^{2\gamma} p}{\partial x^{2\gamma}}. \tag{11.172}$$

$(0 < \alpha \leq 2, 0 < \gamma \leq 1)$ where $\partial^\alpha/\partial t^\alpha$ and $\partial^{2\gamma}/\partial x^{2\gamma}$ are the fractional Caputo and the Riesz-Feller fractional derivatives defined in Chapter 4. Equation (11.172) represents the most general form of fractional Brownian motion and it is termed the space-time fractional diffusion equation. However, if $\gamma \neq 1$ any process represented by (11.172) has no finite moments (besides the first one) which greatly restricts its use in practical applications.

When $\gamma = 1$ we have the so-called time-fractional diffusion equation

$$\frac{\partial^\alpha p}{\partial t^\alpha} = \frac{1}{2} D \frac{\partial^2 p}{\partial x^2}, \tag{11.173}$$

which if $0 < \alpha < 1$ represents subdiffusion and if $1 < \alpha < 2$ superdiffusion. In Sect. 4.6 of Chapter 4 we have obtained the solution to Eq. (11.173) in terms of Mainardi's functions (cf. Eq. (4.97)). This solution depends on the initial conditions attached to the time-fractional equation which, in turn, depend on the range of values taken by the fractional index α. Thus, when $0 < \alpha \le 1$ the initial condition is the usual one $p(x, 0) = \delta(x - x_0)$.[13] In the case $1 < \alpha \le 2$ the initial conditions are

$$p(x, 0|x_0) = \delta(x - x_0), \qquad \left.\frac{\partial p(x, t|x_0)}{\partial t}\right|_{t=0} = 0. \qquad (11.174)$$

We will now address the level crossing problem for the fractional Brownian motion and start with the escape problem.[14]

11.8.1 *The escape problem*

Let $X(t)$ be the time-fractional Brownian motion defined in an interval (a, b) with transition density $p(x', t|x)$ where $x = X(0) \in (a, b)$ is the initial value. We want to characterize when such process escapes out of (a, b). This characterization is determined by the survival probability $S_{ab}(t|x)$ which, as we know, is given in terms of $p(x', t|x)$ by

$$S_{ab}(t|x) = \int_a^b p(x', t|x) dx'. \qquad (11.175)$$

The transition density $p(x', t|x)$ satisfies the fractional equation

$$\frac{\partial^\alpha p}{\partial t^\alpha} = \frac{1}{2} D \frac{\partial^2 p}{\partial x'^2},$$

$(0 < \alpha < 2)$ with initial conditions given by Eq. (11.174) with the replacements $x \to x'$ and $x_0 \to x$. Let us note that the solution to this initial-value problem is a function of $x' - x$, i.e., $p(x', t|x) = p(x' - x, t)$. Hence, $\partial p/\partial x' = -\partial p/\partial x$ and $\partial^2 p/\partial x'^2 = \partial^2 p/\partial x^2$ and $p(x', t|x)$ also obeys the backward fractional equation

$$\frac{\partial^\alpha p(x', t|x)}{\partial t^\alpha} = \frac{1}{2} D \frac{\partial^2 p(x', t|x)}{\partial x^2}, \qquad (11.176)$$

with initial conditions

$$p(x', 0|x) = \delta(x' - x), \qquad \left.\frac{\partial p(x', t|x)}{\partial t}\right|_{t=0} = 0. \qquad (11.177)$$

---

[13]In Chapter 4 we have assumed (without loss of generality) that $x_0 = 0$. For the following discussion on first-passage problems is more convenient to assume a non-zero initial value.

[14]The subject is still under current research and many information is scattered in the literature. A very incomplete sample of references is given by Barkai (2001), Condamin *et al.* (2007), Rangarajan and Ding (2000), Sanders and Amjörnsson (2012) and Yuste and Lindenberg (2005).

Integrating Eq. (11.176) with respect to x' on the interval (a, b), interchanging the integrals with derivatives and finally using Eq. (11.175) and the initial conditions (11.177), we see that the survival probability obeys the fractional equation

$$\frac{\partial^\alpha S_{ab}(t|x)}{\partial t^\alpha} = \frac{1}{2} D \frac{\partial^2 S_{ab}(t|x)}{\partial x^2}, \qquad (11.178)$$

with initial conditions

$$S_{ab}(0|x) = 1, \qquad \left.\frac{\partial S_{ab}(t|x)}{\partial t}\right|_{t=0} = 0, \qquad (11.179)$$

where $x \in (a, b)$. Like the ordinary Brownian motion, survival is impossible if the fractional motion starts at each end of the interval (a, b), then the survival probability must obey the following boundary conditions

$$S_{ab}(t|a) = S_{ab}(t|b) = 0. \qquad (11.180)$$

We can solve the initial and boundary value problem posed by Eqs. (11.178)–(11.180) in Laplace space,

$$\hat{S}_{ab}(s|x) = \int_0^\infty e^{-st} S_{ab}(t|x) dt.$$

In effect, taking the Laplace transform of Eq. (11.178) and recalling that (cf. Eq. (4.81) of Chapter 4 and Eq. (11.179))

$$\mathcal{L}\left\{\frac{\partial^\alpha S_{ab}(t|x)}{\partial t^\alpha}\right\} = s^\alpha \hat{S}_{ab}(s|x) - s^{\alpha-1},$$

we get the simpler problem

$$\frac{1}{2} D \frac{d^2 \hat{S}_{ab}}{dx^2} - s^\alpha \hat{S}_{ab} = -s^{\alpha-1} \quad \text{and} \quad \hat{S}_{ab}(s|a) = \hat{S}_{ab}(s|b) = 0.$$

As we can see by direct substitution the solution of the differential equation is

$$\hat{S}_{ab}(s|x) = \frac{1}{s} + \hat{A}(s) \cosh \sqrt{2s^\alpha/D} + \hat{B}(s) \sinh \sqrt{2s^\alpha/D},$$

where $\hat{A}(s)$ and $\hat{B}(s)$ are found by using boundary conditions. As the reader can easily check, the final result reads

$$\hat{S}_{ab}(s|x) = \frac{1}{s}\left[1 - \frac{\cosh \sqrt{2s^\alpha/D}[x - (a+b)/2]}{\cosh \sqrt{2s^\alpha/D}(a-b)/2}\right]. \qquad (11.181)$$

Instead of trying to invert this expression and get $S_{ab}(t|x)$, which results in a cumbersome and rather impractical expression (see, for instance

Rangarajan and Ding, 2000) we will obtain an asymptotic expression for the survival probability valid as $t \to \infty$. To this end we first obtain the form taken by $\hat{S}_{ab}(s|x)$ when $s \to 0$ and then use Tauberian theorems to get the asymptotic expression of $S_{ab}(t|x)$ when $t \to \infty$.

Since $\cosh z = 1 + z^2/2! + O(z^4)$, simple manipulations yield

$$\frac{\cosh \sqrt{2s^\alpha/D}[x - (a+b)/2]}{\cosh \sqrt{2s^\alpha/D}(a-b)/2} = 1 + s^\alpha (x-a)(x-b)/D + O(s^{2\alpha}).$$

Substituting into (11.181) we finally get

$$\hat{S}_{ab}(s|x) = s^{\alpha-1}(x-a)(b-x)/D + O(s^{2\alpha-1}). \tag{11.182}$$

We know by Tauberian theorems that the behavior of $S_{ab}(t|x)$ as $t \to \infty$ is given by the Laplace inversion of the expression for the Laplace transform $\hat{S}_{ab}(s|x)$ as $s \to 0$ (cf. Sect. 10.2.2 of Chapter 10). Thus, when $0 < \alpha < 1$ we use the inversion formula (Roberts and Kaufman, 1966)

$$\mathcal{L}^{-1}\left\{\frac{1}{s^\beta}\right\} = \frac{t^{\beta-1}}{\Gamma(\beta)}, \qquad (\beta > 0),$$

and transforming back Eq. (11.182) we obtain the asymptotic behavior

$$S_{ab}(t|x) \sim \frac{t^{-\alpha}}{D\Gamma(1-\alpha)}(x-a)(b-x), \qquad (t \to \infty), \tag{11.183}$$

valid when $0 < \alpha < 1$.

Let us note that when $\alpha = 1$ we deal with the ordinary Brownian motion and Eq. (11.182) reads $\hat{S}_{ab}(s|x) = (x-a)(b-x)/D + O(s^{2\alpha-1})$ and, as $s \to 0$, we recover the well known expression of the mean exit time

$$T_{ab}(x) = \hat{S}_{ab}(0|x) = \frac{1}{D}(x-a)(b-x).$$

11.8.2 The first passage problem

We will next discuss the first passage problem to some threshold x_c. We know that this problem is equivalent to the exit problem out of a semi-infinite interval. We have also shown in Sect. 11.2, that the equation satisfied by the survival probability $S_c(t|x)$ is the same than that of the escape problem. That is to say, $S_c(t|x)$ is the finite solution to the equation

$$\frac{\partial^\alpha S_c}{\partial t^\alpha} = \frac{1}{2}D\frac{\partial^2 S_c}{\partial x^2},$$

$(0 < \alpha < 2)$ with initial and boundary conditions

$$S_c(0|x) = 1, \qquad \left.\frac{\partial S_c(t|x)}{\partial t}\right|_{t=0} = 0, \qquad S_c(t|x_c) = 0.$$

Proceeding as in the escape problem we easily see that, in the Laplace space, we have to find the finite solution to the boundary value problem

$$\frac{1}{2}\frac{d^2 \hat{S}_c}{dx^2} - s^\alpha \hat{S}_c = -s^{\alpha-1}, \qquad \hat{S}_c(s|x_c) = 0. \tag{11.184}$$

As the reader can readily check by direct substitution, the solution to this problem that is finite as $x \to \pm\infty$ reads

$$\hat{S}_c(s|x) = \frac{1}{s}\left[1 - e^{-|x-x_c|\sqrt{2s^\alpha/D}}\right]. \tag{11.185}$$

Let us now proceed to obtain the Laplace inversion of this expression. Since $\mathcal{L}^{-1}\{1/s\} = 1$, we write

$$S_c(t|x) = 1 - \mathcal{L}^{-1}\left\{\frac{1}{s}e^{-|x-x_c|\sqrt{2s^\alpha/D}}\right\}.$$

According to the inversion formula,

$$\mathcal{L}^{-1}\left\{e^{-|x-x_c|\sqrt{2s^\alpha/D}}\right\} = \frac{1}{2\pi i}\int_{\mathrm{Br}}\frac{ds}{s}e^{st-|x-x_c|\sqrt{2s^\alpha/D}},$$

where Br denotes the Browmwich path. Setting $\sigma = st$ we have

$$\mathcal{L}^{-1}\left\{e^{-|x-x_c|\sqrt{2s^\alpha/D}}\right\} = \frac{1}{2\pi i}\int_{\mathrm{Br}}e^{\sigma-z\sigma^{\alpha/2}}\frac{d\sigma}{\sigma},$$

where

$$z = |x - x_c|\sqrt{\frac{2}{Dt^\alpha}},$$

In terms of the Wright function (Erdelyi, 1953; Gorenflo *et al.*, 1999), which in the Browmwich representation is defined as [15]

$$\phi(\rho, \beta, z) = \frac{1}{2\pi i}\int_{\mathrm{Br}}e^{\sigma+z\sigma^{-\rho}}\frac{d\sigma}{\sigma^\beta}, \tag{11.186}$$

where $\rho > -1$, $\beta > 0$, we see that

$$\mathcal{L}^{-1}\left\{e^{-|x-x_c|\sqrt{2s^\alpha/D}}\right\} = \phi(-\alpha/2, 1, -z),$$

$(0 < \alpha < 2)$. We therefore obtain the following exact expression for the survival probability

$$S_c(t|x) = 1 - \phi\left(-\frac{\alpha}{2}, 1, -|x - x_c|\sqrt{\frac{2}{Dt^\alpha}}\right), \tag{11.187}$$

---

[15]Wright function is usually defined by the Hankel representation, which is similar to Eq. (11.186) but with the Browmwich contour replaced by the Hankel contour. Mainardi (1996) proved that both representations are equivalent.

$(0 < \alpha < 2)$.

Wright function is often defined by the power series (Erdelyi, 1953; Gorenflo *et al.*, 1999)

$$\phi(\rho, \beta, z) = \sum_{n=0}^{\infty} \frac{z^n}{n!\Gamma(n\rho + \beta)}, \qquad (11.188)$$

$(\rho > -1, \beta > 0)$. We can use this power series for obtaining an asymptotic expression for the survival probability valid for long values of time. Indeed, using (11.188) in (11.187) we readily get

$$S_c(t|x) = \frac{\sqrt{2/D}|x - x_c|}{\Gamma(1 - \alpha/2)} t^{-\alpha/2} + O(t^{-\alpha}), \qquad (11.189)$$

$(0 < \alpha < 2)$. Note that, as $t \to \infty$, $S_c(t|x)$ decays slower than t^{-1} which shows that, like in the ordinary motion, the mean first-passage time of fractional Brownian motion is infinite.

Chapter 12

First-passage and extremes in socio-economic systems

Despite the long and standing tradition of applications of level-crossing methods in physical sciences and engineering, the practical use of these problems to socio-economic sciences has traditionally been scarce. A situation that started to change around 2005.

It is clear that first-passage and extreme-value problems have a great interest in socio-economic systems, ranging from population dynamics to numerous economic settings. From the point of view of finance and, beyond the obvious relation to the classical ruin problem, there are many situations in which level-crossings are a fundamental constituent. This is, among others, the case of the so-called "leverage certificates" which are financial derivatives offering a nonzero payoff only if the underlying asset does not escape from a pre-established domain over a certain time window.

On the other hand, the explosion during recent years of what has been called "big data" has substantially increased the data available in a large list of socio-economic phenomena which, in turn, has allowed a better and more reliable statistical characterization of extreme events.

In the study of any socio-economic system one inevitably deals with time series. In order to model these series and extract some regularities that may be hidden in them, a useful hypothesis is to assume that the series correspond to some realization of a continuous random process, usually a diffusion process. This is, for instance, the case of continuous-time finance where speculative prices are supposed to follow the geometric Brownian motion. In such an approach, if $x_1, x_2, \ldots, x_n, \ldots$ is a given time series, one assumes the existence of a random process in continuous time $X(t)$ –as, for instance, a diffusion process– such that $X(t_n) = x_n$ at discrete times t_n $(n = 1, 2, 3, \ldots)$. Note, however, that in taking this approach one is assuming that between two adjacent data points x_n and x_{n+1} there exist infinitely many registers corresponding to $X(t)$ with $t_n < t < t_{n+1}$, in other words, we are assuming that the actual trajectories of $X(t)$ are continuous.

If, on the other hand, we know with certainty that time is a discrete variable and that between x_n and x_{n+1} there are no intermediate points (for example in tic-by-tic financial time series) then the diffusion approach may not be the most convenient. In such cases a powerful tool is provided by the Continuous Time Random Walk formalism.

In this chapter we shall address the first-passage problem using both approaches –diffusions and random walks– and using the tools developed in previous chapters with most challenging issues in reference to socio-economical contexts and afterwards consider both the continuous and discrete time approaches with some models which are enough sophisticated to improve the basic Gaussian dynamics and solve first-passage and extreme value problems.

12.1 Financial diffusion models

As we have discussed in Chapter 9 (Sect. 9.1), within the framework of financial time series in which $P(t)$ denotes the (random) price of a given asset, the first diffusion model was proposed by the French mathematician Louis Bachelier who in 1900 assumed that prices are themselves a Brownian motion. In other words, $P(t)$ was a diffusion process with constant drift and constant diffusion coefficient. This simple model suffers severe shortcomings being the most critical one that $P(t)$ may become negative with certainty which is contrary to limited liability, perhaps the most basic principle in finance.

Since prices cannot be negative, Bachelier model became unrealistic and it was replaced during the nineteen fifties by the geometric Brownian motion in which prices are the exponential of an ordinary Brownian motion (Osborne, 1959). The model assumes that[1]

$$P(t) = P_0 e^{X(t)}, \tag{12.1}$$

where $P_0 = P(t_0)$ is an initial price and $X(t)$ is the return which is described by an ordinary Brownian motion with constant drift m and constant diffusion coefficient k^2, that is to say, by the following stochastic differential equation

$$dX(t) = m dt + k dW(t), \tag{12.2}$$

---

[1]cf Eqs. (9.1) and (9.2) of Chapter 9 with the change of notation $S(t) \to P(t)$, $r(t) \to X(t)$ and $\sigma \to k$.

where $W(t)$ is the standard Wiener process with zero mean and unit standard deviation. Since the average of $dW(t)$ is zero we see that $m = \langle X(t) \rangle$ is the average return, while $k = \sigma$ is the standard deviation, $\sigma^2 = \langle [X(t) - m]^2 \rangle$, which in finance is usually called "volatility" (see Sect. 12.3).

We have shown in Chapter 9 that using the Itö convention the geometric Brownian motion can be defined by the multiplicative stochastic differential equation

$$\frac{dP(t)}{P(t)} = (m + k^2/2)dt + kdW(t). \tag{12.3}$$

We have also proved that the solution to this equation is given by

$$P(t) = P_0 \exp\Big\{ m(t - t_0) + k\big[W(t) - W(t_0)\big] \Big\}, \tag{12.4}$$

which, by virtue of the normal distribution of $W(t)$, shows the log-normal character of the price $P(t)$.

The log-normal model is rather simple and neglects possible different dynamics in price formation. One possible generalization assumes that the return obeys a more general diffusion process,

$$dX(t) = f(X(t))dt + \sqrt{D(X(t))}dW(t), \tag{12.5}$$

where $f(x)$ is the drift setting the dynamics of the return and, hence, ruling the dynamics of price formations, $D(x)$ is the diffusion coefficient and determines the volatility of prices.

Two particular diffusion processes are widely used in the modeling of many socio-economic problems, both models have linear drift, meaning that there is a linear force which tends to restore the return $X(t)$ to some "equilibrium" value. One model is the Ornstein-Uhlenbeck process for which, besides $f(x)$ being linear, the diffusion coefficient $D(x)$ is constant. We have thoroughly studied the level-crossing properties of this process in the previous chapter.

Another random process that has been found most useful, not only in modeling socio-economic systems, but also in theoretical biology such as population dynamics and neuron firing processes, it is the *Feller process* which is a diffusion process with linear drift and linear diffusion coefficient vanishing at the origin. The time evolution of the process is governed by

$$dX(t) = [-\alpha X(t) + \beta]dt + k\sqrt{X(t)}dW(t), \tag{12.6}$$

where $\alpha > 0, \beta > 0$ and $k > 0$ are constant parameters.

Note that the linear drift $f(x) = -\alpha x + \beta$ with $\alpha > 0$ results in a restoring force which, in the absence of noise, makes the process decaying toward the value β/α. Indeed, when $k = 0$ the solution to (12.6) with initial condition $X(t_0) = X_0$ is

$$X(t) = (X_0 - \beta/\alpha)\,e^{-\alpha(t-t_0)} + \beta/\alpha,$$

which tends to β/α as $t \to \infty$.

On the other hand, the state dependent coefficient, $D(x) = k^2 x$, for large values of x enhances the effect of noise while when x goes to zero the effect of noise vanishes. Therefore, when the process reaches the origin noise disappears and the drift drags $X(t)$ toward β/α. Hence, since α and β are positive, the process, starting at some positive value, cannot reach the negative region with the overall result that the Feller process remains always positive.

The fact that $X(t)$ never attains negative values makes the process an ideal candidate for modeling a number of phenomena in natural and social sciences. Thus, for instance, in financial markets the process was introduced in 1985 to represent the term structure of interest rates –under the name of Cox, Ingersoll, and Ross (CIR) model. The process is also being considered to provide a random character for the volatility of a given stock (see below).

12.2   First-passage problems in the Feller process *

We see from Eq. (12.6) that the drift and the diffusion coefficient of the Feller process are

$$f(x) = -(\alpha x - \beta), \qquad D(x) = kx.$$

Hence, the equation for the first-passage probability, $W_c(t|x)$, to some threshold x_c is given by (cf. Eqs. (11.28) and (11.29) of Chapter 11)

$$\frac{\partial W_c}{\partial t} = -(\alpha x - \beta)\frac{\partial W_c}{\partial x} + \frac{1}{2}k^2 x \frac{\partial^2 W_c}{\partial x^2}, \tag{12.7}$$

with initial and boundary conditions

$$W_c(0|x) = 0, \qquad W_c(t|x_c) = 1. \tag{12.8}$$

The definition of dimensionless variables

$$t' = \alpha t, \qquad x' = \left(\frac{2\alpha}{k^2}\right)x,$$

turns Eq. (12.7) into (we drop the primes)

$$\frac{\partial W_c}{\partial t} = -(x - \theta)\frac{\partial W_c}{\partial x} + x\frac{\partial^2 W_c}{\partial x^2}, \tag{12.9}$$

where

$$\theta = \frac{2\beta}{k^2} > 0, \tag{12.10}$$

is called "saturation parameter" or "normal level" and plays a key role in the behavior of the Feller process.[2]

The time Laplace transform of the first-passage probability,

$$\hat{W}_c(s|x) = \int_0^\infty e^{-st} W_c(t|x) dt,$$

turns the partial differential equation (12.9) into the ordinary differential equation

$$x \frac{d^2 \hat{W}_c}{dx^2} - (x - \theta) \frac{d\hat{W}_c}{dx} - s\hat{W}_c = 0, \tag{12.11}$$

with boundary condition

$$\hat{W}_c(s|x_c) = \frac{1}{s}. \tag{12.12}$$

We must also bear in mind that since W_c is a probability, it is necessary that any solution of the problem (12.11) and (12.12) must be finite and nonnegative for all $x \geq 0$.

The ordinary differential equation (12.11) is a Kummer equation whose general solution is given by a linear combination of the Kummer functions $F(s, \theta, x)$ and $U(s, \theta, x)$ (Magnus *et al.*, 1966)

$$\hat{W}_c(s|x) = AF(s, \theta, x) + BU(s, \theta, x),$$

where A and B are arbitrary constants to be found through the boundary condition (12.12) and the finiteness of W_c. Proceeding as in Sect. 11.4 of the previous chapter it is not difficult to get the final result (Masoliver and Perelló, 2012)

$$\hat{W}_c(s|x) = \begin{cases} \dfrac{F(s, \theta, x)}{sF(s, \theta, x_c)}, & x < x_c, \\[2ex] \dfrac{U(s, \theta, x)}{sU(s, \theta, x_c)}, & x > x_c. \end{cases} \tag{12.13}$$

In general these expressions for the Laplace transform of the hitting probability cannot be inverted exactly in an analytical manner and one has to resort to numerical inversion. There are, nonetheless, some instances in which we can obtain analytical expressions in real time. This is the case of

---

[2]For more information and some technical details on the material presented in this section we refer the reader to Masoliver and Perelló, 2012 (see also Masoliver and Perelló, 2014a, for some erratum corrections).

hitting the origin ($x_c = 0$) which has a significant interest in many practical applications as, for example, in financial markets where reaching the origin may be related to bankruptcy or in population dynamics where it means extinction.

We denote by $W_0(t|x)$ the first-passage probability to threshold $x_c = 0$, we have shown elsewhere that (Masoliver and Perelló, 2012)

$$W_0(t|x) = \begin{cases} \frac{1}{\Gamma(1-\theta)}\Gamma\left(1-\theta, \frac{2\alpha x e^{-\alpha t}}{k^2(1-e^{-\alpha t})}\right), & \theta < 1, \\ 0, & \theta > 1, \end{cases}$$

where $\Gamma(a, z)$ is the incomplete Gamma function. Let us incidentally note that when $\theta > 1$ the probability of reaching the origin is zero. In other words, when $\theta > 1$ the Feller process cannot reach the origin. Therefore, the estimation of the parameter θ is crucial to determine whether the process hits the threshold $x_c = 0$.

We know that in terms of the Laplace transform of the hitting probability, $\hat{W}_c(s|x)$, the mean first-passage time to level x_c is given by

$$T_c(x) = \lim_{s \to 0}\left[\frac{1}{s} - \hat{W}_c(s|x)\right]$$

which after using Eq. (12.13) yields

$$T_c(x) = \lim_{s \to 0}\left[\frac{F(s, \theta, x_c) - F(s, \theta, x)}{sF(s, \theta, x_c)}\right] \qquad (x < x_c) \qquad (12.14)$$

and

$$T_c(x) = \lim_{s \to 0}\left[\frac{U(s, \theta, x_c) - U(s, \theta, x)}{sU(s, \theta, x_c)}\right] \qquad (x > x_c). \qquad (12.15)$$

On the other hand, as we have proved elsewhere (Masoliver and Perelló, 2012), the expansions as $s \to 0$ of the Kummer functions F and U are

$$F(s, \theta, x) = 1 + \frac{s}{\theta}\int_0^x F(1, 1 + \theta, z)dz + O(s^2), \qquad (12.16)$$

and

$$U(s, \theta, x) = 1 - s\left[\psi(1-\theta) + \int^x U(1, 1 + \theta, z)dz\right] + O(s^2), \qquad (12.17)$$

where $\psi(z) = \Gamma'(z)/\Gamma(z)$ is the digamma function. Plugging Eqs. (12.16) and (12.17) into Eqs. (12.14) and (12.15) yields

$$T_c(x) = \frac{1}{\theta}\int_x^{x_c} F(1, 1 + \theta, z)dz, \qquad (x < x_c), \qquad (12.18)$$

and

$$T_c(x) = \int_{x_c}^{x} U(1, 1 + \theta, z) dz, \qquad (x > x_c). \qquad (12.19)$$

For the Feller process one can solve the escape problem as well. In such a case the Laplace transform of the exit probability out of the interval (a, b) also obeys Kummer's equation

$$\frac{1}{2} kx \frac{d^2 \hat{W}_{ab}}{dx^2} - (\alpha x - \beta) \frac{d\hat{W}_{ab}}{dx} - s\hat{W}_{ab} = 0,$$

with the boundary conditions

$$\hat{W}_{ab}(s|a) = \hat{W}_{ab}(s|b) = 1/s.$$

The exact solution is again given by combinations of the confluent hypergeometric functions F and U and we refer the interested reader to our previous work for a complete account of either W_{ab} and the mean exit time T_{ab} (Masoliver and Perelló, 2012).

12.3 Random diffusions and stochastic volatility models

The dynamics of particles in random media may be used for a large variety of phenomena in statistical mechanics and condensed matter physics (see, for instance, Ben-Avraham and Havlin, 2000). One possible approach to address the problem of motion in random media consists in assuming that the diffusion coefficient of the random process under study is itself random. In such an approach the function $D(x)$ appearing in Eq. (12.5),

$$dX(t) = f(X(t))dt + \sqrt{D(X(t))}dW(t),$$

is a stochastic quantity which in many instances is supposed to be another diffusion process. In such a case one deals with a two-dimensional random process

$$\mathbf{X}(t) = (X(t), Y(t)),$$

such that

$$\phi(Y(t)) = \sqrt{D(X(t))},$$

where $\phi(\cdot)$ is an arbitrary function and $Y(t)$ is a diffusion process with drift $h(y)$ and noise intensity $g(y)$. The bidimensional process is thus governed by two coupled stochastic differential equations one representing the usual diffusion process and the other "the diffusion of diffusion":

$$dX(t) = f(X(t))dt + \phi(Y(t))dW_1(t),$$
$$dY(t) = h(Y(t))dt + g(Y(t))dW_2(t), \qquad (12.20)$$

where the Wiener processes W_1 and W_2 may, or may not, be correlated, although, in what follows, we suppose that they are uncorrelated, that is,

$$\langle dW_1(t)dW_2(t)\rangle = 0, \qquad \langle dW_1(t)^2\rangle = \langle dW_2(t)^2\rangle = dt.$$

Note that the set given by Eq. (12.20) can be written as a vector equation

$$d\mathbf{X}(t) = \mathbf{f}(\mathbf{X}) + \mathbf{g}(\mathbf{X})d\mathbf{W}(t)$$

where $\mathbf{X} = (X,Y)$ is the bidimensional diffusion process and $\mathbf{W} = (W_1, W_2)$. The drift is given by the vector function

$$\mathbf{f}(\mathbf{x}) = (f(x), h(y)),$$

where $\mathbf{x} = (x, y)$, and the noise matrix by

$$\mathbf{g}(\mathbf{x}) = \begin{pmatrix} \phi(y) & 0 \\ 0 & g(y) \end{pmatrix}.$$

As we have seen in Chapter 7 (Sect. 8.1) this corresponds to a bidimensional diffusion process with diffusion matrix given by $\mathbf{D} = \mathbf{g}\mathbf{g}^T$. That is,

$$\mathbf{D}(\mathbf{x}) = \begin{pmatrix} \phi^2(y) & 0 \\ 0 & g^2(y) \end{pmatrix}$$

Let us now address the level crossing problem for the vector process $\mathbf{X}(t)$ out of some bidimensional region $\mathbf{R}$ with boundary $\mathbf{B}$. Let us denote by $S(t|\mathbf{x})$ the probability that starting at $t = 0$ at some point $\mathbf{x}$ inside the region $\mathbf{R}$, the process has not reached at time t, or during any previous time, the boundary $\mathbf{B}$. We have seen in Chapter 11 (Sect. 11.5) that the survival probability $S(t|\mathbf{x})$ satisfies the Fokker-Planck equation

$$\frac{\partial S}{\partial t} = f(x)\frac{\partial S}{\partial x} + h(y)\frac{\partial S}{\partial y} + \frac{1}{2}\phi^2(y)\frac{\partial^2 S}{\partial x^2} + \frac{1}{2}g^2(y)\frac{\partial^2 S}{\partial y^2}, \qquad (12.21)$$

with initial and boundary conditions given respectively by

$$S(0|\mathbf{x}) = \begin{cases} 1, & \text{if } \mathbf{x} \in \mathbf{R} \\ 0, & \text{if } \mathbf{x} \notin \mathbf{R}, \end{cases} \qquad (12.22)$$

and

$$S(t|\mathbf{x}) = 0, \qquad (\mathbf{x} \in \mathbf{B}). \qquad (12.23)$$

12.3.1 *Stochastic volatility*

In socio-economic systems one also encounters random diffusions. This is the case of the so-called stochastic volatility (SV) models appearing in financial analysis for which the volatility $\sigma(t)$, defined as the standard deviation of the return $X(t)$,

$$\sigma^2(t) = \left\langle \left[X(t) - \langle X(t) \rangle \right]^2 \right\rangle,$$

is a random quantity. There is a growing empirical evidence that supports this assumption and financial engineering have started considering this approach since the 1987 crash. The industry in this way improved the geometric Brownian motion description with the aim of providing more precise evaluation of financial option prices. Empirical evidence –embodied in the so-called stylized facts– strongly suggests that the assumption of constant volatility does not properly account for important features of markets such as fat tails in price fluctuations, negative skewness in price changes, long-range volatility autocorrelation and return-volatility asymmetric cross-correlation.[3]

The starting point of any stochastic volatility model is the geometric Brownian motion described above (cf. Eqs. (12.1)-(12.3)). In what follows is more convenient to work with the zero-mean return (also called detrended return) defined as

$$X'(t) = X(t) - \langle X(t) \rangle,$$

so that the evolution of $X'(t)$ is given by the simple equation (cf. Eq. (12.2))

$$dX(t) = \sigma dW(t),$$

where we have dropped the prime in the return.

We next assume that the volatility is itself a random quantity and set

$$\sigma = \phi(Y(t)), \tag{12.24}$$

where $Y(t)$ is a random processes which is generally assumed to be a diffusion process. The general model is, therefore, given by the bidimensional process $(X(t), Y(t))$,

$$dX(t) = \phi(Y(t))dW_1(t),$$
$$dY(t) = h(Y(t))dt + g(Y(t))dW_2(t). \tag{12.25}$$

---

[3]For a mathematical introduction to stochastic volatility see Fouque *et al.* (2000); for a complete review on stylized facts see Cont (2001).

In other words, stochastic volatility models are special cases of the random diffusion models discussed above.

There is a wide consensus that volatility seems to return to some average value, the so-called mean reversion. This means that there exists a normal level to which the volatility eventually goes back. For the general model given in Eq. (12.25), the existence of mean reversion implies restrictions on the form of the drift coefficient $h(Y)$. A simple way of incorporating this empirical fact into the model is to assume a linear restoring force (as in the harmonic oscillator) which drags the process toward the normal level:

$$h(Y) = -\alpha(Y - \mu), \tag{12.26}$$

where $\alpha > 0$ and μ are given parameters of the model. Before proceeding further let us prove that μ is the normal level of the process $Y(t)$, that is to say, $Y(t) \to \mu$ as $t \to \infty$. In effect, we can easily check by direct substitution that the formal solution to the second equation of the set (12.25) when $h(Y)$ given by Eq. (12.26), reads

$$Y(t) = Y_0 e^{-\alpha(t-t_0)} + \mu\big(1 - e^{-\alpha(t-t_0)}\big)$$
$$+ \int_{t_0}^{t} e^{-\alpha(t-t')} g(Y(t'))dW_2(t'), \tag{12.27}$$

where t_0 is some initial time and $Y_0 = Y(t_0)$.[4] From Eq. (12.27) we clearly see that $Y(t) \to \mu$ as $t \to \infty$ and the volatility reverts to (cf. Eq. (12.24))

$$\lim_{t\to\infty} \sigma(Y(t)) = \phi(\mu), \tag{12.28}$$

where $\phi(\mu) \equiv \sigma_s$ is the normal level of volatility.

Apart from the linearity of the volatility drift, the complete dynamics of the volatility has not been conclusively attached to any specific model. We next detail some of the most common stochastic volatility models.

(a) The Ornstein-Uhlenbeck model

In this case $\phi(Y) = Y$, that is,

$$\sigma = Y(t)$$

and $Y(t)$ is the Ornstein-Uhlenbeck process with linear drift given by Eq. (12.26) and $g(Y) = k$ $(k > 0)$. The model is thus specified by

$$dX(t) = Y(t)dW_1(t) \tag{12.29}$$
$$dY(t) = -\alpha(Y(t) - m) + kdW_2(t), \tag{12.30}$$

---

[4]Even though we call (12.27) as the "formal solution" to the differential equation for $Y(t)$, it is in fact an integral equation for $Y(t)$ equivalent to Eq. (12.25).

where we have now written $\mu = m$ as the normal level of volatility.

Unfortunately this model is not fully satisfactory because volatility – which is, by definition, a positive quantity– may attain negative values in the course of time. Despite this inconsistence, the model has been used a number of times because of its simplicity (it is completely linear).

The next two models repair this irregularity because in them the volatility turns out to be a positive quantity.

(b) The Heston model

In this model $\phi(Y) = \sqrt{Y}$, that is,

$$\sigma = \sqrt{Y(t)} \tag{12.31}$$

and $Y(t)$ is the Feller process discussed above. The model is thus specified by

$$dX(t) = \sqrt{Y(t)}dW_1(t) \tag{12.32}$$

$$dY(t) = -\alpha(Y(t) - m^2) + k\sqrt{Y(t)}dW_2(t). \tag{12.33}$$

Since the Feller process never attains negative values, the volatility (12.31) is real and positive. Note that in this case $\mu = m^2$ represents the normal level of volatility.

The Heston model is undoubtedly one of the most popular models of stochastic volatility. Another useful model in which volatility is always positive is the exponential Ornstein-Uhlenbeck (expOU) model.

(c) The exponential Ornstein-Uhlenbeck model

In this model $\phi(Y) = me^Y$ $(m > 0)$, so that the volatility

$$\sigma(t) = me^{Y(t)} \tag{12.34}$$

is always positive regardless the values of $Y(t)$. The process $Y(t)$ is taken to be an Ornstein-Uhlenbeck process with zero mean. The complete expOU model is thus specified by

$$dX(t) = me^{Y(t)}dW_1(t) \tag{12.35}$$

$$dY(t) = -\alpha Y(t) + kdW_2(t). \tag{12.36}$$

Let us denote by $\sigma_s = \lim_{t\to\infty} \sigma(t)$ the normal level of volatility. We see from the solution of Eq. (12.36) (see Eq. (12.27) with $\mu = 0$ and $g(x) = k$)

$$Y(t) = Y_0 e^{-\alpha(t-t')} + k \int_{t_0}^{t} e^{-\alpha(t-t')}dW_2(t'),$$

so that $Y(t) \to 0$ as $t \to \infty$ and hence (cf. Eq. (12.34)) $m = \sigma_s$ is the normal level of volatility.

12.4   The first-passage problem in the Heston model *

As an illustration of the formalism presented above we will review an example of first-passage problem for the Heston model (12.32)–(12.33). We want to treat the first-passage to a certain critical value x_c of the zero-mean return $X(t)$ regardless the value of the volatility.

Recall that in the Heston model the volatility variable $Y(t)$ is the Feller process,

$$dY(t) = -\alpha(Y(t) - m^2) + k\sqrt{Y(t)}dW_2(t).$$

Hence the level crossing problem, including first-passage and escape, for the volatility alone has been already treated in Sect. 12.2. Let us remark that we will study here the level crossing problem (specifically the first-passage problem) for the return $X(t)$. Since return depends on volatility through the pair of stochastic differential equations (12.33) we will deal with a two dimensional first-passage problem

We therefore denote by $S(t|x,y)$ the probability that return $X(t)$ initially at $X(0) = x$ with volatility variable $Y(0) = y$ has never attained the value x_c before time t. When $x_c \leq x$ –in such a case x_c is a level of losses– S coincides with the survival probability for the joint process $(X(t), Y(t))$ to be, at time t, still inside the infinite bidimensional region $\mathbf{R}$ defined by

$$\mathbf{R} = \{(x,y) \in \mathbb{R}^2 | x_c \leq x < \infty, 0 < y < \infty\}.$$

Note that the boundary $\mathbf{B}$ of $\mathbf{R}$ is the straight line $x = x_c$. We also note that the probability of first reaching x_c, i.e., the probability of having a loss labeled by x_c (which as $x_c \to -\infty$ becomes the "default probability") is given by

$$W(t|x,y) = 1 - S(t|x,y). \tag{12.37}$$

When $x_c \geq x$ –now x_c is a profit level– the bidimensional region $\mathbf{R}$ is

$$\mathbf{R} = \{(x,y) \in \mathbb{R}^2 | -\infty < x \leq x_c, 0 < y < \infty\},$$

and the boundary is again the straight line $x = x_c$. Note that the probability of having a profit labeled by x_c is obviously given by Eq. (12.37).

Since in the Heston model (12.32)–(12.33), we do not consider any bias, both situations –loss and profit– are symmetrical.

The survival probability will be the solution to the problem (12.21)-(12.23) with the replacements (compare Eq. (12.25) with Eq. (12.33))

$$f(x) = 0, \quad \phi(y) = \sqrt{y}, \quad h(y) = -\alpha(y - m^2), \quad g(y) = k\sqrt{y}.$$

That is,

$$\frac{\partial S}{\partial t} = -\alpha(y - m^2)\frac{\partial S}{\partial y} + \frac{1}{2}y\frac{\partial^2 S}{\partial x^2} + \frac{1}{2}k^2 y\frac{\partial^2 S}{\partial y^2}, \tag{12.38}$$

with initial and boundary conditions given by

$$S(0|x, y) = 1, \qquad S(t|x_c, y) = 0, \quad \text{for all } y > 0. \tag{12.39}$$

The symmetry between losses and profits just mentioned is apparent because the problem (12.38)–(12.39) remains invariant under the change of variable

$$z = |x - x_c|. \tag{12.40}$$

Indeed, if $x_c > x$ then $z = x_c - x$ and $\partial S/\partial x = -\partial S/\partial z$ which implies $\partial^2 S/\partial x^2 = \partial^2 S/\partial z^2$ and the same expression holds when $x_c < x$ and $z = x - x_c$. Since in Eq. (12.38) the only dependence on x is through the second derivative we see that the equation is invariant under the change of variable (12.40). For both cases the boundary condition is given by

$$S(t|z = 0, y) = 0.$$

If in addition to the change (12.40) we define the dimensionless variables

$$\tau = \alpha t, \qquad v = y/\alpha, \tag{12.41}$$

the problem (12.38)–(12.39) can be written as

$$\frac{\partial S}{\partial \tau} = -\alpha(v - \theta)\frac{\partial S}{\partial y} + \frac{1}{2}\beta^2 v\frac{\partial^2 S}{\partial v^2} + \frac{1}{2}v\frac{\partial^2 S}{\partial z^2}, \tag{12.42}$$

where

$$\theta = m^2/\alpha \quad \text{and} \quad \beta = k/\alpha \tag{12.43}$$

are respectively the dimensionless normal level and noise intensity. The initial and boundary conditions are

$$S(0|z, v) = 1, \qquad S(\tau|0, v) = 0. \tag{12.44}$$

The exact solution to the problem can be obtained by means Fourier transform methods. Since $z \geq 0$ and $S(\tau|0, v) = 0$, we will use the Fourier sine transform, $\tilde{S} = \mathcal{F}_s\{S\}$, defined as

$$\tilde{S}(\tau|\omega, v) = \int_0^\infty S(\tau|z, v)\sin\omega z dz. \tag{12.45}$$

The inverse transform is

$$S(\tau|z, v) = \frac{2}{\pi}\int_0^\infty \tilde{S}(\tau|\omega, v)\sin\omega z d\omega. \tag{12.46}$$

Let us observe that written in terms of its Fourier sine transform, the survival probability automatically satisfies the boundary condition

$$S(\tau|z = 0, v) = 0.$$

We thus proceed to transform Eq. (12.42) which, after taking into account the standard property

$$\mathcal{F}_s \left\{ \frac{\partial^2 S}{\partial z^2} \right\} = -\omega^2 \tilde{S},$$

yields the simpler equation

$$\frac{\partial \tilde{S}}{\partial \tau} = -\alpha(v - \theta)\frac{\partial \tilde{S}}{\partial v} + \frac{1}{2}\beta^2 v \frac{\partial^2 \tilde{S}}{\partial v^2} - \frac{1}{2}\omega^2 v \tilde{S}, \qquad (12.47)$$

with the initial condition

$$\tilde{S}(0|\omega, v) = \mathcal{P}(1/\omega), \qquad (12.48)$$

where

$$\mathcal{P}(1/\omega) = \int_0^\infty \sin \omega z \, dz = 1/\omega \qquad (\omega \neq 0)$$

is the Cauchy principal value.[5]

As can be easily seen by direct substitution the solution to the initial problem (12.47)–(12.48) is furnished by [6]

$$\tilde{S}(\tau|\omega, v) = \mathcal{P}(1/\omega) \exp\{-A(\omega, \tau) - 2B(\omega, \tau)v/\beta^2\}, \qquad (12.49)$$

where

$$A(\omega, \tau) = (2\theta/\beta^2) \int_0^\tau B(\omega, \tau')d\tau', \qquad (12.50)$$

and $B(\omega, \tau)$ obeys the Riccati equation

$$\dot{B} = -B - B^2 + (\beta\omega^2/2)^2, \qquad (12.51)$$

---

[5] See Erderlyi (1954) and also Vladimirov (1984) for the Cauchy principal value.

[6] In effect, substituting (12.49) into (12.47) and some rearranging of terms yields (dots refer to τ derivatives)

$$\dot{A} - (2\theta/\beta^2)B + (2/\beta^2)[\dot{B} + B + B^2 - \beta^2\omega^2/4]v = 0.$$

Since this equation holds for all values of v, we conclude that the unknown functions $A(\omega, \tau)$ and $B(\omega, \tau)$ must obey the equations

$$\dot{A} - (2\theta/\beta^2)B = 0, \qquad \dot{B} + B^2 - \beta^2\omega^2/4 = 0,$$

from which Eqs. (12.50) and (12.51) follow. The initial condition (12.48) is satisfied by imposing the initial conditions $A(\omega, 0) = B(\omega, 0) = 0$.

with initial condition $B(\omega, 0) = 0$. Equation (12.51) can readily solved by setting $B = \dot{Z}/Z$ which transforms Riccati equation for B into a second-order linear equation with constant coefficients for Z.

We have shown elsewhere (see Masoliver and Perelló, 2009, for details on the calculations of this section) that the final expression for the survival probability $S(t|x, y)$ (after substituting Eq. (12.49) into Eq. (12.46) and using the original variables (12.40)–(12.41)) reads

$$S(t|x, y) = \frac{2}{\pi} \int_0^\infty \frac{d\omega}{\omega} \sin \omega |x - x_c| \tag{12.52}$$

$$\times \left[\frac{\Delta(\omega) e^{-yB(\omega,t)/\alpha\theta}}{\mu_+(\omega) + \mu_-(\omega) e^{-\Delta(\omega)\alpha t}} \right]^{2\theta/\beta^2} e^{-(2\theta\alpha t/\beta^2)\mu_-(\omega)},$$

where

$$\Delta(\omega) = \sqrt{1 + (\beta\omega)^2}, \qquad \mu_\pm(\omega) = \frac{1}{2}[\Delta(\omega) \pm 1] \tag{12.53}$$

and

$$B(\omega, t) = \mu_-(\omega) \frac{1 - e^{-\Delta(\omega)\alpha t}}{1 + [\mu_-(\omega)/\mu_+(\omega)]e^{-\Delta(\omega)\alpha t}}. \tag{12.54}$$

Equation (12.52) provides the complete solution to the first-passage problem for the Heston model which can be used to get numerical values for the survival and hitting probabilities. It is not, however, the most convenient form of $S(t|x, y)$ for uncovering general properties of the first-passage problem. We will, therefore, obtain asymptotic approximations that reveal general and qualitative properties which are very useful in many practical situations.

12.4.1 Long-time behavior and the mean first-passage time

We first obtain an approximate expression to the survival probability that is valid not only for long times but also for large (initial) volatilities. For times such that

$$\alpha t \gg 1,$$

we can use the saddle-point method (Erdelyi, 1956) to get an approximate expression for the integral in Eq. (12.52). The saddle-point method is based on the fact that as $t \to \infty$ the exponential in Eq. (12.52),

$$e^{-(2\theta\alpha t/\beta^2)\mu_-(\omega)},$$

decreases so fast that the only relevant values for the integral come from the vicinity of the minimum value of $\mu_-(\omega)$. We easily see from Eq. (12.53) that this minimum corresponds to $\omega = 0$. Taking into account (cf. (12.53) and (12.54))

$$\mu_-(\omega) = (\beta\omega/2)^2 + O(\omega^4), \quad \mu_+(\omega) = 1 + (\beta\omega/2)^2 + O(\omega^4) \quad (12.55)$$

and

$$\Delta(\omega) = 1 + (\beta\omega)^2 + O(\omega^4), \quad B(\omega, t) = (\beta\omega/2)^2(1 - e^{-\alpha t}) + O(\omega^4), \quad (12.56)$$

we have

$$\left[\frac{\Delta(\omega)e^{-yB(\omega,t)/\alpha\theta}}{\mu_+(\omega) + \mu_-(\omega)e^{-\Delta(\omega)\alpha t}} \right]^{2\theta/\beta^2}$$

$$= \left[\frac{\Delta(\omega)}{\mu_+(\omega) + \mu_-(\omega)e^{-\Delta(\omega)\alpha t}} \right]^{2\theta/\beta^2} e^{-(2y/\alpha\beta^2)B(\omega,t)}$$

$$= [1 + O(\omega^2)] \exp\left[-\frac{\omega^2 y}{2\alpha}(1 - e^{-\alpha t}) \right].$$

Substituting this into Eq. (12.52) and bearing in mind that $\mu_-(\omega) = (\beta\omega/2)^2 + O(\omega^2)$ we get

$$S(t|x,y) \simeq \frac{2}{\pi} \int_0^\infty \frac{\sin\omega|x - x_c|}{\omega} \exp\left[-\frac{1}{2}\lambda(t,y)\omega^2 \right] d\omega, \quad (12.57)$$

where

$$\lambda(t,y) = \theta\alpha t + y(1 - e^{-\alpha t})/\alpha \simeq \theta\alpha t + y/\alpha, \quad (\alpha t \gg 1). \quad (12.58)$$

Let us also note that for small to moderate values of the volatility y, the long-time behavior of $\lambda(t,y)$ is simply given by

$$\lambda(t,y) \simeq \theta\alpha t. \quad (12.59)$$

The integral in Eq. (12.57) can be performed exactly after using the result (Erdelyi, 1954)

$$\int_0^\infty e^{-a\omega^2} \frac{\sin(\omega x)}{\omega} d\omega = \frac{\pi}{2}\mathrm{erf}\left(\frac{x}{2\sqrt{a}} \right),$$

where

$$\mathrm{erf}(z) = \frac{2}{\sqrt{\pi}} \int_0^x e^{-\xi^2} d\xi$$

is the error function. The approximation (12.57) is thus given by the Gaussian probability

$$S(t|x,y) \simeq \mathrm{erf}\left(\frac{|x - x_c|}{2\sqrt{\lambda(t,y)}} \right), \quad (12.60)$$

$(\alpha t \gg 1)$, where $\lambda(t, y)$ is given by Eq. (12.58) (or by Eq. (12.59) for moderate values of the volatility).

In this approximation the first-passage probability $W(t|x, y)$, related to the survival probability by

$$W(t|x, y) = 1 - S(t|x, y),$$

reads

$$W(t|x, t) \simeq \mathrm{Erfc}\left(\frac{|x - x_c|}{2\sqrt{\lambda(t, y)}}\right),\qquad(12.61)$$

$(\alpha t \gg 1)$, where

$$\mathrm{Erfc}(z) = 1 - \mathrm{erf}(z)$$

is the complementary error function.

We shall briefly comment on the mean first-passage time. We know that this statistic is given by the time integral of the survival probability (cf. (10.29) of Chapter 10) which we denote

$$T(x, y) = \int_0^\infty S(t|x, y)dt.\qquad(12.62)$$

Mathematical analysis tells us that in order for the integral to be finite it is necessary that $S(t|x, y)$ decays faster than $1/t$ as $t \to \infty$. For the Heston model the behavior of the survival probability at long times is given by Eq. (12.60). Thus looking at Eq. (12.60), we see that the behavior of S is determined by the behavior of $\mathrm{erf}(z)$ as $z \to 0$ and since (Magnus *et al.*, 1966)

$$\mathrm{erf}(z) = \frac{2z}{\sqrt{\pi}}\left[1 + O(z^2)\right],\qquad(12.63)$$

we see that the survival probability of the Heston model decays when $t \to \infty$ as [7]

$$S(t|x, y) = \frac{2|x - x_c|}{\sqrt{2\theta\alpha t}}[1 + O(1/\alpha t)].\qquad(12.64)$$

Therefore, the survival probability falls off as $1/\sqrt{t}$ which is slower than $1/t$ implying that the integral in (12.62) diverges. Hence, the mean first-passage time of the Heston model is infinite.

$$T(x, y) = \infty.\qquad(12.65)$$

---

[7]Note that as $t \to \infty$, $\lambda(t, y) \simeq \theta\alpha t$.

12.4.2 *Large volatility and default*

Other situations where we can find simpler approximations for the survival probability refer to large values of the volatility. Let us note that, as it appears in the right-hand side of Eq. (12.52), the dependence of the survival probability on the volatility y is that of a linear exponential. This suggests, similarly to the case $\alpha t \gg 1$, that as long as y/α is not small –i.e., for moderate to large volatilities– we may use again the saddle-point method and perform an approximate evaluation of Eq. (12.52) valid when $y \gg \alpha$. We have proved elsewhere (Masoliver and Perelló, 2009) that the case of large volatility turns out to be completely analogous to the long-time approximation just discussed. In other words, approximation (12.60), valid for long times, is also valid for large volatilities. That is to say

$$S(t|x,y) \simeq \operatorname{erf}\left(\frac{|x-x_c|}{\sqrt{\lambda(t,y)}}\right) \qquad (y \gg \alpha), \qquad (12.66)$$

in addition of being true when $t \gg 1/\alpha$.

Therefore, as long as neither the volatility nor the time are small, we can safely use (12.66) to obtain the behavior of the survival probability as $|x_c| \to \infty$. This is, perhaps, one of the most interesting aspects of the first-passage problem to look at, because that limit is intimately related to extreme risks and, ultimately, to default. Using the following asymptotic expression of the error function $\operatorname{erf}(z)$ as $z \to \infty$ (Magnus *et al.*, 1966)

$$\operatorname{erf}(z) \sim 1 - \frac{e^{-z^2}}{\sqrt{\pi}z}[1 + O(1/z^2)], \qquad (12.67)$$

we have

$$S(t|x,y) \sim 1 - \frac{\sqrt{\lambda(t,y)}}{\sqrt{\pi}|x-x_c|}e^{-|x-x_c|^2/\lambda(t,y)}\left[1 + O(1/|x-x_c|^2)\right]. \quad (12.68)$$

The first-passage probability to the level x_c, $W(t|x,y)$ is given by Eq. (12.37). The default probability will be obtained by assuming $x_c \to -\infty$.[8] In this way we find the following exponential decay:

$$W \sim \frac{e^{-x_c^2/\lambda}}{|x_c|} \qquad (x_c \to \pm\infty). \qquad (12.69)$$

Let us take a closer look at the behavior of $S(t|x,y)$ with increasing values of the volatility. Observe that Eq. (12.66) is appropriate for this

---

[8]Note that, due to the dependence on the absolute value $|x-x_c|$, the default probability is identical to the "uprising" probability when $x_c \to \infty$.

purpose because it is true for large values of the volatility. Intuition tells us that for high volatility when $y \to \infty$, the survival probability should tend to zero and this is indeed readily seen from Eq. (12.58) since when $y \to \infty$ then $\lambda(t,y) \to \infty$ and $\text{erf}(0) = 0$.

Let us elucidate how this limit works. From Eq. (12.58) we first note that as $y \gg \alpha$

$$\frac{1}{\sqrt{\lambda(t,y)}} = \frac{1}{\sqrt{\theta \alpha t + (1 - e^{-\alpha t})(y/\alpha)}}$$

$$= \frac{1}{\sqrt{(1 - e^{-\alpha t})(y/\alpha)}} \frac{1}{\sqrt{1 + \dfrac{\theta \alpha t}{1 - e^{-\alpha t}}(\alpha/y)}}$$

$$= \frac{1}{\sqrt{(1 - e^{-\alpha t})(y/\alpha)}} \left[1 + \frac{\theta \alpha t}{2(1 - e^{-\alpha t})}(\alpha/y) + O((\alpha/y)^2) \right].$$

Merging this equation with the expansion (12.63) into Eq. (12.66)

$$S(t|x,y) \simeq \frac{2|x - x_c|}{\sqrt{\pi(1 - e^{-\alpha t})(y/\alpha)}} \left[1 + O(\alpha/y) \right]. \tag{12.70}$$

Therefore, when volatility y increases the survival probability decreases as $y^{-1/2}$.

12.4.3 *Volatility fluctuations*

Besides the approximate solutions to the first-passage problem depending on large values of time or volatility, there are other interesting and useful approximations based on the magnitude of the fluctuations of the volatility. The key parameter on which such an approximate analysis is performed, is the dimensionless noise intensity parameter defined in Eq. (12.43):

$$\beta = k/\alpha \tag{12.71}$$

This parameter measures the strength of volatility fluctuations, given by the noise intensity k (also called "vol of vol") relative to the deterministic pull α, the latter driving volatility towards the normal level m^2 (cf. Eq. (12.33)). Large values of this dimensionless parameter ($\beta \gg 1$) imply intense volatility fluctuations (k large) but also weak reversion to the normal level (α small). When $\beta \ll 1$ we will have the reverse situation in which fluctuations are mild and reversion to the mean strong. One can, therefore, expect different patterns in the first-passage problem according to whether β is large or small.

Small fluctuations

We first treat the case of weak fluctuations, so that the vol of vol is much smaller than the pull toward normal level, $k \gg \alpha$; that is

$$\beta \ll 1.$$

We return to the exact expression of the survival probability given in Eq. (12.52). A glance at Eq. (12.53) suffices to realize that the assumption of $\beta \to 0$ is equivalent to assuming $\omega \to 0$. But the latter is precisely the limit we have taken above, in the asymptotic time and volatility limits, as a consequence of the saddle-point approximation. We therefore expect that the case of weak fluctuations will lead to the same approximation than that of long times or large volatilities. Let us briefly show that this is certainly the case.

In effect, we start from the exact survival probability, Eq. (12.52), which we rewrite in the form

$$S(t|x,y) = \frac{2}{\pi} \int_0^\infty \frac{\sin \omega |x - x_c|}{\omega} \left[\frac{\Delta(\omega)}{\mu_+(\omega) + \mu_-(\omega)e^{-\Delta(\omega)\alpha t}} \right]^{2\theta/\beta^2} \tag{12.72}$$
$$\times \exp\left\{ - (2/\beta^2) \left[\theta\mu_-(\omega)\alpha t + B(\omega,\tau)y/\alpha \right] \right\} d\omega.$$

If $\beta \to 0$ the integral can be approximately evaluated by the saddle-point method. Indeed in this case $2/\beta^2 \to \infty$ and the exponential term in (12.72) decays so quickly that the only significant contribution to the integral comes from around the minimum value of the factor

$$\theta\mu_-(\omega)\alpha t + B(\omega,\tau)y/\alpha,$$

which, as seen above, is located at $\omega = 0$. Expanding the exponential we have

$$\exp\left\{ -\left[\theta\mu_-(\omega)\alpha t + B(\omega,\tau)y/\alpha \right] \right\}$$
$$= \exp\left\{ \left[\theta\alpha t + y\left(1 - e^{-\alpha t}\right)/\alpha \right](\beta\omega/2)^2 + O((\beta\omega)^4) \right\}.$$

Moreover (see Eq. (12.55))

$$\left[\frac{\Delta(\omega)}{\mu_+(\omega) + \mu_-(\omega)e^{-\Delta(\omega)\alpha t}} \right]^{2\theta/\beta^2} = 1 + O((\beta\omega)^2),$$

and Eq. (12.72) yields

$$S(t|x,y) \simeq \frac{2}{\pi} \int_0^\infty \frac{\sin \omega |x - x_c|}{\omega} \exp\left\{ -\frac{1}{2}[\theta\alpha t + y(1 - e^{-\alpha t})/\alpha]\omega^2 \right\}$$

which coincides exactly with Eq. (12.57). Hence, we have again

$$S(t|x,y) \simeq \text{erf}\left(\frac{|x - x_c|}{\sqrt{\lambda(t,y)}}\right) \qquad (\beta \ll 1), \qquad (12.73)$$

with $\lambda(t,y)$ defined in Eq. (12.58).

Before proceeding further let us notice an apparent contradiction in the above statements on the coincidence of the approximate expressions for the survival probability between the cases of large volatility and reduced volatility fluctuations, because, at first sight, one would expect the opposite, i.e., that high volatility and enhanced volatility fluctuations would be equivalent. Let us remind, however, that a large value of the volatility solely means that the initial volatility is large and this, by any means, does not imply that volatility fluctuations must be intense and vice versa.

12.4.3.1 *Large fluctuations*

We now suppose intense volatility fluctuations, for which the noise intensity is much higher than the deterministic pull towards the normal level, that is $k \gg \alpha$. Hence

$$\beta \gg 1.$$

As in the case of small fluctuations described above, the starting point of the asymptotic analysis is the exact expression of the survival probability in the form given by Eq. (12.72) in which we now use the approximations (cf. Eq. (12.53))

$$\Delta(\omega) \equiv \sqrt{1 + (\beta\omega)^2} = \beta\omega\sqrt{1 + \frac{1}{(\beta\omega)^2}}$$

$$= \beta\omega\left[1 + O\left(\frac{1}{\beta^2}\right)\right],$$

and

$$\mu_{\pm}(\omega) \equiv \frac{1}{2}[\Delta(\omega) \pm 1] = \frac{1}{2}\left\{\pm 1 + \beta\omega\left[1 + O\left(\frac{1}{\beta^2}\right)\right]\right\}$$

$$= (\beta\omega/2)\left[1 \pm \frac{1}{\beta\omega} + O\left(\frac{1}{\beta^2}\right)\right].$$

Hence (cf. Eq. (12.54))

$$e^{-\Delta(\omega)\alpha t} \simeq e^{-\beta\omega\alpha t},$$

which, except for very small values of time, is an exponentially small term (this certainly excludes the initial stages of the process as will become clearer soon).[9] Hence (cf. Eq. (12.54))

$$B(\omega, t) = \mu_-(\omega) \frac{1 - e^{-\Delta(\omega)\alpha t}}{1 + [\mu_-(\omega)/\mu_+(\omega)]e^{-\Delta(\omega)\alpha t}} \simeq (\beta\omega/2)\frac{1 - e^{-\beta\omega\alpha t}}{1 + e^{-\beta\omega\alpha t}},$$

whence,

$$B(\omega, t) \sim \beta\omega/2. \tag{12.74}$$

Moreover

$$\left[\frac{\Delta(\omega)}{\mu_+(\omega) + \mu_-(\omega)e^{-\Delta(\omega)\alpha t}}\right]^{2\theta/\beta^2} \sim 1 + O(1/\beta^2),$$

where we have proceeded as in Eq. (12.74) by neglecting exponentially small terms. Collecting these results into Eq. (12.72) we find

$$S(t|x, y) \sim \int_0^\infty \frac{1}{\omega} \sin\omega|x - x_c| \exp\left[-(\theta\alpha t + y/\alpha)(\omega/\beta)\right] d\omega.$$

The integral can be done exactly (Gradshteyn and Ryzhik, 1994) giving the non-Gaussian form[10]

$$S(t|x, y) \sim \frac{2}{\pi} \arctan\left(\frac{\beta|x - x_c|}{m^2 t + y/\alpha}\right) \qquad (\beta \gg 1). \tag{12.75}$$

Let us observe that this asymptotic form satisfies the boundary condition $S_c(t|x_c, y) = 0$, but not the initial condition, since $S(0|x, y) \neq 1$ for $y \neq 0$. Consequently, the approximation (12.75) will work only after an initial period of time, representing transient effects, has elapsed.

The most striking difference between the above asymptotic forms of the first-passage probability –Eqs. (12.73) and (12.75), appears when one considers the extreme risk of default. In such a case $x_c \to -\infty$ and the hitting probability, $W_c(t|x, y)$, corresponding to the Gaussian form (12.73) quickly decays by following the decreasing quadratic exponential shown in Eq. (12.69):

$$W \sim \frac{e^{-x_c^2/\lambda}}{|x_c|} \qquad (|x_c| \to \infty, \ \beta \ll 1). \tag{12.76}$$

On the other hand, as volatility fluctuations increase, the non-Gaussian approximation given by Eq. (12.75) implies a much slow decay since now default is ruled by the hyperbolic law:

$$W \sim \frac{1}{\beta|x_c|}, \qquad (|x_c| \to \infty, \ \beta \gg 1). \tag{12.77}$$

In other words, we have *a significant increase of risk when the fluctuations of the volatility soar.*

---

[9]Indeed for $\beta\omega \gg 1$ the exponent $\beta\omega\alpha t \gg 1$ except when $t \to 0$. This means, in particular, that $1 \pm e^{-\beta\omega\alpha t} \sim 1$.

[10]In writing Eq. (12.75) we take into account that $\theta\alpha = m^2$ (cf. Eq. (12.43)).

12.5   The Continuous Time Random Walk

In many practical situations the time series of socio-economic systems represent the complete information on the underlying random process. This is the case of tic-by-tic data in finance which give the microscopic description of the market. In such a situation, diffusion models –which have, by definition, continuous trajectories– only represent, at most, a "mesoscopic"[11] approximation to the actual process. A more accurate description is provided by Continuous Time Random Walks (CTRW's) where the inherent discontinuous nature of any microscopic description is taken into account from the start. This kind of random walk was introduced more than five decades ago by Elliott Montroll and George H. Weiss and a rather complete account of the formalism and its application to financial problems may be found in Masoliver *et al.* (2008).[12] We shall briefly summarize the main traits of CTRW's before addressing the level-crossing problem.

As we have mentioned at the beginning of this chapter, if $P(t)$ denotes a financial price or the value of an index at time t, the return of the investment from some initial time $t_i \leq t$ is given by $\ln\big[P(t+t_i)/P(t_i)\big]$. In general the return has non-zero average and since this introduces a systematic trend in its evolution it turns out to be more convenient to work with the zero-mean return $X(t)$ (also called detrended return) defined by

$$X(t) = \ln\big[P(t+t_i)/P(t_i)\big] - \Big\langle \ln\big[P(t+t_i)/P(t_i)\big] \Big\rangle.$$

The object of the CTRW in finance is to study the time evolution of the zero-mean return.

12.5.1   *A short report*

We next present an introduction to the CTRW which is not restricted to financial applications but applies to any time series.

As its name suggests the CTRW generalizes simple random walk models. Although the term "random walk" was coined by Karl Pearson in 1905, the formalism had been devised in the seventeenth century in the context of gambling games such as the probability of ruin after betting n times in a coin tossing game. In this case, the sum of gains minus losses in n trials is equal to the state of a player's fortune. Let us thus note that the ordinary random walk is based on the assumption that changes are made at equal

---

[11]Somewhere between microscopic and macroscopic description of markets.

[12]See also Kutner and Masoliver, 2017, for a recent and rather complete outlook to the CTRW in many branches of science, not only finance.

time intervals. However this is a first approximation and the CTRW relaxes this restriction since it assumes that time intervals between consecutive changes are not constant but random. In fact, as we shall see next, in the CTRW there are two sources of randomness: one coming from the waiting times and another from the amplitudes of random jumps.

In the most common version of the CTRW a given random process $X(t)$ shows a series of random increments or jumps at random times $\ldots, t_{-1}, t_0, t_1, t_2, \ldots, t_n, \ldots$ remaining constant between these jumps. We will first assume that the time intervals between successive jumps, which we call sojourns or waiting times,

$$\tau_n = t_n - t_{n-1},$$

are independent and identically distributed random variables with a probability density function, $\psi(t)$, defined as

$$\psi(t)dt = \text{Prob}\{t < \tau_n \le t + dt\}.$$

At the end of a given sojourn the zero-mean return $X(t)$ suffers a random change equal to

$$\Delta X_n = X(t_n) - X(t_{n-1})$$

whose probability density function is defined as

$$h(x)dx = \text{Prob}\{x < \Delta X_n \le x + dx\}.$$

We now define an auxiliary function $\rho(x,t)$ such that $\rho(x,t)dxdt$ is the joint density that an increment of return is added whose magnitude, ΔX_n, is between x and $x + dx$ and that the time interval τ_n between successive jumps is between t and $t + dt$; that is,

$$\rho(x,t)dxdt = \text{Prob}\{x < \Delta X_n \le x + dx, t < \tau_n \le t + dt\}. \qquad (12.78)$$

It is implicit in this definition that the probability for any jump ΔX_n to take the value x does not depend on the value of previous jumps and the probability for any time interval between jumps, τ_n, to take the value t does not depend on previous waiting times either. This restriction can be relaxed by assuming some kind of memory between different sojourns (Montero and Masoliver, 2007b, 2017).

It is frequently assumed that $\rho(x,t)$ is an even function of x so that there is no net drift in the evolution of $X(t)$. We observe that if waiting times are independents of jumps then $\rho(x,t) = h(x)\psi(t)$. In any other situation one has to specify a functional form of $\rho(x,t)$ that is compatible with the observed data. Moreover, taking into account that jump and waiting time densities, $h(x)$ and $\psi(t)$, are the marginal densities of the joint density, any proposed form of $\rho(x,t)$ must satisfy:

$$h(x) = \int_0^\infty \rho(x,t)dt, \qquad \psi(t) = \int_{-\infty}^\infty \rho(x,t)dx. \qquad (12.79)$$

Let us note that a direct consequence of the unbiassed assumption, i.e.: $\rho(-x,t) = \rho(x,t)$, is that $h(x)$ is also symmetric around the origin

$$h(-x) = h(x), \qquad (12.80)$$

so that its odd moments are equal to zero.

Two key statistics are associated with densities $\psi(t)$ and $h(x)$. One of them is the average sojourn time, also called mean waiting time between consecutive jumps,

$$\langle \tau \rangle = \int_0^\infty t\psi(t)dt.$$

The second statistic is the variance of jump sizes (recall that due to the unbiassed assumption the mean jump size is zero)

$$\sigma^2 = \int_{-\infty}^\infty x^2 h(x)dx.$$

Regarding the trajectories of the return process, we assume that $X(t)$ consists of a series of step functions. In other words, the return evolves discontinuously and during any sojourn the value of return remains constant. In consequence, the value of $X(t)$ at time t will be given by the height of the step at time t (see Fig. 12.1).

The main objective of the formalism consists in obtaining the probability density function of the process $X(t)$. This function, sometimes called propagator, is defined as

$$p(x,t)dx = \mathrm{Prob}\{x < X(t) \le x + dx\}.$$

From this definition and Fig. 12.1 we see that the propagator prior to the first jump, which we denote by $p_0(x,t)$, is equal to

$$p_0(x,t) = \Psi(t)\delta(x), \qquad (12.81)$$

where we have assumed that the initial jump occurred at $t = 0$ while $X(0) = 0$ and $\Psi(t)$ is the probability that the time between two consecutive jumps is greater than t. In other words, $\Psi(t)$ is the cumulative probability associated with the density $\psi(t)$

$$\Psi(t) = \int_t^\infty \psi(t')dt'. \qquad (12.82)$$

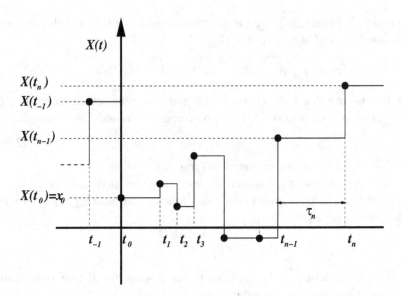

Fig. 12.1   Schematic representation of the return process. The dots mark the value $X(t_n)$ of the return after each sojourn. $\tau_n = t_n - t_{n-1}$ is the time increment of the n-th sojourn.

In terms of p_0 and ρ, the complete density $p(x,t)$ satisfies the following integral equation

$$p(x,t) = p_0(x,t) + \int_0^t dt' \int_{-\infty}^{\infty} \rho(x-x',t-t')p(x',t')dx'. \qquad (12.83)$$

This equation is derived from the following renewal argument. Indeed, if a jump occurs at time t, it must be either the first one (after the initial one at $t = 0$) this described by the first term on the right hand side of this equation, or else an earlier jump occurred at time $t' < t$ when the process had the value x' and no further jump occurred during the time interval $t - t'$. All this integrated over all possible positions $-\infty < x' < \infty$ and intermediate times $0 < t' < t$.

We can solve Eq. (12.83) in terms of the joint Fourier-Laplace transform:

$$\hat{\tilde{p}}(\omega,s) = \int_0^{\infty} dt e^{-st} \int_{-\infty}^{\infty} e^{i\omega x} p(x,t)dx.$$

In effect, the convolution theorem for both the Laplace and Fourier transforms yields

$$\mathcal{L}\left\{\mathcal{F}\left\{\int_0^t dt' \int_{-\infty}^{\infty} \rho(x-x',t-t')p(x',t')dx'\right\}\right\} = \hat{\tilde{\rho}}(\omega,s)\hat{\tilde{p}}(\omega,s),$$

and the integral equation (12.83) transforms into the algebraic equation

$$\hat{\tilde{p}}(\omega, s) = \hat{\tilde{p}}_0(\omega, s) + \hat{\tilde{\rho}}(\omega, s)\hat{\tilde{p}}(\omega, s),$$

whose solution is

$$\hat{\tilde{p}}(\omega, s) = \frac{\hat{\tilde{p}}_0(\omega, s)}{1 - \hat{\tilde{\rho}}(\omega, s)}, \tag{12.84}$$

where $\hat{\tilde{p}}_0(\omega, s)$ and $\hat{\tilde{\rho}}(\omega, s)$ are respectively the joint Fourier-Laplace transforms of the functions $p_0(x, t)$ and $\rho(x, t)$.

Taking into account that the Fourier transform of Dirac's delta function $\delta(x)$ is the unity, we immediately get from Eq. (12.81)

$$\hat{\tilde{p}}_0(\omega, s) = \hat{\Psi}(s), \tag{12.85}$$

where $\hat{\Psi}(s)$ is the Laplace transform of the cumulative probability $\Psi(t)$. Using standard properties of the Laplace transform [Roberts and Kaufman (1966)]

$$\mathcal{L}\left\{\int_t^\infty \psi(t')dt'\right\} = [1 - \hat{\psi}(s)]/s,$$

where $\hat{\psi}(s)$ is the Laplace transform of the waiting time density $\psi(t)$, we see from Eqs. (12.82) and (12.85) that

$$\hat{\tilde{p}}_0(\omega, s) = \frac{1}{s}\left[1 - \hat{\psi}(s)\right]. \tag{12.86}$$

In many situations jumps and waiting times are independent random variables so that the joint density factorizes

$$\rho(x, t) = \psi(t)h(x). \tag{12.87}$$

In such a case Eq. (12.84) results in the celebrated Montroll-Weiss equation:

$$\hat{\tilde{p}}(\omega, s) = \frac{[1 - \hat{\psi}(s)]/s}{1 - \hat{\psi}(s)\tilde{h}(\omega)}, \tag{12.88}$$

where $\tilde{h}(\omega)$ is the Fourier transform of the jump density $h(x)$.

Let us finally recall that moments $\langle X^n(t)\rangle$ are related to derivatives of the characteristic function $\tilde{p}(\omega, t)$ with respect to ω evaluated at $\omega = 0$. Therefore, the Laplace transform of moments can be written as

$$\mathcal{L}\{\langle X^n(t)\rangle\} = i^{-n} \left. \frac{\partial^n \hat{\tilde{p}}(\omega, s)}{\partial \omega^n}\right|_{\omega=0}.$$

Using Eq. (12.88) we can obtain the Laplace transform of the moments of the CTRW in terms of the moments of the waiting time density $\psi(t)$ and the jump density $h(x)$. In terms of derivatives of the Laplace and Fourier transform, $\hat{\psi}(s)$ and $\tilde{h}(\omega)$, these moments are given by

$$\langle \tau^n \rangle \equiv \int_0^\infty t^n \psi(t)dt = (-1)^n \hat{\psi}^{(n)}(0), \tag{12.89}$$

and

$$\mu_n \equiv \int_{-\infty}^\infty x^n h(x)dx = i^{-n}\tilde{h}^{(n)}(0). \tag{12.90}$$

We illustrate this procedure for the first two moments of the CTRW. Taking into account that $\tilde{h}(0) = 1$ (which corresponds to the normalization of $h(x)$), the first derivative with respect to ω of Eq. (12.88) leads to

$$\mathcal{L}\left\{\langle X(t) \rangle\right\} = \frac{\mu_1 \hat{\psi}(s)}{s[1 - \hat{\psi}(s)]}. \tag{12.91}$$

Let us recall that for an unbiased walk $h(x) = h(-x)$ (cf. Eq. (12.80)) and $\mu_1 = 0$. Hence $\mathcal{L}\{\langle X(t) \rangle\} = 0$ which implies that the first moment of the CTRW is also zero (the same applies to all odd moments). The second derivative finally yields

$$\mathcal{L}\left\{\langle X^2(t) \rangle\right\} = \frac{2\mu_1^2 \hat{\psi}^2(s)}{s[1 - \hat{\psi}(s)]} + \frac{\mu_2 \hat{\psi}(s)}{s[1 - \hat{\psi}(s)]}. \tag{12.92}$$

For the unbiased walk $\mu_1 = 0$, $\mu_2 = \sigma^2$ is the variance of jumps and we have

$$\mathcal{L}\left\{\langle X^2(t) \rangle\right\} = \frac{\sigma^2 \hat{\psi}(s)}{s[1 - \hat{\psi}(s)]}. \tag{12.93}$$

12.5.2   *Fractional diffusion**

The CTRW provides a derivation of the main properties of the fractional Brownian motion discussed in Chapters 4 and 11. In particular the formalism allows for a derivation of the fractional diffusion equation

$$\frac{\partial^\alpha p}{\partial t^\alpha} = \frac{1}{2}D\frac{\partial^{2\gamma}p}{\partial x^{2\gamma}} \tag{12.94}$$

in the case when $0 < \alpha \leq 1$ and $0 < \gamma \leq 1$ as we will see below.[13]

As mentioned above, the CTRW essentially differs from the ordinary random walk in considering random waiting times and random jump sizes. The overall walk can be handled even if these times and jumps have no

---

[13]The superdiffusive case $1 < \alpha \leq 2$ is more difficult to handle and, attending the elementary character of this introduction, we will not address this case here.

finite moments which permits an unified treatment of both fractional and non-fractional random motion. Thus in the so-called "fluid limit" –that is to say, large times and distances– one assumes, as $s \to 0$ and $\omega \to 0$, that [14]

$$\hat{\psi}(s) = 1 - (Ts)^\alpha \ldots, \qquad \tilde{h}(\omega) = 1 - (L\omega)^{2\gamma} \ldots, \qquad (12.95)$$

where T and L are positive constant parameters, T sets a characteristic time and L a characteristic length.[15]

In Eq. (12.95) we have assumed $0 < \alpha \le 1$ and $0 < \gamma \le 1$. Hence, $\hat{\psi}'(s) = -\alpha s^{\alpha-1} \cdots \to -\infty$ as $s \to 0$ which by virtue of Eq. (12.89) implies that the mean waiting time $\langle \tau \rangle$ is infinite (except for the case $\alpha = 1$ where $\langle \tau \rangle = T$ is finite).[16] We have excluded the range of values $1 < \alpha \le 2$ because from Eqs. (12.89) and (12.95) we see that for this range of values the average sojourn time is zero, $\langle \tau \rangle = 0$, which is rather unphysical. If $1 < \alpha < 2$ and in order to allow for a nonzero mean sojourn time, the expansion (12.95) needs to be modified, for instance as (Weiss, 1994)

$$\hat{\psi}(s) = 1 - \langle \tau \rangle s + (Ts)^{1+\beta} \ldots, \qquad (0 < \beta < 1).$$

However, this may result in another type of fractional diffusion equation different than the standard one and we will not treat this case here.

Within the fluid-limit approximation (12.95) we have

$$1 - \hat{\psi}(s)\tilde{h}(\omega) = (Ts)^\alpha + (L\omega)^{2\gamma} \ldots, \qquad 1 - \hat{\psi}(s) = (Ts)^\alpha \ldots,$$

and Montroll-Weiss equation (12.88) reads

$$\hat{\tilde{p}}(\omega, s) = \frac{s^{\alpha-1}}{s^\alpha + (L^{2\gamma}/T^\alpha)\omega^{2\gamma}}. \qquad (12.96)$$

We distinguish two cases: (i) $\alpha = \gamma = 1$ and (ii) $0 < \alpha < 1$, $0 < \gamma < 1$. In case (i) waiting times and jumps have finite average values. Indeed in this case from Eq. (12.95) we see that $\hat{\psi}'(0) = -T$, $\tilde{h}'(0) = 0$, and $\tilde{h}''(0) = -L^2$ and, since

$$\hat{\psi}'(0) = -\int_0^\infty t\psi(t)dt = -\langle \tau \rangle, \qquad \tilde{h}''(0) = -\int_{-\infty}^\infty x^2 h(x)dx = -\sigma^2,$$

---

[14]Recall that due to Tauberian theorems large times and distances, $t \to \infty$ and $|x| \to \infty$, correspond to small Laplace and Fourier variables, $s \to 0$ and $\omega \to 0$. See Chapter 10, Sect. 10.2.2.

[15]The combinations Ts and $L\omega$ are dimensionless. Therefore, since s has units of t^{-1} and ω of x^{-1}, T has units of time and L of length. The fluid limit approximation means times greater than T and lengths greater than L, both parameters determined by the physical characteristics of the problem at hand.

[16]The same applies to jump moments where, due to the assumed symmetry of $h(x)$, the first moment vanishes and the variance, $\sigma^2 = \mu_2$, is infinite (cf. Eq. (12.90)).

we conclude that the characteristic time T and the characteristic length L are in this case respectively given by the mean waiting time and the variance of jumps, $T = \langle \tau \rangle$ and $L = \sigma$. The transformed density (12.96) now reads

$$\hat{\tilde{p}}(\omega, s) = \frac{1}{s + D\omega^2/2},$$

where $D = 2L^2/T$ is the diffusion coefficient. The joint Fourier-Laplace inversion of this equation yields the Gaussian distribution of the Brownian motion:

$$p(x,t) = \frac{1}{\sqrt{2\pi Dt}}e^{-x^2/2Dt}. \tag{12.97}$$

Therefore, when jumps and waiting times have finite moments, at large times and distances, the CTRW reduces to the ordinary Brownian motion.

(ii) For fractional values of the exponents α and γ one ends up with a fractal probability distribution of the Lévy type for which there is neither finite mean waiting time nor finite jump variance. In Sect. 4.6 of Chapter 4, we have thoroughly studied this case and transformed back Eq. (12.96) to obtain $p(x,t)$ in terms of Manardi functions (cf. Eq. (4.97)) (see also Sect. 11.8 of Chapter 11 for the first passage problem of the fractional motion).

We will now prove that the equation satisfied by $p(x,t)$ is indeed the fractional diffusion equation (12.94). In other words, from Montroll-Weiss equation (12.96) we will derive Eq. (12.94). To this end we start writing Eq. (12.96) in the form

$$\left[s^\alpha + \frac{1}{2}D\omega^{2\gamma} \right]\hat{\tilde{p}}(\omega, s) = s^{\alpha-1},$$

where $D = 2L^{2\gamma}/T^\alpha$ is the fractional diffusion coefficient. Fourier inverting this equation and taking into account the definition of the Riesz-Feller derivative (cf. Eq. (4.83), Chapter 4)

$$\mathcal{F}^{-1}\left\{ |\omega|^{2\gamma}\hat{\tilde{p}}(\omega, s) \right\} = -\frac{\partial^{2\gamma}\tilde{p}(x, s)}{\partial x^{2\gamma}},$$

we get [17]

$$\left[s^\alpha - \frac{1}{2}D\frac{\partial^{2\gamma}}{\partial x^{2\gamma}} \right]\tilde{p}(x, s) = s^{\alpha-1}\delta(x),$$

---

[17]We use the Fourier inversion

$$\mathcal{F}^{-1}\{s^{\alpha-1}\} = s^{\alpha-1}\mathcal{F}^{-1}\{1\} = s^{\alpha-1}\frac{1}{2\pi}\int_{-\infty}^{\infty}e^{-i\omega x}d\omega = s^{\alpha-1}\delta(x).$$

where $\hat{p}(x, s)$ is the Laplace transform of the density function. We rewrite the last expression as

$$s^\alpha \hat{p}(x, s) - s^{\alpha-1}\delta(x) = \frac{1}{2}D\frac{\partial^{2\gamma}\hat{p}}{\partial x^{2\gamma}}. \tag{12.98}$$

We next proceed to Laplace inversion. Recall that when $0 < \alpha \leq 1$ the Laplace transform of the Caputo derivative is given by Eq. (4.82) of Chapter 4,

$$\mathcal{L}\left\{\frac{\partial^\alpha p(x, t)}{\partial t^\alpha}\right\} = s^\alpha \hat{p}(x, s) - s^{\alpha-1}p(x, 0).$$

But $p(x, 0) = \delta(x)$ and Laplace inversion yields

$$\mathcal{L}^{-1}\left\{s^\alpha \hat{p}(x, s) - s^{\alpha-1}\delta(x)\right\} = \frac{\partial^\alpha p(x, t)}{\partial t^\alpha}, \tag{12.99}$$

$(0 < \alpha \leq 1)$. Finally the Laplace inversion of Eq. (12.98) and the use of Eq. (12.99) prove that the probability density function $p(x, t)$ of the fractional walk obeys the fractional diffusion equation (12.94):

$$\frac{\partial^\alpha p}{\partial t^\alpha} = \frac{1}{2}D\frac{\partial^{2\gamma}p}{\partial x^{2\gamma}}.$$

We have therefore shown that the CTRW becomes the foundation for anomalous (i.e., dispersive, non-Gaussian) transport which, in the last two decades, has opened a most fashionable field in modern statistical mechanics, condensed and soft-matter physics and also socio-economic sciences that goes beyond the traditional Boltzmann-Gibbs statistical mechanics (see, for instance, Masoliver, 2016, for a short review).

12.6 First-passage problems in the CTRW formalism

We will now address the level-crossing problem within the framework of CTRW's. From a mathematical point of view, and contrary to diffusion models where all probabilities related to first-passage problems obey differential equations, in CTRW's they satisfy integral equations as otherwise occurs with the unrestricted walk of the previous section.

12.6.1 *The escape problem*

We first address the escape problem out of an interval (a, b). Let $S_{ab}(t|x)$ be the survival probability of the random walker within the interval (a, b). That is to say, $S_{ab}(t|x)$ is the probability that the walker, having started at some point $x \in (a, b)$ inside the interval, has not left that interval at time t or before. In order to obtain $S_{ab}(t|x)$ we will use renewal techniques –as we have done in the previous section for the unrestricted walk– but now the limits of the resulting integral equations will be determined by the boundaries a and b of the interval.

As in the unrestricted walk, the key quantity is the joint density $\rho(x,t)$ defined in Eq. (12.78) which, recall, gives the joint distribution of jump sizes and time intervals in every sojourn. We shall now derive an integral equation for the survival probability in terms of ρ.

Let us first notice that due to the discontinuous character of the trajectory (see Fig. 12.1) the process $X(t)$ will leave the interval (a,b) as long as a sufficiently large jump occurs.

The escape out of an interval (a,b) may take place at the end of the first sojourn or after it, and these two possibilities are mutually exclusive. We can, therefore, write the survival probability as the sum

$$S_{ab}(t|x) = S_{ab}^{(0)}(t|x) + S_{ab}^{(1)}(t|x),$$

where $S_{ab}^{(0)}(t|x)$ is the probability of remaining inside the interval conditioned on the event that no jump occurred, whereas $S_{ab}^{(1)}(t|x)$ takes into account that one or more jumps occurred but the walker is still inside (a,b). In other words, $S_{ab}^{(1)}(t|x)$ is the survival probability under the condition that at least one jump occurred between 0 and t.

Notice that, since the initial position $X(0) = x \in (a,b)$ lies inside the interval, $S_{ab}^{(0)}(t|x)$ will be given by the probability that no jump has occurred up to time t which is the cumulative distribution $\Psi(t)$ of the sojourn time density $\psi(t)$ (cf. Eq. (12.82)), that is

$$S_{ab}^{(0)}(t|x) = \Psi(t).$$

On the other hand, suppose that the walker has not left the interval (a,b) during the first sojourn. Thus, since the duration of the first sojourn is the time interval $t' - t_0 = t' < t$ (recall that the initial time is $t_0 = 0$), this means that the process has attained at time t' some value x' still inside (a,b) (i.e., the first jump has an amplitude $x' - x$). Note that the probability density of this joint event is $\rho(x' - x, t')$. From the new initial position x' the survival probability –for the time interval up to time t– is given by $S_{ab}(t - t'|x')$. Hence, the conditional probability that, at time t, the process is still inside the interval and at least one jump occurred is given by

$$S_{ab}^{(1)}(t|x) = \int_0^t dt' \int_a^b \rho(x' - x, t') S_{ab}(t - t'|x') dx'.$$

Summing $S_{ab}^{(0)}$ and $S_{ab}^{(1)}$ we find that the survival probability obeys the following integral equation:

$$S_{ab}(t|x) = \Psi(t) + \int_0^t dt' \int_a^b \rho(x' - x, t') S_{ab}(t - t'|x') dx'. \quad (12.100)$$

As we will see below the escape probability $W_{ab}(t|x)$ obeys a similar integral equation. However, it turns out to be more convenient to calculate W by means of its relation in terms of the survival probability,

$$W_{ab}(t|x) = 1 - S_{ab}(t|x), \quad (12.101)$$

once we know S. Indeed, The integral equation for $W_{ab}(t|x)$ is readily obtained after substituting Eq. (12.101) into the integral equation (12.100). The result is

$$W_{ab}(t|x) = \chi(t|x) + \int_0^t dt' \int_a^b \rho(x' - x, t') W_{ab}(t - t'|x') dx', \quad (12.102)$$

where

$$\chi(t|x) = 1 - \Psi(t) - \int_0^t dt' \int_a^b \rho(x' - x, t') dx'.$$

It is rather obvious that the integral equation for W_{ab} is more involved than Eq. (12.100) for the survival.

The difficulty in solving Eq. (12.100) is reduced by taking the Laplace transform,

$$\hat{S}_{ab}(s|x) = \int_0^\infty e^{-st} S_{ab}(t|x) dt,$$

which turns Eq. (12.100) into

$$\hat{S}_{ab}(s|x) = \hat{\Psi}(s) + \int_a^b \hat{\rho}(x' - x, s) \hat{S}_{ab}(s|x') dx', \quad (12.103)$$

where

$$\hat{\rho}(x, s) = \int_0^\infty e^{-st} \rho(x, t) dt, \qquad \hat{S}_{ab}(s|x') = \int_0^\infty e^{-st} S_{ab}(t|x) dt \quad (12.104)$$

are the Laplace transforms of ρ and S_{ab}.

Recall from Chapter 10 (cf. Eq. (10.16)) that in terms of $\hat{S}_{ab}$ the mean exit time $T_{ab}(x)$ is given by

$$T_{ab}(x) = \hat{S}_{ab}(0|x). \quad (12.105)$$

Setting $s = 0$ in Eq. (12.103) we get the following equation for the mean exit time

$$T_{ab}(x) = \hat{\Psi}(0) + \int_a^b \hat{\rho}(x' - x, 0) T_{ab}(x') dx'. \quad (12.106)$$

Let us rewrite Eq. (12.106) in a simpler and more intuitive form. We first note from Eq. (12.104) that

$$\hat{\rho}(x' - x, 0) = \int_0^\infty \rho(x' - x, t)dt,$$

hence (cf. Eq, (12.79))

$$\hat{\rho}(x' - x, 0) = h(x' - x), \tag{12.107}$$

is the jump density.

On the other hand, we expand $\hat{\psi}(s)$ in powers of s,

$$\hat{\psi}(s) = \hat{\psi}(0) + s \left. \frac{d\hat{\psi}(s)}{ds} \right|_{s=0} + O(s^2),$$

and bearing in mind the definition of the Laplace transform and the normalization of the density $\psi(t)$ we easily see that

$$\hat{\psi}(0) = 1, \qquad \left. \frac{d\hat{\psi}(s)}{ds} \right|_{s=0} = -\langle \tau \rangle,$$

where

$$\langle \tau \rangle = \int_0^\infty t\psi(t)dt,$$

is the mean waiting time between consecutive jumps. Therefore, as long as $\langle \tau \rangle$ is finite, we have the expansion

$$\psi(s) = 1 - \langle \tau \rangle s + O(s^2). \tag{12.108}$$

Now, since $\hat{\Psi}(s)$ is given by Eq. (12.86) we see that

$$\hat{\Psi}(0) = \lim_{s \to 0} \frac{1}{s} \left[1 - \hat{\psi}(s) \right],$$

which after using Eq. (12.108) yields

$$\hat{\Psi}(0) = \langle \tau \rangle. \tag{12.109}$$

Substituting Eqs. (12.107) and (12.109) into Eq. (12.106) we have

$$T_{ab}(x) = \langle \tau \rangle + \int_a^b h(x' - x)T_{ab}(x')dx' \tag{12.110}$$

which shows that the mean exit time only depends on the jump density $h(x)$ and the first moment of $\psi(t)$, that is, the average time interval between consecutive jumps.

12.6.2 The first-passage problem

We next briefly address the first-passage or hitting problem within the CTRW formalism. We recall from Chapter 10 that the first-passage probability, $W_c(t|x)$, to some threshold x_c is a special cases of the escape probability. For example, if the threshold is above the starting point ($x_c > x$) the hitting probability W_c is given by the exit probability $W_{ab}(t|x)$ after setting $a = -\infty$ and $b = x_c$, while when threshold is below x ($x_c < x$) substitutions are $a = x_c$ and $b = \infty$, and analogously for the survival probability $S_c(t|x)$. Focusing on this latter we have

$$S_c(t|x) = \begin{cases} S_{-\infty x_c}(t|x), & x < x_c, \\ S_{x_c\infty}(t|x), & x > x_c. \end{cases} \qquad (12.111)$$

We therefore see from Eq. (12.100) that the integral equation for the survival probability $S_c(t|x)$ to threshold x_c is

$$S_c(t|x) = \Psi(t) + \int_0^t dt' \int_{-\infty}^{x_c} \rho(x' - x, t')S_c(t - t'|x')dx', \qquad (12.112)$$

when $x < x_c$, or

$$S_c(t|x) = \Psi(t) + \int_0^t dt' \int_{x_c}^{\infty} \rho(x' - x, t')S_c(t - t'|x')dx', \qquad (12.113)$$

when $x > x_c$.

Similarly to the escape problem, the time Laplace transform of the survival probability $\hat{S}_c(s|x)$ obeys the simpler equations

$$\hat{S}_c(s|x) = \hat{\Psi}(s) + \int_{-\infty}^{x_c} \hat{\rho}(x' - x, s)\hat{S}_c(s|x')dx' \quad (x < x_c) \qquad (12.114)$$

or

$$\hat{S}_c(t|x) = \hat{\Psi}(s) + \int_{x_c}^{\infty} \hat{\rho}(x' - x, s)\hat{S}_c(s|x')dx' \quad (x > x_c). \qquad (12.115)$$

Recall that the mean first-passage time is given by the Laplace transform of the survival probability evaluated at $s = 0$,

$$T_c(x) = \hat{S}_c(0|x).$$

Thus, taking into account that $\hat{\Psi}(0) = \langle \tau \rangle$ is the average time between jumps and $\hat{\rho}(x, 0) = h(x)$ is the jump density (cf. Eq. (12.107)), we see from Eqs (12.114) and (12.115) that the mean first-passage time then obeys the integral equation

$$T_c(x) = \langle \tau \rangle + \int_{-\infty}^{x_c} h(x' - x)T_c(x')dx', \qquad (x < x_c), \qquad (12.116)$$

or

$$T_c(x) = \langle \tau \rangle + \int_{x_c}^{\infty} h(x' - x)T_c(x')dx', \qquad (x > x_c). \qquad (12.117)$$

12.6.3   *An example: Laplacian jumps * *

As an illustration of the formalism and equations above we will now obtain some explicit expressions of the level crossing problem. To this end we assume that jumps and waiting times are independent, so that

$$\rho(x,t) = h(x)\psi(t),$$

and also that jumps are exponentially distributed according to the Laplace density,

$$h(x) = \frac{1}{2}\gamma e^{-\gamma|x|}, \tag{12.118}$$

where the parameter $\gamma > 0$ is such that γ^{-1} is the average jump size.[18] Despite this particular (but widely used) jump distribution, we will not suppose any special distribution for sojourn times and assume a general $\psi(t)$.

The escape problem

We first treat the escape problem out of the interval (a,b). Our objective is obtaining the expression for the survival probability or, at least, for its Laplace transform. The starting point is the integral equation (12.103) for $\hat{S}_{ab}(s|x)$. In the case of Laplacian jumps this equation reads

$$\hat{S}_{ab}(s|x) = \hat{\Psi}(s) + \frac{1}{2}\gamma\hat{\psi}(s)\int_a^b e^{-\gamma|x'-x|}\hat{S}_{ab}(s|x')dx', \tag{12.119}$$

which, after implementing the absolute value in the exponential inside the integral, can be written more explicitly as

$$\hat{S}_{ab}(s|x) = \hat{\Psi}(s) + \frac{1}{2}\gamma\hat{\psi}(s)\left[e^{-\gamma x}\int_a^x e^{\gamma x'}\hat{S}_{ab}(s|x')dx'\right.$$
$$\left. + e^{\gamma x}\int_x^b e^{-\gamma x'}\hat{S}_{ab}(s|x')dx'\right]. \tag{12.120}$$

Let us see that this integral equation is equivalent to an ordinary differential equation. The derivative of Eq. (12.120) with respect to x and some

---

[18] In effect, the mean jump size is given by

$$\langle\Delta X\rangle \equiv \int_{-\infty}^{\infty} xh(x)dx = \gamma\int_0^{\infty} xe^{-\gamma x}dx = 1/\gamma.$$

cancellations yield

$$\frac{d\hat{S}_{ab}}{dx} = \frac{\gamma^2\hat{\psi}(s)}{2}\left[-e^{-\gamma x}\int_a^x e^{\gamma x'}\hat{S}_{ab}(s|x')dx'\right.$$
$$\left. + e^{\gamma x}\int_x^b e^{-\gamma x'}\hat{S}_{ab}(s|x')dx'\right]. \qquad (12.121)$$

Taking the second derivative we have

$$\frac{d^2\hat{S}_{ab}}{dx^2} = \gamma\frac{\gamma^2\hat{\psi}(s)}{2}\left[e^{-\gamma x}\int_a^x e^{\gamma x'}\hat{S}_{ab}(s|x')dx'\right.$$
$$\left. +e^{\gamma x}\int_x^b e^{-\gamma x'}\hat{S}_{ab}(s|x')dx'\right] - \gamma^2\hat{\psi}(s)\hat{S}_{ab}(s|x), \quad (12.122)$$

but from Eq. (12.120) we see that

$$\frac{1}{2}\gamma\hat{\psi}(s)\left[e^{-\gamma x}\int_a^x e^{\gamma x'}\hat{S}_{ab}(s|x')dx'\right.$$
$$\left. +e^{\gamma x}\int_x^b e^{-\gamma x'}\hat{S}_{ab}(s|x')dx'\right] = \gamma^2\left[\hat{S}_{ab}(s|x) - \hat{\Psi}(s)\right]. \quad (12.123)$$

Substituting (12.123) into (12.122) we see that the survival probability satisfies the following inhomogeneous ordinary differential equation of second order with constant coefficients

$$\frac{d^2\hat{S}_{ab}}{dx^2} - \gamma^2\left[1 - \hat{\psi}(s)\right]\hat{S}_{ab} = \gamma^2\hat{\Psi}(s). \qquad (12.124)$$

Two boundary conditions accompany this differential equation. Setting $x = a$ in (12.120) we get

$$\hat{S}_{ab}(s|a) = \hat{\Psi}(s) + \frac{1}{2}\gamma\hat{\psi}(s)e^{\gamma a}\int_a^b e^{-\gamma x'}\hat{S}_{ab}(s|x')dx'.$$

From Eq. (12.121) with $x = a$ we see that

$$\frac{1}{\gamma}\hat{S}'_{ab}(s|a) = \frac{1}{2}\gamma\hat{\psi}(s)e^{\gamma a}\int_a^b e^{-\gamma x'}\hat{S}_{ab}(s|x')dx',$$

where the prime denotes the derivative with respect to x. Subtracting these two expressions we get the first boundary condition:

$$\hat{S}'_{ab}(s|a) = \gamma\left[\hat{S}_{ab}(s|a) - \hat{\Psi}(s)\right]. \qquad (12.125)$$

Setting $x = b$ in (12.120) and (12.121) and proceeding analogously we obtain the second boundary condition

$$\hat{S}'_{ab}(s|b) = -\gamma\Big[\hat{S}_{ab}(s|b) - \hat{\Psi}(s)\Big]. \qquad (12.126)$$

Solving the differential equation for $\hat{S}_{ab}(s|x)$ is straightforward. One can easily see by direct substitution that the general solution to Eq. (12.124) is

$$\hat{S}_{ab}(s|x) = \hat{A}(s)\cosh\hat{\lambda}(s)x + \hat{B}(s)\sinh\hat{\lambda}(s)x - \frac{1}{s}, \qquad (12.127)$$

where $\hat{A}(s)$ and $\hat{B}(s)$ are arbitrary constants (i.e., independent of x although they may depend on the Laplace variable s) and

$$\hat{\lambda}(s) = \gamma\sqrt{1 - \hat{\psi}(s)}. \qquad (12.128)$$

We incidentally note that in writing Eq. (12.127) we have taken into account that $\hat{\Psi}(s) = [1 - \hat{\psi}(s)]/s$ (cf. Eq. (12.86)).

The explicit expressions for $\hat{A}(s)$ and $\hat{B}(s)$ are obtained using the boundary conditions by the standard procedure. Thus, substituting (12.127) into (12.125) and (12.126) results in a linear set of two algebraic equations which determines the integration constants. As the reader may easily check, the final solution reads

$$\hat{S}_{ab}(s|x) = \frac{1}{s}\left[1 - \frac{\gamma\hat{\psi}(s)\cosh\big[\hat{\lambda}(s)(x - (a+b)/2)\big]}{\hat{\lambda}(s)\sinh\big[\hat{\lambda}(s)L/2\big] + \gamma\cosh\big[\hat{\lambda}(s)L/2\big]}\right], \qquad (12.129)$$

where

$$L = b - a$$

is the length of the interval. The Laplace transform of the escape probability,

$$\hat{W}_{ab}(s|x) = \frac{1}{s} - \hat{S}_{ab}(s|x),$$

then reads

$$\hat{W}_{ab}(s|x) = \frac{1}{s}\,\frac{\gamma\hat{\psi}(s)\cosh\big[\hat{\lambda}(s)(x - (a+b)/2)\big]}{\hat{\lambda}(s)\sinh\big[\hat{\lambda}(s)L/2\big] + \gamma\cosh\big[\hat{\lambda}(s)L/2\big]}. \qquad (12.130)$$

From Eq. (12.105) we know that the mean exit time is given by the Laplace transform of the survival probability evaluated at $s = 0$. Thus setting $s = 0$ in Eqs (12.124), (12.125) and (12.126) and recalling that

$\hat{\psi}(0) = 1$ and $\hat{\Psi}(0) = \langle \tau \rangle$ ($\langle \tau \rangle$ is the mean sojourn time) we see that the mean exit time for Laplacian jumps obeys the differential equation

$$\frac{d^2 T_{ab}}{dx^2} = -\gamma^2 \langle \tau \rangle, \qquad (12.131)$$

with boundary conditions

$$T'_{ab}(a) = \gamma [T_{ab}(a) - \langle \tau \rangle], \qquad T'_{ab}(b) = -\gamma [T_{ab}(b) - \langle \tau \rangle]. \qquad (12.132)$$

The solution to this problem is straightforward and reads

$$T_{ab}(x) = \frac{\langle \tau \rangle}{2} \left[1 + \left(1 + \frac{\gamma L}{2} \right)^2 - \gamma^2 \left(x - \frac{a+b}{2} \right)^2 \right], \qquad (12.133)$$

and the mean exit time is a quadratic function not only of the initial position x but also of the length $L = b - a$ of the interval.

It is very illustrative to compare the above expression of the mean exit time with that of the ordinary random walk. If the price follows a random walk then, in the continuous limit, the zero-mean return is the Wiener process, $X(t) = \sigma W(t)$, where $\sigma > 0$ is the volatility and $W(t)$ is the Wiener process with zero mean and unit variance. In this case the mean exit time out of an interval (a, b) is given by Eq. (11.45) of Chapter 11

$$T_{ab}^{(\mathrm{RW})}(x) = \frac{1}{\sigma^2}(x - a)(b - x),$$

which is also a quadratic function of the position. However, boundary conditions now read

$$T_{ab}^{(\mathrm{RW})}(a) = T_{ab}^{(\mathrm{RW})}(b) = 0,$$

which are quite different than those of the CTRW given in Eq. (12.132). In fact, from Eq. (12.133) we easily see

$$T_{ab}(a) = T_{ab}(b) = \langle \tau \rangle,$$

and the mean exit time from any boundary equals the mean sojourn time, as otherwise one may expect from an intuitive point of view.

The hitting problem

Let us briefly comment on the first-passage problem for Laplacian jumps. We know that the hitting problem to some threshold x_c equals the escape problem out of an interval when one of the boundaries is x_c while the other boundary tends to $\pm\infty$. Specifically, first-passage and escape probabilities are related by (cf. Eq. (12.111))

$$W_c(t|x) = \begin{cases} W_{-\infty x_c}(t|x), & x < x_c, \\ W_{x_c\infty}(t|x), & x > x_c. \end{cases} \tag{12.134}$$

For Laplacian jumps the Laplace transform of the escape probability is given by Eq. (12.130). Since $L = b - a$, the limits $a \to -\infty$ and $b \to \infty$ are equivalent to $L \to \infty$ and

$$\sinh(\hat{\lambda}(s)L/2) \sim \cosh(\hat{\lambda}(s)L/2) \sim \frac{1}{2}e^{\hat{\lambda}(s)L/2},$$

as $L \to \infty$. We have

$$\begin{aligned}\hat{W}_{ab}(s|x) &\sim \frac{\gamma\hat{\psi}(s)}{s[\gamma + \hat{\lambda}(s)(s)]} \frac{e^{\hat{\lambda}(s)[x-(a+b)/2]} + e^{-\hat{\lambda}(s)[x+(a+b)/2]}}{e^{\hat{\lambda}(s)L/2}} \\ &= \frac{\gamma\hat{\psi}(s)}{s[\gamma + \hat{\lambda}(s)(s)]}\left[e^{\hat{\lambda}(s)(x-b)} + e^{\hat{\lambda}(s)(a-x)}\right].\end{aligned}$$

Thus, when $a \to -\infty$ or $b \to \infty$, we get

$$\hat{W}_{-\infty b}(s|x) = \frac{\gamma\hat{\psi}(s)}{s[\gamma + \hat{\lambda}(s)(s)]}e^{-\hat{\lambda}(s)(b-x)},$$

or

$$\hat{W}_{a\infty}(s|x) = \frac{\gamma\hat{\psi}(s)}{s[\gamma + \hat{\lambda}(s)]}e^{-\hat{\lambda}(s)(x-a)}.$$

From Eq. (12.134) we conclude that the Laplace transform of the first-passage probability to threshold x_c for Laplacian jumps is

$$\hat{W}_c(s|x) = \frac{\gamma\hat{\psi}(s)}{s[\gamma + \hat{\lambda}(s)]}e^{-\hat{\lambda}(s)|x-x_c|}. \tag{12.135}$$

It is instructive to compare this probability with that of the simplest diffusion process, the free Brownian motion. As we have shown in Chapter 11 the hitting probability for the Wiener process of variance σ^2 is (cf. Eq. (11.41) (with $D = \sigma^2$))

$$\hat{W}_c^{(\mathrm{RW})}(s|x) = \frac{1}{s}e^{-\sqrt{2s}|x-x_c|/\sigma}, \tag{12.136}$$

whose Laplace inversion yields (see Eq. (11.43) and recall that $W_c = 1 - S_c$)

$$W_c^{(\mathrm{RW})}(t|x) = \mathrm{Erfc}\left[\frac{|x-x_c|}{\sigma\sqrt{2t}}\right], \tag{12.137}$$

where $\mathrm{Erfc}(z)$ is the complementary error function.

As the reader may easily check, the two expressions, (12.135) and (12.136), coincide for small values of s as long as the waiting time density $\psi(t)$ has finite moments. In this case, as we have shown above, $\hat{\psi}(s) = 1 - sT + \ldots$ ($s \to 0$) and (cf. Eq. (12.128)) as

$$\lim_{s \to 0} \frac{\gamma \hat{\psi}(s)}{\gamma + \hat{\lambda}(s)} = 1, \qquad \hat{\lambda}(s) = \gamma T^{1/2} \sqrt{s}, \qquad (s \to 0),$$

and Eq. (12.135) tends as $s \to 0$ to Eq. (12.136) with $\sigma = 2/\gamma\sqrt{T}$. That is to say, for large timescales ($t \to \infty$) the non-Gaussian probability tends to a Gaussian function as in Eq. (12.137) which constitutes a restatement of the Central Limit Theorem.

As we have seen above, the mean first-passage time $T_c(x)$ to threshold x_c is obtained from the mean exit time, $T_{ab}(x)$, by setting $a = -\infty$ and $b = x_c$ if $x < x_c$ or $b = \infty$ and $a = x_c$ if $x > x_c$. From Eq. (12.133) we see that $T_{ab}(x) \to \infty$ in both cases when $a \to -\infty$ and $b \to \infty$. Consequently,

$$T_c(x) = \infty,$$

and, like in the Wiener process, the mean first-passage time does not exist

12.7   Risk control *

Risk control is one of the most central issues in any financial setting which is a direct consequence of the random evolution of speculative prices. It returns attain considerable negative values this implies significant drops in prices and the risk of loosing investor's wealth. It is, therefore, clear that this question is related to first passages and level crossings.

Let us consider a set of different assets corresponding to different economic sectors, what is called a "portfolio". Certainly such a set reduces risk but, yet, the risk is not completely removed. The problem was addressed during the 1950's by Harry Markowitz with his theory of portfolio allocation which has had a deep and standing influence not only in economic theory but also in financial practice.

The standard way of addressing the risk of a given asset (and, by extension, of a given portfolio) is by evaluating the "Value at Risk" (VaR). This is a number giving the maximum negative return, that is to say, the minimum return that the investor can bear during a certain period of time $t - t_0$.[19] Clearly this value is a random quantity. In this way, and in order

---

[19]In what follows we will set $t_0 = 0$ since we make the usual assumption that prices and returns are stationary random processes that are invariant under time translations.

to get an operative definition of the VaR, we should specify the level of confidence we want. This level of confidence, denoted by α, is a probability and, therefore, a number between 0 and 1.

The VaR can be formally defined as follows: Given a confidence level $\alpha \in (0, 1)$, the Value at Risk (at a confidence level α and for a time interval t) of a portfolio with return $X(t)$ is the smallest negative number $x < 0$ such that the probability that $X(t)$ exceeds x is at least α. In other words, VaR is the negative of the α-quantile (see, for instance, McNeil et al., 2005, for more information).

Let us denote the value at risk by a somewhat cumbersome but traditional notation $\text{VaR}_{\alpha t}$ indicating that this value depends on both the confidence level α and the time interval t. Note that from the definition above the VaR can be defined implicitly as

$$\text{Prob}\left\{ X(t) \geq \text{VaR}_{\alpha t} \right\} = \alpha. \tag{12.138}$$

Note that the lower VaR is, the higher α becomes. In fact when $\text{VaR} = -\infty$ then $\alpha = 1$. In practice the usual values of α range from 0.95 to 0.99.

In a more operative way Eq. (12.138) can be written as

$$\int_{\text{VaR}_{\alpha t}}^{\infty} p(x, t)dx = \alpha, \tag{12.139}$$

where $p(x, t)$ is the probability density function of $X(t)$. Let us note that the VaR can also be defined through the distribution function of the return. Indeed, since the distribution function $F(x, t)$ is related to the probability density function by

$$F(x, t) = \int_{-\infty}^{x} p(x', t)dx',$$

we see from Eq. (12.139), after recalling the normalization of $p(x, t)$, that

$$F(\text{VaR}_{\alpha t}, t) = 1 - \alpha. \tag{12.140}$$

The value at risk as defined above is the standard measure of risk in the financial sector and it is one of the most important parameters of any portfolio. However, it ignores the instantaneous risk aversion of investors. Indeed, although VaR gives the worst return one can get at the end of a fixed time interval for a given confidence level α, this does not mean that $X(t)$ could not have reached smaller and dangerous values before time t. In other words, VaR neglects the fact that investors may not assume all paths leading to the same final return at time t, because some paths, while ending at the given point, may have gone through any intermediate region with much higher risk.

This would have not been the case had we used the survival probability, instead of the unrestricted probability, for quantifying risk. Indeed, the use of the survival probability ensures, within the desired level of confidence, that any value of the return $X(t)$ has never reached the risk measure at time t. We call this safer measure of risk *survival probability risk level.*

We denote by

$$S_c(t - t_0|x) = S_{x_c,\infty}(t - t_0|x) \qquad (12.141)$$

the survival probability that $X(t)$ starting initially at $x > x_c$ has never crossed x_c during the time interval $t - t_0$. We now adapt this definition to our risk measure. To this end, and using again a cumbersome but apparent notation, we denote by $\text{SpR}_{\alpha t}$ the critical value that we wish return has never attained at time t or before with a level of confidence $\alpha \in (0,1)$. Since we have assumed that $t_0 = 0$ and the starting (i.e., initial) return x is, by definition, equal to zero. Then, and looking at Eq. (12.138), we define this critical value, the survival probability risk level $\text{SpR}_{\alpha t}$, by replacing the unrestricted probability by the survival probability:

$$S_{risk}(t|0) = \alpha, \qquad (12.142)$$

where (cf. (12.141))

$$S_{risk}(t|0) = S_{\text{SpR}_{\alpha t},\infty}(t|0). \qquad (12.143)$$

We incidentally note that implicit in this definition is the fact that this risk measure is a negative number, $\text{SpR}_{\alpha t} < 0$ (recall the the initial return is zero).

Example: The Wiener process

In order to illustrate both procedures, VaR and SpR, we assume that the value -that is, the price– of the portfolio evolves following a geometric Brownian motion, so that its return $X(t)$ obeys the Wiener process which, recall, is the simplest but otherwise most widely used model (cf. Eqs. (12.1) and (12.2)). In what follows, and for the sake of simplicity, we will assume that the return has zero mean (the case of non vanishing average only requires minor modifications). In this case the probability density function of $X(t)$ is

$$p(x,t) = \frac{1}{\sqrt{2\pi\sigma^2 t}} e^{-x^2/2\sigma^2 t}.$$

and substituting into Eq. (12.139) we have

$$\alpha = \frac{1}{\sqrt{2\pi\sigma^2 t}} \int_{\text{VaR}_{\alpha t}}^{\infty} e^{-x^2/2\sigma^2 t} dx = \frac{1}{\sqrt{2\pi}} \int_{\text{VaR}_{\alpha t}/\sigma\sqrt{t}}^{\infty} e^{-z^2/2} dz$$

$$= 1 - \frac{1}{\sqrt{2\pi}} \int_{-\infty}^{\text{VaR}_{\alpha t}/\sigma\sqrt{t}} e^{-z^2/2} dz.$$

In terms of the Gaussian distribution function (also called probability integral)

$$\Phi(u) = \frac{1}{\sqrt{2\pi}} \int_{-\infty}^{u} e^{-z^2/2} dz \qquad (12.144)$$

we can then write

$$\alpha = 1 - \Phi\left(\frac{\text{Var}_{\alpha t}}{\sigma\sqrt{t}}\right),$$

and we have

$$\text{VaR}_{\alpha t} = \sigma\sqrt{t}\,\Phi^{-1}(1-\alpha), \qquad (12.145)$$

where $\Phi^{-1}(\cdot)$ is the inverse of the probability integral.

We next obtain the survival probability risk level for the Wiener return. From Eqs. (12.142) and (12.143) we have

$$\alpha = S_{\text{risk}}(t|0).$$

For the Wiener process we have seen in Chapter 11 (cf. Eq. (11.43)) that

$$S_c(t|x) = 1 - \text{Erfc}\left[\frac{|x - x_c|}{\sqrt{2Dt}}\right], \qquad (12.146)$$

where $\text{Erfc}(z)$ is the complementary error function,

$$\text{Erfc}(u) = \frac{2}{\sqrt{\pi}} \int_{u}^{\infty} e^{-z^2} dz. \qquad (12.147)$$

The expression for $S_{\text{risk}}(t|0)$ is then given by (12.146) with $x_c = \text{SpR}_{\alpha t}$ and $x = 0$. Hence,

$$\alpha = 1 - \text{Erfc}\left[\frac{|\text{SpR}_{\alpha t}|}{\sqrt{2\sigma^2 t}}\right]. \qquad (12.148)$$

However, from (12.144) and (12.147) we easily see that the complementary error function is related to the probability integral by

$$\text{Erfc}(z) = 2\Phi(-\sqrt{2}z), \qquad (12.149)$$

and we write

$$\alpha = 1 - 2\Phi\left(\frac{-|\text{SpR}_{\alpha t}|}{\sqrt{2\sigma^2 t}}\right).$$

The survival probability risk level is a negative number, $|\mathrm{SpR}_{\alpha t}| = -\mathrm{SpR}_{\alpha t}$, and we have

$$\Phi\left(\frac{\mathrm{SpR}_{\alpha t}}{\sqrt{2\sigma^2 t}}\right) = \frac{1-\alpha}{2}.$$

Finally

$$\mathrm{SpR}_{\alpha t} = \sigma\sqrt{t}\,\Phi^{-1}\left(\frac{1-\alpha}{2}\right). \tag{12.150}$$

Let us finally compare the values of VaR and SpR. From Eqs. (12.145) and (12.150), we write

$$1 - \alpha = \Phi\left(\frac{\mathrm{VaR}_{\alpha t}}{\sigma\sqrt{t}}\right), \qquad \frac{1-\alpha}{2} = \Phi\left(\frac{\mathrm{SpR}_{\alpha t}}{\sigma\sqrt{t}}\right).$$

Hence,

$$\Phi\left(\frac{\mathrm{VaR}_{\alpha t}}{\sigma\sqrt{t}}\right) > \Phi\left(\frac{\mathrm{SpR}_{\alpha t}}{\sigma\sqrt{t}}\right),$$

and since $\Phi(z)$ is an increasing function of z,[20] we conclude that

$$\mathrm{VaR}_{\alpha t} > \mathrm{SpR}_{\alpha t}$$

for all confidence levels and all $t \geq 0$, meaning that, compared with SpR, *VaR systematically underestimates risk*. This may have significant practical consequences since VaR is one of the most widely used measures of risk and the above analysis shows that SpR improves the efficiency of more traditional methods based on VaR (Bouchaud and Potters, 2011; Montero and Masoliver, 2007a).

---

[20]Indeed, taking the derivative of Eq. (12.144) we see that $\Phi'(u) = e^{-u^2/2}/\sqrt{2\pi} > 0$ for all $u \in \mathbb{R}$.

Bibliography

Abramowitz, M. and Stegun, I. A. (eds.) (1965). *Handbook of Mathematical Functions* (Dover, New York).

Arnold, L. (1974). *Stochastic Differential Equations* (J. Wiley, New York).

Arnold, V. I. (1978). *Ordinary Differential Equations* (MIT Press, Cambridge, MA).

Bachelier, L. (1900). Téorie de la spéculation, *Ann. Sci. École Norm. Sup* **17**, pp. 21–86.

Balescu, R. (2005). *Aspects of Anomalous Transport in Plasmas* (Taylor and Francis, London).

Balescu, R. (2007). V- langevin, continuous time random walks and fractional diffusion, *Chaos Solitons and Fractals* **34**, pp. 62–80.

Barkai, E. (2001). Fractional fokker-planck equation, solution and application, *Phys. Rev. E* **63**, p. 046118.

Baxter, M. and Rennie, A. (1998). *Financial Calculus* (Cambridge University Press, Cambridge, UK).

Ben-Avraham, D. and Havlin, S. (2000). *Diffusion and reactions in Fractals and Disordered Systems* (Cambridge University Press, Cambridge, UK).

Berman, S. M. (1992). *Sojourn and Extremes of Stochastic Processes* (Wadsworth and Brooks/Cole, Belmont, CA).

Black, F. and Sholes, M. (1973). The pricing of options and corporate liabilities, *J. Pol. Econ.* **81**, pp. 637–659.

Bouchaud, J.-P. and Georges, A. (1990). Anomalous diffusion in disordered media: statistical mechanics, models and applications, *Phys. Rep* **195**, pp. 127–293.

Bouchaud, J.-P. and Potters, M. (2011). *Theory of Financial risk and Derivative Pricing: From Statistical Physics to Risk Management* (Cambridge University Press, Cambridge, UK).

Brigo, D. and Mercurio, F. (2006). *Interest Rate Models: Theory and Practice* (Springer-Verlag, Berlin).

Burkhardt, T. (2014). First-passage of a random accelerated particle, in R. Metzler, G. Oshanin and S. Redner (eds.), *First-passage Phenomena and their Applications* (World Scientific, Singapore).

Carslaw, S. and Jaeger, J. C. (1990). *Conduction of Heat in Solids* (Oxford Science

Publications, Oxford).

Condamin, S., Bénichou, O., Tejedor, V. and Klafter, J. (2007). First-passage times in complex scale-invariant media, *Nature* **450**, pp. 77–80.

Cont, R. (2001). Empirical properties of asset returns: stylized facts and statistical issues, *Quant. Fin.* **1**, pp. 223–236.

Cootner, P. H. (ed.) (1964). *The Random Character of Stock Market Prices* (MIT Press, Cambridge, MA).

Courant, R. and Hilbert, R. (1991). *Methods of Mathematical Physics: Partial Differential Equations* (J. Wiley, New York).

Doob, J. L. (1942). The brownian movement and stochastic equations, *Annals of Mathematics* **43**, pp. 351–369.

Doob, J. L. (1966). *Stochastic Processes* (J. Wiley, New York).

Duffie, D. (2001). *Asset Pricing Theory*, 3rd edn. (Princeton University Press, Princeton, NJ).

Eliazar, I. I. and Shlesinger, M. F. (2013). Fractional motions, *Phys. Rep* **527**, pp. 101–129.

Erdelyi, A. (ed.) (1953). *Higher Transcendental Functions (Vol. 3)* (McGraw-Hill, New York).

Erdelyi, A. (ed.) (1954). *Table of Integral Transforms (Vol. 1)* (McGraw-Hill, New York).

Erdelyi, A. (1956). *Asymptotic Expansions* (Dover, New York).

Farmer, J. D., Geanakoplos, J., Masoliver, J., Montero, M. and Perelló, J. (2015). Value of the future: discounting in random environments, *Phys. Rev. E* **91**, p. 052816.

Feller, W. (1991). *An Introduction to Probability Theory and its Applications (Vols. 1 and 2)* (J. Wiley, New York).

Fouque, J.-P., Papanicolau, G. and Sircar, K. R. (2000). *Derivatives in Financial Markets with Stochastic Volatility* (Cambridge University Press, Cambridge, UK).

Fouque, J.-P., Papanicolau, G., Sircar, K. R. and Solna, K. (2011). *Multiscale Stochastic Volatility for Equity, Interest Rate and Credit Derivatives* (Cambridge University Press, Cambridge, UK).

Gardiner, C. W. (1986). *Handbook of Stochastic Methods* (Springer-Verlag, Berlin).

Gelfand, I. M. and Shilov, G. E. (2016a). *Generalized Functions, Vol. 1* (AMS Chelsea Publishing).

Gelfand, I. M. and Shilov, G. E. (2016b). *Generalized Functions, Vol. 4* (AMS Chelsea Publishing).

Ghiman, I. F. and Skorohod, A. V. (1972). *The Theory of Stochastic Processes* (Springer-Verlag, Berlin).

Gitterman, M. (2012). *The Noisy Oscillator* (World Scientific, Singapore).

Gnedenko, B. V. (1976). *The Theory of Probability* (Mir Publishers, Moscow).

Gorenflo, R., Luchko, Y. and Mainardi, F. (1999). Analytical properties and applications of the wright function, *Fractional Calc. Appl. Anal.* **2**, pp. 383–414.

Gorenflo, R., Mainardi, F. and Vivoli, A. (2007). Continuous time random walks and parametric subordination in fractional diffusion, *Chaos Solitons and Fractals* **34**, pp. 87–103.

Gradshteyn, I. S. and Ryzhik, I. M. (1994). *Tables of Integrals Series and Products* (Academic Press, San Diego).

Grimmett, G. R. and Stirzaker, D. R. (1992). *Probability and Random Processes*, 2nd edn. (Clarendon Press, Oxford, UK).

Gumbel, E. J. (2004). *Statistics of Extremes* (Dover, New York).

Hamilton, J. D. (1994). *Time series Analysis* (Princeton University Press, Princeton, NJ).

Handelsman, R. A. and Lew, J. S. (1974). Asymptotic expansions of laplace convolutions for large arguments and tail densities for certain sums of random variables, *SIAM J. Math. Anal.* **5**, pp. 425–451.

Hull, J. C. (2010). *Fundamentals of Futures and Option Markets* (Prentice-Hall, New Jersey).

Klafter, J. and Sokolov, I. (2005). Anomalous diffusion spreads its wings, *Physics World* **August**, pp. 29–32.

Kutner, R. and Masoliver, J. (2017). The ctrw still trendy: fifty-year history, state of the art and outlook, *Eur. Phys. J. B* **90**, p. 50.

Laha, R. G. and Rohatgi, V. K. (1979). *Probability Theory* (J. Wiley, New York).

Langer, R. E. (ed.) (1960). *On a Unified Theory of Boundary Problems for Elliptic-Parabolic Equations of Second Order* (University of Wisconsin Press).

LePage, W. R. (1990). *Complex Variables and Laplace Transforms for Engineers* (Dover, New York).

Lukacs, E. (1970). *Characteristic Functions*, 2nd edn. (Griffin, London).

Magnus, W., Oberhettinger, F. and Soni, R. P. (1966). *Formulas and Theorems for the Special Functions of Mathematical Physics* (Springer-Verlag, Berlin).

Mainardi, F. (1996). The fundamental solution of the fractional diffusion-wave equation, *Appl. Math. Lett* **9**, pp. 23–28.

Mainardi, F., Luchko, F. and Pugnini, G. (2001). The fundamental solutions of the space-time fractional diffusion equation, *Fractional Calculus and Applications* **4**, pp. 153–192.

Mantegna, M. R. and Stanley, H. E. (2007). *Introduction to Econophysics: Correlations and Complexity in Finance* (Cambridge University Press, Cambridge, UK).

Masoliver, J. (2014a). Extreme values and the level-crossing problem: An application to the feller process, *Phys. Rev. E* **89**, p. 042106.

Masoliver, J. (2014b). The level-crossing problem: first-passage, escape and extremes, *Fluct. Noise. Lett.* **13**, p. 14300001.

Masoliver, J. (2016). Fractional telegrapher's equation from fractional persistent random walks, *Phys. Rev. E* **93**, p. 052107.

Masoliver, J., Montero, M., Perelló, J. and Weiss, G. H. (2008). The continuous time random walk formalism in financial markets, *J. Econ. Behavior and Org.* **61**, pp. 577–598.

Masoliver, J. and Perelló, J. (2009). First-passage and risk evaluation under stochastic volatility, *Phys. Rev. E* **80**, p. 016108.

Masoliver, J. and Perelló, J. (2012). First-passage and escape problems in the feller process, *Phys. Rev. E* **86**, p. 041116.

Masoliver, J. and Perelló, J. (2014a). Erratum: First-passage and escape problems in the feller process, *Phys. Rev. E* **89**, pp. 029902–1.

Masoliver, J. and Perelló, J. (2014b). First-passage and extremes in socioeconomic systems, in R. Metzler, G. Oshanin and S. Redner (eds.), *First-Passage Phenomena and their Applications* (World Scientific, Singapore).

Masoliver, J. and Porrà, J. (1995). Exact solution to the mean exit time problem for free inertial processes driven by gaussian white noise, *Phys. Rev. Lett* **75**, pp. 189–192.

Masoliver, J. and Porrà, J. (1996). Exact solution to the exit problem for an undamped free particle driven by gaussian white noise, *Phys. Rev. E* **53**, pp. 2243–2256.

Mattos, T., Mejía-Monasterio, C., Metzler, R. and Oshamin, G. (2012). First passages in bounded domains: When is the mean first passage time meaningful? *Phys. Rev. E* **86**, p. 031143.

McNeil, A., Frey, R. and Embrechts, P. (2005). *Quantitative Risk Management: Concepts, Techniques and Tools* (Princeton University Press, Princeton, NJ).

Merton, R. (1973). Theory of rational option pricing, *Bell J. Econ. Manage Sci* **4**, pp. 141–183.

Merton, R. (1976). Option pricing when underlying stock returns are discontinuous, *J. Fin. Econ.* **3**, pp. 125–144.

Metzler, R. and Klafter, J. (2000). The random walk's guide to anomalous diffusion: a fractional diffusion approach, *Phys. Rep* **339**, pp. 1–77.

Metzler, R. and Klafter, J. (2004). The restaurant at the end of the walk: recent developments in the description of anomalous transport by fractional dynamics, *J. Phys. A* **37**, pp. 161–208.

Metzler, R., Oshanin, G. and Redner, S. (eds.) (2014). *First-Passage Phenomena and their Applications* (World Scientific, Singapore).

Middleton, D. (1960). *An Introduction to Statistical Communication Theory* (Princeton University Press, Princeton, NJ).

Montero, M. and Masoliver, J. (2007a). Mean exit time and survival probability within the ctrw formalism, *Eur. Phys. J. B* **57**, pp. 181–185.

Montero, M. and Masoliver, J. (2007b). Nonindependent continuos-time random walks, *Phys. Rev. E* **76**, p. 061115.

Montero, M. and Masoliver, J. (2017). Continuos-time random walks with memory and financial distributions, *Eur. Phys. J. B* **90**, p. 207.

Oksendal, B. (2010). *Stochastic Differential Equations: An Introduction with Applications* (Springer-Verlag, Berlin).

Oppenheim, I., Shuler, K. E. and Weiss, G. H. (1977). *Stochastic Processes in Chemical Physics: The Master Equation* (MIT Press, Cambridge, MA).

Osborne, M. F. M. (1959). Brownian motion in stock markets, *Operations Research* **7**, pp. 145–173.

Papoulis, A. (1977). *Signal Analysis* (McGraw-Hill, New York).

Papoulis, A. (1984). *Probability, Random Variables and Stochastic Processes*

(McGraw-Hill, New York).

Pitt, H. R. (1958). *Tauberian Theorems* (Oxford University Press, Oxford, UK).

Podbury, I. (1999). *Fractional Differential Equations* (Academic Press, San Diego).

Porter, D. and Stirling, D. S. G. (1990). *Integral Equations* (Cambridge University Press, Cambridge, UK).

Rangarajan, G. and Ding, M. (2000). Anomalous diffusion and the first passage problem, *Phys. Rev. E* **62**, pp. 120–133.

Redner, S. (2007). *A Guide to First-Passage Processes* (Cambridge University Press, Cambridge, UK).

Risken, H. (1987). *The Fokker-Planck Equation* (Springer-Verlag, Berlin).

Roberts, G. E. and Kaufman, H. (1966). *Table of Integral Transforms* (W. B. Saunders, Philadelphia).

Rytov, S. M., Kravtsov, Y. and Tatarski, V. I. (1987). *Principles of Statistical Radiophysics (Vol. 1)* (Springer-Verlag, Berlin).

Sanders, L. P. and Amjornsson, T. (2012). First passage times for a tracer particle in single file diffusion and fractional brownian motion, *J. Chem. Phys.* **136**, p. 175103.

Smith, C. W. (1976). Option pricing: A review, *J. Fin. Econ.* **3**, pp. 3–51.

Stratonovich, R. L. (1966). A new representation for stochastic integrals and equations, *SIAM J. Control* **4**, pp. 362–371.

Stratonovich, R. L. (1967). *Topics in the Theory of Random Noise (Vol. 1)* (Gordon and Breach, new York).

van Kampen, N. G. (1992). *Stochastic Processes in Physics and Chemistry* (North-Holland, Amsterdam).

Vasicek, O. (1977). An equilibrium characterization of the term structure, *J. Fin. Econ.* **5**, pp. 177–188.

Vladimirov, V. S. (1984). *Equations of Mathematical Physics* (Mir Publishers, Moscou).

Wax, N. (ed.) (1954). *Selected Papers on Noise and Stochastic Processes* (Dover, New York).

Weiss, G. H. (1994). *Aspects and Applications of the Random Walk* (North-Holland, Amsterdam).

West, B. J. (2014). Colloquium: Fractional view of complexity: a tutorial, *Rev. Mod. Phys.* **86**, pp. 1169–1184.

West, B. J. (2016). *Fractional Calculus View of Complexity: Tomorrow Science* (CRC, Boca Raton, Florida).

West, B. J., Bologna, M. and Grigolini, P. (2003). *Physics of the Fractal operators* (Springer-Verlag, Berlin).

Wilmott, P. (1998). *Derivatives* (J. Wiley, New York).

Wilmott, P., Dewynne, J. and Howison, S. (1995). *Option Pricing: Mathematical Models and Computation* (Oxford Financial Press, Oxford).

Wio, H. S., Deza, R. R. and López, J. M. (2012). *An Introduction to Stochastic Processes and Nonequilibrium Statistical Physics* (World Scientific, Singapore).

Wong, E. and Zakai, M. (1965). On the convergence of ordinary integrals to stochastic integrals, *Ann. Math. Statist.* **36**, pp. 1560–1564.

Yaglom, A. M. (1973). *An Introduction to the Theory of Random Functions* (Dover, New York).

Yuste, S. and Lindenberg, K. (2005). Trapping reactions with subdiffusive traps and particles characterized by different diffusion exponents, *Phys. Rev. E* **72**, p. 061103.

Zaslavsky, G. M. (2002). Chaos, fractional kinetics, and anomalous transport, *Phys. Rep* **371**, pp. 411–580.

Index

Printed in the United States
By Bookmasters